NATO ASI Series

Advanced Science Institutes Series

A series presenting the results of activities sponsored by the NATO Science Committee, which aims at the dissemination of advanced scientific and technological knowledge, with a view to strengthening links between scientific communities.

The Series is published by an international board of publishers in conjunction with the NATO Scientific Affairs Division

A Life Sciences **B Physics**	Plenum Publishing Corporation London and New York
C Mathematical and Physical Sciences **D Behavioural and Social Sciences** **E Applied Sciences**	Kluwer Academic Publishers Dordrecht, Boston and London
F Computer and Systems Sciences **G Ecological Sciences** **H Cell Biology** **I Global Environmental Change**	Springer-Verlag Berlin Heidelberg New York London Paris Tokyo Hong Kong Barcelona Budapest

PARTNERSHIP SUB-SERIES

1. Disarmament Technologies	Kluwer Academic Publishers
2. Environment	Springer-Verlag
3. High Technology	Kluwer Academic Publishers
4. Science and Technology Policy	Kluwer Academic Publishers
5. Computer Networking	Kluwer Academic Publishers

The Partnership Sub-Series incorporates activities undertaken in collaboration with NATO's Cooperation Partners, the countries of the CIS and Central and Eastern Europe, in Priority Areas of concern to those countries.

NATO-PCO DATABASE

The electronic index to the NATO ASI Series provides full bibliographical references (with keywords and/or abstracts) to about 50000 contributions from international scientists published in all sections of the NATO ASI Series. Access to the NATO-PCO DATABASE compiled by the NATO Publication Coordination Office is possible in two ways:

- via online FILE 128 (NATO-PCO DATABASE) hosted by ESRIN, Via Galileo Galilei, I-00044 Frascati, Italy.
- via CD-ROM "NATO Science & Technology Disk" with user-friendly retrieval software in English, French and German (© WTV GmbH and DATAWARE Technologies Inc. 1992).

The CD-ROM can be ordered through any member of the Board of Publishers or through NATO-PCO, Overijse, Belgium.

2. Environment – Vol. 8

Springer
Berlin
Heidelberg
New York
Barcelona
Budapest
Hong Kong
London
Milan
Paris
Santa Clara
Singapore
Tokyo

Urban Air Pollution

Monitoring and Control Strategies

Edited by

Ivo Allegrini
Franco De Santis

CNR, Istituto Inquinamento Atmosferico
Via Salaria Km 29, 300, CP 10
I-00016 Monterotondo Stazione, Italy

Springer

Published in cooperation with NATO Scientific Affairs Division

Proceedings of the NATO Advanced Research Workshop "Monitoring and Control Strategies for Urban Air Pollution", held in Erice, Italy, October 9–15, 1994

Cataloging-in-Publication Data applied for

Die Deutsche Bibliothek - CIP-Einheitsaufnahme

Urban air pollution : monitoring and control strategies ; [proceedings of the NATO Advanced Research Workshop "Monitoring and Control Strategies for Urban Air Pollution", held in Erice, Italy, October 9 - 15, 1994] / ed. by Ivo Allegrini ; Franco DeSantis. Publ. in cooperation with NATO Scientific Affairs Division. - Berlin ; Heidelberg ; New York ; Barcelona ; Budapest ; Hong Kong ; London ; Milan ; Paris ; Santa Clara ; Singapore ; Tokyo : Springer, 1996
(NATO ASI series : Sub-ser. 2, Environment ; Vol. 8)

ISBN-13: 978-3-642-64703-1 e-ISBN-13: 978-3-642-61120-9

DOI: 10.1007/978-3-642-61120-9

NE: Allegrini, Ivo [Hrsg.]; Advanced Research Workshop Monitoring and Control Strategies for Urban Air Pollution <1994, Erice>; NATO: NATO ASI series / 02

Softcover reprint of the hardcover 1st edition 1996

Typesetting: Camera-ready by authors/editors
SPIN: 10475231 31/3137 – 5 4 3 2 1 0 – Printed on acid-free paper

PREFACE

This book presents the proceedings of a NATO Advanced Research Workshop which was also financially supported by the National Research Council of Italy. The Workshop was held from October 9 to 15, 1994, at the Centro Ettore Maiorana in Erice, Italy. Over 40 researchers from a wide variety of fields attended the Workshop, which brought to attention the ongoing research on various phenomena related to urban air pollution.

The presence of high levels of atmospheric pollutants in the air of several urban centres of developed and developing countries causes a great concern among authorities and public opinion. Some 20% of the European population live in cities of more than 500,000 inhabitants and about 40% in cities of more than 50,000. Since exceedance of the Air Quality Guidelines has been observed to occur worldwide, a great effort has been addressed to the control of primary pollutants, but many problems related to secondary pollutants such as nitrogen containing species (nitrogen oxides, nitric and nitrous acid, nitrates) and photochemical oxidants (ozone, PAN and others) are far from being solved. The importance of atmospheric chemistry in understanding the processes occurring in urban atmospheres has been well recognised, thus there is a strong need to exchange experiences and results from urban centres in different Countries. Indeed, atmospheric pollution is very much dependent on the type of emissions which are very different according to the economic development of the urban centre under consideration. Drastic improvement in energy efficiency and use of less polluting energy systems are needed to reduce the impact of pollutant emission while allowing continued economic growth. It is evident that the implementation of the best technologies in Eastern Europe requires an intense exchange of know-how, resources and capital investment between Western and Eastern Europe.

The workshop identified a number of items which are still a matter of speculation and which are relevant keys in understanding physico-chemical processes. For instance, high nitrous acid concentration has been detected in several cities, but the mechanism for its formation and for the initiation of photochemical reactions to produce photooxidants is not very clear. The Chemical Mass Balance technique may give important information on the relative contribution to individual volatile organic components in the formation of ozone and other photooxidants. This includes aromatic hydrocarbons for which the chemical mechanisms in ozone formation are still under investigation. In the field of meteorology, a clear gap was identified in the evaluation of the dynamics of the boundary layer which is known to play the most relevant role in the definition of pollution properties.

Experience with the measurements of Radon was found very useful.

Attention was also called to the very complex area of emission inventories.

Chemistry to describe complex processes is very much related to photochemistry and the development of comprehensive models for the description of chemical transformation was one of the key issues. However, attention was drawn to applying models in locations, such as the Mediterranean area, characterised by complex meteorological processes at mesoscale level. In addition, chemical parameters need to be defined in laboratory studies or in simulated smog chambers where a more realistic picture may be obtained.

These were just some of the issues which were discussed extensively during the Workshop. The objective were to review the scientific basis of urban pollution, to exchange information among participants and to indicate the way towards appropriate collaborative and integrated research programmes, a prerequisite condition to achieve any further progress.

The Directors of the ARW

Ivo Allegrini and Jacek Walczewski

TABLE OF CONTENTS

Urban atmospheric pollution: review of the state-of-the-art

Ivo Allegrini
CNR-Ist. Inquinamento Atmosferico
Via Salaria Km. 29,300 - CP10
00016 Monterotondo Stazione (Roma), Italy

Introduction

The problem of urban atmospheric pollution is one of the most challenging for modern society. Indeed, when adverse meteorological conditions occur in densely populated locations, high levels of primary pollutants should be expected. In addition, since most primary emitted compounds may react in the atmosphere with radicals, sites experiencing intense solar irradiation may also experience high concentration levels of secondary pollutants. This makes the atmospheric environment incompatible with accepted standards therefore, proper actions should be strongly solicited by public opinion from those responsible for air quality.

The emission of pollutants in urban environments is mainly due to traffic and domestic heating in addition to, the eventuality of industrial sources in settlements not far from the city. Thus, most of these pollutants are emitted at ground level where severe pollution levels might be found. This is quite different from pollution by industrial sources where most of the problems are related to fumigation processes caused by emissions from elevated sources. However, the transformation of primary into secondary pollutants requires an amount of time which is dependent upon reaction rates, thus secondary pollution is much more diffused and uniformly distributed. In other words, citizens are usually exposed to peak levels of primary pollutants whereas they are continuously exposed to secondary pollutants. This makes the description of the possible effects very complex and likewise the strategy for a possible reduction of exposition is equally complex.

As the result of such simple premises, urban air pollution management requires knowledge of the parameters related to three important aspects which are the basis of the workshop. They are:

NATO ASI Series, Partnership Sub-Series, 2. Environment – Vol. 8
Urban Air Pollution
Edited by Ivo Allegrini and Franco De Santis

1) the air quality monitoring network, which provides measurements and data relevant to several pollutants, their occurrence in time, their spatial distribution and, eventually, episodes exceeding the quality standards.

2) emission inventories, which provide information about the emitting source and their temporal evolution.

3) numerical models, which provide information about the prediction of air quality from emission data.

Of course these components are interrelated as it is shown in figure 1. Monitoring data are used to validate models. Inventory data are used as input for models which are able to provide numerical prediction. Results from the model application and inventory data, might suggest the formulation of proper control strategies. This is clearly the most important problem for those responsible for air pollution management.

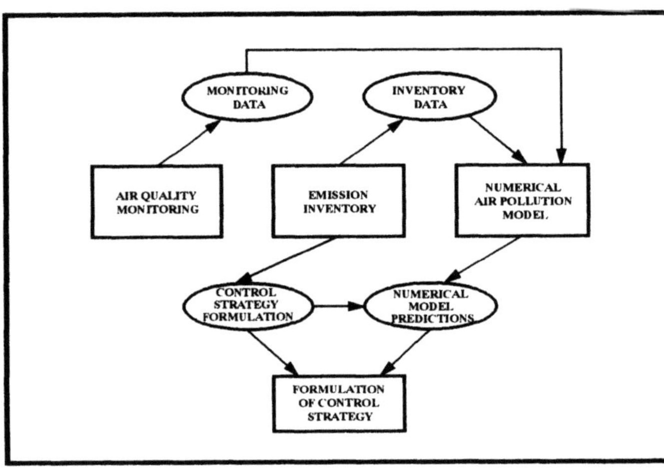

FIGURE 1 - Scheme for air pollution management process in an urban environment.

In the operating scheme reported in Figure 1, major importance must be given to monitoring aspects and to models. In fact, procedures for the compilation of emission inventories are well established. Even models have nowadays reached a sufficient degree of accuracy. However, even though models usually appear to be sufficiently accurate, some very unusual boundary

conditions require additional studies, .
The purpose of this presentation is to give some indication about the state of the art relevant to monitoring atmospheric pollutants in urban centres, pointing out the need for a better understanding of the processes responsible for atmospheric pollution. This include the development of techniques, the proper siting of monitoring stations, quality control of the data and the acquisition of non-conventional parameters such as those related to meteorology. In addition, topics related to atmospheric chemitry are presented and discussed.

Monitoring urban atmospheric pollution

Due to the importance of monitoring techniques several efforts have been made to improve them. Monitoring air quality is usually carried out through monitoring networks, which are made up of many measuring stations. They are composed of many chemical and physico-chemical sensors which provide the concentration of selected compounds. Unfortunately, only a few species can be measured automatically and, if a complete picture of atmospheric processes is needed, manual chemical procedures must be employed. These are required for inorganic as well as organic components.
When monitoring networks are taken into consideration, monitoring techniques should be discussed in terms of possible automation, sensitivity, accuracy, lack of interference and time resolution. An exhaustive list of the techniques which can be used for air pollution monitoring is given in Table I. This shows that the state of the art of monitoring techniques is sufficiently advanced and that many components in gas phase and particulate matter, as well as several organic components may be measured. Some species may be measured by means of a large variety of methods, thus allowing the required time resolution and sensitivity.
For example, sulphur dioxide can be measured through a variety of methods which range from the acidimetric method to standard methods based on the flame photometry or the Differential Optical Absorption Spectrometer (DOAS). For nitrogen dioxide,

chemiluminescence is commonly used, but DOAS may also be employed. The latter technique, may also be used for ozone.

Pollutant	**Technique**
SO_2	☐ Hydrogen peroxide acidimetric and colorimetric
	☐ Colorimetric pararosaniline
	☐ Gas phase fluorescence
	☐ Differential optical absorption spectroscopy (DOAS)
	☐ Flame photometry
NO_2	☐ Chemiluminescence
	☐ Christie Arsenite
	☐ Diffusion tube
	☐ Differential optical absorption spectroscopy (DOAS)
O_3	☐ Ultra-violet photometry
	☐ Chemiluminescence
	☐ Differential optical absorption spectroscopy (DOAS)
CO	☐ Infra-red absorption
	☐ Electrochemical cell
Particulate Matter	☐ Hi-volume sampler
	☐ Smoke shade reflectance
	☐ Beta-gauge
PAN	☐ Gas chromatography/electron capture
	☐ Gas chromatography/Chemiluminescence
Benzene	☐ Gas chromatography
	☐ Differential optical absorption spectroscopy (DOAS)
VOCs	☐ Total hydrocarbon/non methane hydrocarbons analyser
	☐ Specific hydrocarbons/halocarbons/Oxygenates
PAH	☐ High performance liquid chromatography (HPLC)
Lead and Cadmium	☐ Atomic absorption spectroscopy
	☐ X-ray fluorescence
Acidity in gases and particles	☐ Diffusion tubes/ion chromatography

TABLE 1 Analytical methods employed in urban pollution monitoring

Moreover, automated chromatography for the measurement of benzene and other organic components is available. The list of available methods ends with the denuder technique which has found several applications especially for the monitoring of secondary pollutants.

In conclusion, the state of the art of monitoring atmospheric pollutants seems to be adequate for solving most problems encountered in the evaluation of the species defining the air quality and the characteristics of the atmospheric environment. However, significant steps are expected in the near future towards the development of area sensors, i.e., sensors which yield responses integrated over a path and remote sensors. In addition, there is a strong need to develop instruments and systems suitable for the monitoring of species which, although not interesting from an environmental point of view, may be significant for understanding the physical and chemical processes which are going on in the atmosphere.

Siting of monitoring stations

Although present technologies are sufficient to give certain information to the air quality managers, the problem now shifts to a different direction, that is the lack of clear strategies prior to monitoring. It is indisputable that atmospheric pollutant concentrations need to be measured in order to compare the data measured on site with standards and limits which have been fixed by the national or international regulation, but unfortunately this is not enough. According to current opinion in the scientific community, clear indications of control strategies should be given, bearing in mind that the main purpose of a monitoring network is to understand the basic processes which are responsible for atmospheric pollution. This means that the proper compounds, in the proper place, with proper instrumentation, at the proper time and with proper quality control and quality assurance procedures should be measured.

This implies that a monitoring network cannot be considered as simply made up of a sum of stations which are measuring the same components (usually the criteria pollutants) at different sites, but it is necessary to differentiate the kind of chemical species which are observed and monitored and how the data are treated. The lack of such a strategy is evident in Italy where most of the monitoring networks are based on stations located just in the centre of the city. In this situation, there is no way of understanding what happens in the outskirts where many people might be exposed to rather large concentration levels of atmospheric pollutants. In addition, since air masses are mixed up, it is very difficult to understand the processes which transport pollutants from the centre and is very difficult to differentiate polluted air masses coming from sources which are in the same area. This is

the case of industrial settlements which are near cities.
An example of how the optimisation of the monitoring strategy in urban centres may be carried out, is that provided by the Italian legislation on the deployment of monitoring network which is shown in Figure 2

Station Type	Site	Measured Pollutants
A	Parks	Primary and Secondary
B	Residential	Primary and Secondary
C	Heavy Traffic	Pollutants from traffic
D	Outskirts	Photochemical Pollutants

FIGURE 2. Definition and equipment of stations a,b,c,d according to the Italian regulation.

As it can be easily observed, the stations are differentiated in equipment and setting. Type A is the main station, located in the centre of the city, but out of the influence of direct traffic. This station contains sensors for automated measurements and samplers suitable for accumulation techniques which collect compounds that are then analysed in the laboratory. B-type stations are stations located in residential sites while C-type stations are located in sites where primary pollution from mobile sources is common occurrence. D-type are stations which are located in the outskirt and, as such, they are particularly suitable for the evaluation of secondary pollution, especially ozone. Although the Italian legislation, as well as most of the legislation enforced in several advanced Countries, distinguishes stations, locations and equipment, there is still a great need to study sampling sites especially according to their ability to represent the space distribution of a pollutant. This can be very poor, especially in sites where significant modulation of the meteorological situation is expected.

Data quality control

Data quality control and quality assurance are aspects which are extremely important since data management are significant only if data are correct or supposed to be so. Very often, incorrect calibration procedures or underestimation of possible problems may be such that the quality of measurement is impaired and control strategies based upon such data sets may not be sufficient.

A good example of quality assurance can be deduced from a practical observation that was carried out for a DOAS calibration for nitrous acid measurements. The instrument response for *HONO*, against a calibration source developed in the laboratory is shown in Figure 3

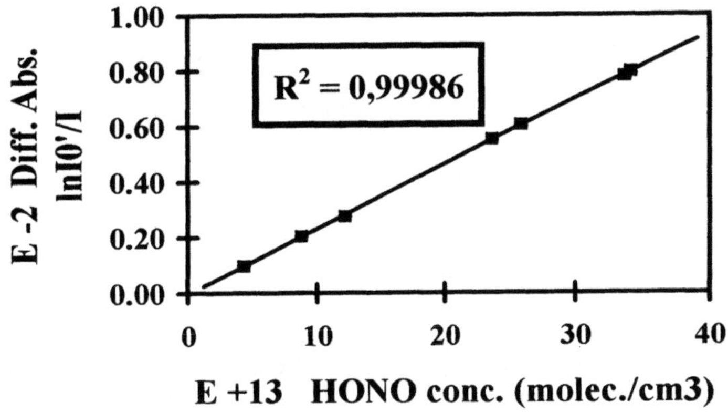

FIGURE 3 - Calibration curve of a DOAS for the long path monitoring of nitrous acid.

As it is shown, the calibration constant is about 30% less than that expected. This means that this instrument is able to measure nitrous acid, but that data obtained so far must be reconciliated with the new calibration constant. Clearly this is the most favourable situation since the instrument has the capability to provide sufficiently precise results. The accuracy can be reconstructed "a posteriori".

However, if another example is taken into consideration, the situation is not so favourable For example, if a comparison of the DOAS response for NO_2 with that obtained from chemilumenescence analyser is carried out, a very bad regression could be obtained. Figure 4a shows a scatter plot which was obtained in Milan during a monitoring campaign in 1993 and which shows that the response of DOAS is very far from the reference chemiluminescence method.

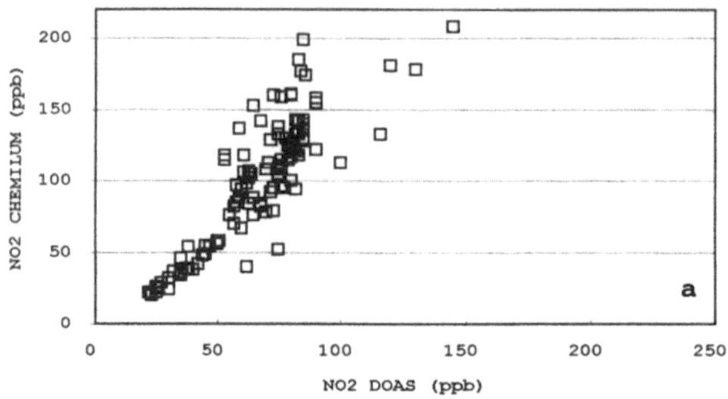

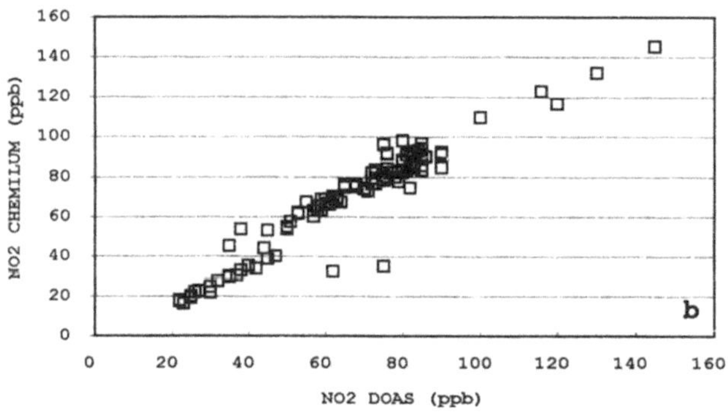

FIGURES 4 - a) Scatter plot of chemiluminescence and DOAS response for the measurement of nitrogen dioxide. b) - Reconstructed scatter plot after correction of calibration parameters of the chemiluminescence analyser.

After a mathematical treatment, which will be presented elsewhere, it was possible to demonstrate that the difference between the chemiluminescence analyser and the DOAS response is a function of the concentration of nitrogen monoxide. This simply means that the chemiluminescence analyser was out of calibration (indeed, it was an unbalanced two-channel instrument). If the chemiluminescence function against the DOAS function is reconstructed taking into account the correct calibration factors for both channels and for both nitrogen monoxide and dioxide (as measured after the campaign) a consistent scatter plot is obtained (See Figure 4b). That means that the two instruments were equivalent.

The most interesting aspect of this experiment lies in the fact that a fairly new technique such as the DOAS, was used to perform a quality assurance on the old chemiluminescence analyser, thus the development of newly developed instruments may also be very important for improving the quality of observation carried out by more conventional analysers. From this standpoint, it is recommended that at least one station in the urban monitoring network takes care of testing for new instruments and methods..

However, deviations between analytical methods (either instrumental or manual) cannot be simply explained in terms of instrumental performances, As an example, it is sufficient to continue the discussion on the results reported in Figures 4. According to the mathematical reconstruction of the chemiluminescence data, the agreement between the DOAS and the conventional analyser is very satisfactory. The residual difference, reported in Figure 5 is modulated and anticorrelates with the concentration of ozone.

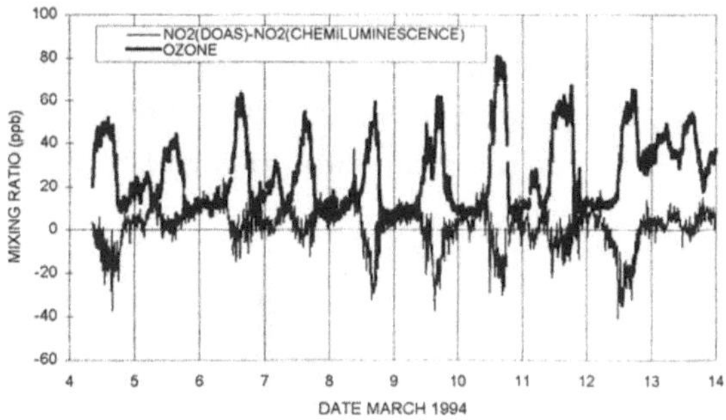

FIGURE 5 - Residual difference of nitrogen dioxide concentrations (DOAS minus chemiluminescence) and ozone concentration.

The functional relationship of the difference of nitrogen dioxide values and the ozone concentration is not simple and so additional studies are needed to fully understand the physical reasons for the observed deviation. Since the ozone reach high values if atmospheric instability is experienced, one possible cause of the deviation has to be attributed to the poor atmospheric homogeneity during unstable periods, i-e-, when the ozone peaks.

Local meteorological effects

Another aspect which is very often overlooked in planning a monitoring campaign or in the interpretation of monitoring data is the evaluation of the boundary layer structure. It is well known that the pollutants distribution in the atmosphere is related to its mixing properties. For example, very poor pollutant mixing ratios are generally associated with a perturbed meteorological situation, while fine weather, due to atmospheric stability, is very often associated with high levels of pollutants.

For measuring atmospheric mixing properties it is possible to exploit the natural radioactivity related to Radon and the radioisotopes coming from its decay and attached to particulate matter. Radon, as it is well known, is emitted at ground level, thus its temporal distribution would be very similar to the distribution of pollutants emitted at ground levels such as those emitted by automobile exhausts. In fact, if the *CO* concentration during time is compared with that of the Radon, a good agreement between the two functions can be found. As an example, Figure 5 shows the pattern of the *CO* and Radon mixing ratio, recorded in Milan during two significant days

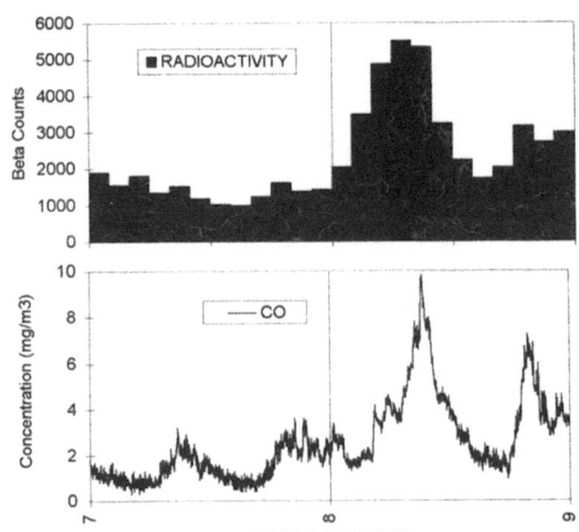

FIGURE 6 - Carbon monoxide and Radon concentrations measured in Milan.

From the figure, it is clear that a good relationship is present, even in the fine structure defined by a strong stability episodes overnight. Clearly, the same consideration is true for other primary pollutants such as poliaromatics hydrocarbons in particulate matter and *NO* as well as for many components relevant for the protection of the urban environment. The direct relationship between mixing properties in the atmosphere and pollution level is limited to primary pollutants. In fact, secondary pollutants are also modulated by the chemical properties of the atmosphere which define the reactions and their rate, converting primary into secondary pollutants. However, the evaluation of mixing properties is still an important measurement because it enables to distinguish the contribution of meteorology and that of chemistry in the formation rates of secondary pollutants.

Although the above considerations are very general, they are particularly pertinent to Mediterranean cities which are characterised by night time stability and day time instability. An example taken in Rome in 1991 is shown in Figure 7.

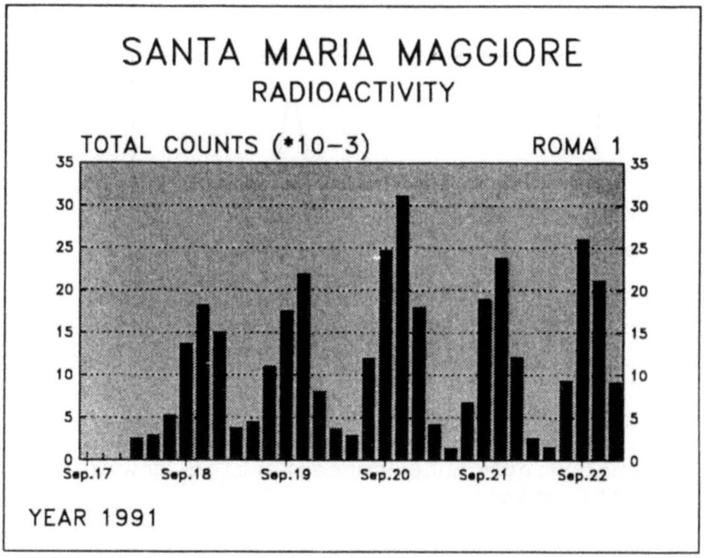

FIGURE 7 - Evolution of the atmospheric mixing properties in Rome as measured through Radon concentration.

Very often, the meteorological situation in Rome is such that night time stability and very strong day time instability is experienced. This means that the pollution in Rome is expected to be quite high in the evening, in the night and in the early morning and drops almost zero during the day. In fact, this is the typical pollution trend in the city of Rome especially during synoptic high pressure conditions. The situation in Milan is slightly different: here there is strong night time stability but moderate stability even during the day. This will prevent pollution dispersion and account sfor most "multiday episodes" where many pollutants accumulate in the course of time.

The possible distinction of meteorology from chemistry will also shed some light into another important environmental issue: the presence of very high ozone concentrations in many urban sites. In fact, ozone is generated by a series of complex photochemical reactions which take place between its precursors (Nitrogen oxides and reactive hydrocarbons). Ozone is formed in the low troposphere as well as in the free troposphere from where it may be transported to ground levels by means of

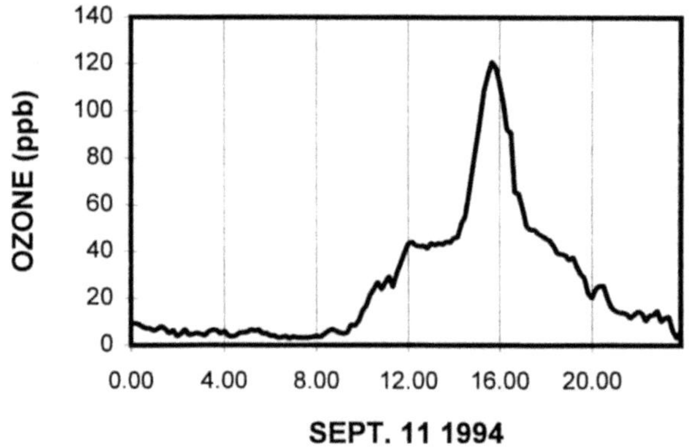

FIGURE 8 - Monitoring ozone concentration downwind of Rome at the Montelibretti station

turbulence. Thus, in a given location close to an urban centre, it is possible to discriminate between the two different sources.

As an example, Figure 8 shows the temporal trend of ozone as measured at the ground station of Montelibretti, about 30 Km N-E of Rome and which is characterised by the same modulation of the boundary layer occurring over the city of Rome. During daytime, as expected, the ozone concentration shows a broad peak which is directly related to the mixing properties characterised by moderate instability. An overlapping strong peak is also observed in these conditions. This is caused by transport of pollutants from the City of Rome as it is possible to demonstrate according to wind speed and direction. During the night, atmospheric stability is such that ozone is almost completely depleted due to the presence of sources of nitrogen monoxide or due to deposition at ground.

Atmospheric chemistry

As said before, atmospheric chemistry plays an important role in the definition of pollution properties of an urban centre since it fully controls the time evolution of secondary

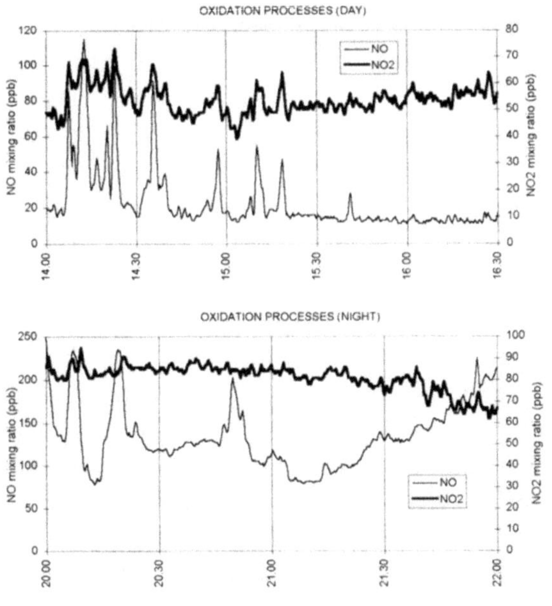

FIGURE 9 - Daytime (a) and nighttime (b) monitoring of nitrogen oxides near a heavy traffic road in Milan.

pollutants such as nitrogen dioxide, ozone and others. In order to demonstrate how pollutant levels recorded in a sampling site may be affected by a meteorological situation and by atmospheric chemistry, it is sufficient to discuss Figure 9 which shows the typical trend of nitrogen oxides in a monitoring station located near a road with heavy traffic in the centre of Milan.

The extent to which radicalic reactions take place in the urban environment can be easily recognised. In fact, the daytime signals related to NO are closely related to those of NO_2. while the data set collected overnight shows that the relation between the temporal gradients of NO and NO_2 are not so evident. Since during these measurements the ozone concentration was very close to zero, radicalic processes are the only explanation for the observed rapid transformation of NO into NO_2. The extent of radicalic reaction may be explained if the "chemical amplification" due to Carbon Monoxide (CO) is also taken into account. In fact, near roads with heavy traffic, the concentration of such species may be sufficiently high to trigger the following reaction chain:

$$RO_2^{\cdot} + NO \rightarrow RO^{\cdot} + NO_2 \qquad [1]$$
$$RO^{\cdot} + O_2 \rightarrow HO_2^{\cdot} + R'CHO \qquad [2]$$
$$HO_2^{\cdot} + NO \rightarrow OH^{\cdot} + NO_2 \qquad [3]$$
$$OH^{\cdot} + CO \rightarrow CO_2 + H^{\cdot} \qquad [4]$$
$$H + O_2 \rightarrow HO_2^{\cdot} \qquad [5]$$

where reaction [5], provides hydroperoxy radicals which oxidise nitrogen monoxide into dioxide and which, in turn, are regenerated by CO through the intermediate hydroxyl radicals. In such a situation, very rapid oxidation takes place with the concurrent formation of many other components. They include aldehydes, generated by reaction [2] and other reaction products of alkoxy radicals $(RO^{\cdot})$ such as ketons or alkyl nitrites. In addition, side chain reactions, such as those leading to the formation of Peroxialkylnitrates or alkyl nitrates, may also generate species which may be very harmful for public health. Clearly this is just a starting model since many additional parameters should be taken into account if a full evaluation of processes leading to the formation of nitrogen dioxide need to be carried out.

Radicalic oxidation processes are very important for urban atmospheric chemistry. In fact they are responsible for the formation of several species which may yet again be extremely harmful to public health. Many processes are well known and have also been successfully used for the modelling of the time and space distribution of atmospheric pollutants in urban sites. However, the extent to which these reactions are known, is still very poor. Several reactions, especially those involving aromatic hydrocarbons are not yet clear, as it is not yet clear what exactly is the true driving force of radicalic processes.

It is widely accepted that atmospheric chemistry has a real pivot in the *OH* radicals which in fact reacts very rapidly with hydrocarbons. One of the major questions still open is how these radicals are generated. Apparently, when *NOx* are very high, such as for instance in urban sites, it has been postulated that most of *OH* are from the photolysis of formaldehyde since the ozone in the mixing layer is completed depleted by reactive substances. However, it seems that a possible route for the formation of radicals in the early morning, which are responsible for triggering intense photochemical episodes, come from the photodissociation of nitrous acid.

Such species, which has not been thoroughly studied in the urban atmosphere, generates nitrous acid when nitrogen dioxide reacts with water through the heterogeneous reaction:

$$NO_2 + H_2O \rightarrow HNO_3 + HNO_2 \qquad [6]$$

yielding nitrous and nitric acid. This reaction has been partly evaluated in laboratory studies and it was shown to be first order in NO_2. It was also shown that its extent is dependent upon the structure of the mixing layer so that, when it is very shallow, high amounts of HNO_2 may be formed. Several studies should be addressed to this reaction since there are premises according to which it may be responsible for most pollution

episodes experienced in coastal cities.
This opens an additional field of investigation because it bring to a completion the study of nitrogen cycle in the atmosphere. Such a cycle, outlined in Figure 10, starts from the emission of nitrogen oxides (most of the mass output as monoxide) and, after a series of reactions which involve ozone and peroxy radicals, nitrogen dioxide is formed. This is, in turn, furtherly oxidised to nitric acid and nitrates. As a result, the amount of nitrogen dioxide present in the

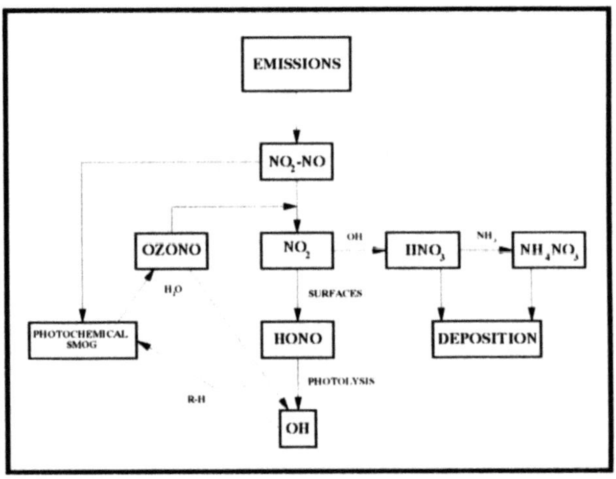

FIGURE 10 - Simplified nitrogen cycle in an urban environment.

atmosphere is a complex function of the extent of radicalic reactions which cause its formation and its removal. Side reactions may give rise to nitrous acid which is, upon photodissociation, able to generate additional radicals. The rapid formation of particulate nitrate, from the neutralisation of nitric acid by ammonia, is demonstrated by several measurements carried out in Milan where sometimes fine ammonium nitrate accounts for 90% of the total particulate nitrates and for about 30% for the total particulate matter load. Very often the concentration of ammonium nitrate may reach levels of about 100 $\mu g/m^3$.
Even though the atmospheric cycle of nitrogen containing compounds has been largely elucidated, many aspects still need to be clarified. Among these, the formation of nitrous acid merit very special attention. Other reactions, such as those

relevant to the formation of nitroderivatives of organic compounds, including *PAN* (Peroxy Acetyl Nitrate), should be furtherly investigated due to the potential harmful effects of such compounds on the human health.

Conclusions

It appears that the present status of the art of urban air pollution research still needs further studies in many basic topics related to the atmospheric environment. As it has been shown, the scientific community needs data which must be precise and accurate enough therefore better analytical methods are required. However, as it was stressed before, a set of data should be taken into consideration only if it is consistent with the objective of the study. In other words, a monitoring activity or a monitoring campaign should be planned in order to obtain data in order to give a positive answer to the questions behind the working hypothesis of the experiment.

The role of meteorology in the definition of urban atmospheric pollution is extremely important and better methods are required to characterise the urban mixed layer and its properties, especially in conjunction with chemistry. Some new techniques may be able to predict the possible evolution of the boundary layer and the mixing properties of the atmosphere, thus predicting the mixing properties of the low troposphere.

As far as the atmospheric chemistry is concerned, there are many additional research topics which merit attention. There are still uncertainties concerning the chemistry of hydrocarbons, especially the aromatics which make up a significant fraction of the total amount emitted in an urban environment. The formation of oxidised organic compounds and their distribution is still a matter of speculation, furthermore the occurrence of secondary pollutants such as nitrogen dioxide and ozone have still left open important questions. Finally, the role of meteorology at microscale or mesocale levels needs further investigation to parametrize chemical models which may be the resolving key to plan effective reduction of primary sources of pollutants.

References

Allegrini I., Cortiello M., Febo A., Perrino C., (1990), "Generation of Standard Atmospheres of Nitrous Acid". In Physico Chemical Behaviour of Atmospheric Pollutants, pp. 140-144. CEC Air Pollution Research Report 23, Kluwer, Dordrecht.

Allegrini I., De Santis F., Di Palo V., Febo A., Perrino C., Possanzini M., Liberti A., (1987) "Annular denuder method for sampling reactive gases and aerosols in the atmosphere." Sci. Total Environ. 67, 1-16.

Allegrini I., Febo A., Pasini A., Schiarini S., (1994), "Monitoring of the Nocturnal mixed Layer by Means of Particulate Radon Progeny Measurement". J. Geophys. Res. 99(D9), 18765-18777.

Allegrini I., Febo A., Perrino C., Masia P. (1994)."Measurement of Atmospheric Nitric Acid in Gas phase and Nitrate in Particulate Matter By means of Annular Denuders". Intern. J. Environ. Anal. Chem. 54, 183 .

Allen A.G., Harrison R.M., Erisman J.W., (1989), "Field Measurement of the Dissociation of Ammonium Nitrate and Ammonium Chloride Aerosols" Atmos. Environ. 23,1591-1599.

Ciccioli P., Possanzini M., Di Palo V., Cecinato A., (1992), "Dynamic Calibnration of Proxyacetylnitrate (PAN) analyser by Annular Denuder and Ion-Chromatographic Techniques", Atmos. Environ. 26A, 1515-1518.

Cox R.A., (1974) "The Photolysis of Gaseous Nitrous Acid". J. Photochem. 3,175-188.

Febo A., C. Perrino and M. Cortiello, (1993) "A Denuder Technique for the Measurement of Nitrous Acid in Urban Atmospheres", Atmos Environ. 27A, 1721-1728.

Febo A. and C. Perrino, (1991) "Prediction and Experimental Evidence for High Air Concentration of Nitrous Acid in Indoor Environment". Atmos. Environ. 25A, 1055-1061.

Finlayson-Pitts B. J. and Pitts Jr. (1986) Atmospheric Chemistry. Wiley, New York.

Goldstein Y.F. And Andrews L.R., (1987) "Peak exposures to nitrogen dioxide and study design to detect their acute health effects" Envir. Int. 13, 285-291.

Hastie D.R., Weissenmayer M., Burrows J.P., Harris G.W., (1991) "Calibrated Chemical Amplifier for Atmospheric ROx Measurements" Anal. Chem. 63, 2048-2057.

Hewitt C.N., (1991), "Spatial Variability in Nitrogen Dioxide Concentration in an Urban Area" Atmos. Environ., 25B, 429-434.

Lalas, D.P., Asimakopoulos, D.N:, Deligiorgi, D.G., Helimis, C.G., (1983), "Sea Breeze Circulation and Photochemical Pollution in Athens, Greece" Atmos. Environ. 17, 1621-1632.

Lightfoot P.D., Cox R.A., Crowley J.N., Destria M., Hayman G.D., Jenkin M.E., Moortgat G.K., Zabel F, (1992) "Organic Peroxy Radicals: Kinetics, Spectroscopy and Tropospheric Chemistry", Atmos. Environ., 26A(10), 1805-1964.

Millan M. And Artinano B. (1992) "Meso meteorological Cycles of Air Pollution in the Iberian Peninsula (MECAPIP)" Final Report of the European Communities Commission EV4V-0097-E(A).

Perrino C., De Santis F., And Febo A., (1990) "Criteria for the Choice of a Denuder Sampling Technique Devoted to the Measurement of Atmospheric Nitric and Nitrous Acids", Atmos. Environ. 24A, 617-626.

Wayne R.P., Barnes I., Biggs P., Burrows J.P., Canosa-Mas C.E., Hjorth J., Le Bras G., Moortgat G.K., Perner D., Poulet G., Restelli G., Sidebottom H., (1991) "The Nitrate Radical: Physics, Chemistry and the Atmosphere". Atmos. Environ. 25A, 1-206.

Atmospheric Chemistry Relevant to Urban Pollution

K.H. Becker and I. Barnes
Bergische Universität - Gesamthochschule Wuppertal
Physikalische Chemie / FB 9
42097 Wuppertal, Germany

Urban pollution is the result of a superposition from large scale, regional and local emissions and the related transformation products due to natural as well as man-made sources, meteorology and interaction with photochemically active light. Persistent inversions, land-sea-breezes or drainage of air masses caused by specific conditions in the surroundings of an urban area will have a significant influence on the extend and the nature of pollution. Vertical mixing resulting in diurnal exchanges of air between the free troposphere and the planetary boundary layer always influences the chemistry and dispersion of polluted air near the surface.

The chemistry on larger scales is mainly controlled by the oxidation processes of CH_4 and CO initiated by OH radicals in the presence of low NO_x concentrations. The contribution of NO_x determines whether these processes are sources or sinks of O_3 in the background atmosphere. On the other hand, the O_3 photolysis in the presence of H_2O is the major OH source.

Taking the urban area as the only man-made emission source the surrounding vegetation, eventually the nearby sea, has also a considerable influence through their biogenic emissions on the chemistry of the local pollution. Atlanta/Georgia can be taken as an example where it was recently possible to show that the largest contribution of volatile organic compounds (VOC) originates from the vegetation. Natural sources of NO_x seem to be insignificant for urban areas but globally they cannot be neglected.

The urban atmospheric system itself can be characterised by chemical processes firstly in the gaseous phase initiated by emissions from sources such as traffic, industry, solvent use and

NATO ASI Series, Partnership Sub-Series, 2. Environment - Vol. 8
Urban Air Pollution
Edited by Ivo Allegrini and Franco De Santis

household activities. A variety of VOC's are emitted: alkanes, alkenes, aromatic and oxygenated organics and also the halogen substituted analogues of these various classes of organic compounds. However, due to their relatively low source strengths compared to the other compounds, the organohalogens have only a minor impact on urban air chemistry. There is also the question as to whether or not reactive chlorine or bromine compounds are released, however, this is a speciality area which will not be discussed further. SO_2 can be considered as a classic emission in urban areas since early industrialisation, in highly developed countries control strategies have resulted a downward trend of these emissions. Of crucial importance besides the VOC and SO_2 emissions are the NO_x ($NO + NO_2$) emissions. To characterise the homogeneous gas-phase chemistry the oxidation processes of alkanes, alkenes, aromatics and aldehydes have to be considered in the presence of different NO_x concentrations. The $[VOC]/[NO_x]$ ratios in space and time very much determine the oxidation pathways leading either to formation of hydroperoxides at low NO_x, to ozone at medium NO_x and nitric acid at high NO_x concentrations for a constant [VOC]. This chemistry is strongly linked to the system of photooxidants with the built-up of high ozone concentrations in urban plumes during the day in summer (photosmog). A certain fraction of SO_2 is also converted by gas-phase reactions to SO_3 and consecutively to H_2SO_4.

Secondly, another side of urban air chemistry is related to heterogeneous or liquid phase chemistry. It is reasonably well known that 60 to 70% of SO_2 is transformed to H_2SO_4 by aqueous phase processes in rain or cloud droplets. However, the oxidation rate of SO_2 is very much dependent on the presence of H_2O_2 and O_3 in the aqueous phase which presents a coupling to the gaseous phase. Another important source of acids is the reaction of NO_2 with O_3 to form NO_3 radicals and eventually N_2O_5 which reacts in water films forming nitric acid. Relatively new is the observation that NO_2 is transformed to nitrous acid in urban atmospheres. The formation seems to be controlled by heterogeneous processes at surfaces within a narrow range of acidity of the surface. The formation of most carboxylic acids is probably also controlled by heterogeneous reactions. The production of acids constitutes a serious problem besides the photooxidant formation in urban and adjacent areas.

The direct formation of particles is a third problem, however, it will not be considered here.

The major concern related to "photosmog" as well as to "acid rain" leads to the question how the oxidation processes in an urban environment are initiated, how the radical chains involved propagate and how they are terminated. There are limiting conditions given by the composition and magnitude of the different emissions. All these problems are related to:

- photooxidant chemistry in general and during episodes of photosmog
- sources and sinks of H_2O_2/hydroperoxide
- formation of O_3, PAN and other oxidant species
- sources and sinks of the acids HNO_2, HNO_3 and H_2SO_4
- the radical budgets for OH, HO_2/RO_2 and NO_3 and the ozonolysis of alkenes

Air quality in urban areas, however, can only be preserved or re-established by controlling the anthropogenic emissions. This requires a reasonably precise knowledge of the impact which the major emissions will have on the urban air chemistry, e.g. the oxidant-precursor relationship has to be known under different meteorological conditions. This should also include knowledge about the limitations for maintaining a sufficiently high oxidation capacity which is necessary to keep the atmosphere clean. In the following sections brief overviews are given on the very recent state of knowledge on the following currently relevant issues concerned with urban air pollution: a) HONO in the atmosphere, b) hydroperoxide formation in the ozonolysis of alkenes and c) the photooxidation of aromatics. All these problems are currently being investigated in the Wuppertal laboratory.

a) Formation of HONO in the Atmosphere

Nitrous acid (HONO) is of particular importance in atmospheric chemistry, because HONO photolysis significantly enhances photooxidant production early in the morning due to the photolysis reaction:

$$HONO + h\nu \longrightarrow OH + NO \qquad (1).$$

The mechanisms by which HONO is formed in the atmosphere are not well established at present. Two reactions have been suggested:

$$NO + NO_2 + H_2O \longrightarrow 2\ HONO \qquad (2)$$

$$2\ NO_2 + H_2O \longrightarrow HONO + HNO_3 \qquad (3).$$

However, the above gas-phase reactions are much too slow to explain the observed nighttime HONO production rates. Evidence of heterogeneous HONO production by the surface reaction of NO_x with water vapour has been found in smog chamber studies. In addition, heterogeneous production of HONO has also been proposed in order to explain the high HONO concentrations which have been measured in the atmosphere.

From laboratory experiments it has been concluded that HONO formation does not require the presence of NO as suggested in reaction (3), whereas results from field measurements tend to favour reaction (2) as source for HONO. However, very recently Harrison and Kitto (1994) concluded from their field experiments that HONO is formed at the surface by heterogeneous reactions of NO_2. In addition, indirect evidence that HONO formation does not require NO was also obtained by field measurements where the presence of high ozone concentrations indicates low NO concentrations or low NO concentrations have been directly measured. The current state of knowledge on HONO formation in the atmosphere has been reviewed very recently by Calvert et al. (1994).

In a recent study in this laboratory the heterogeneous conversion of NO_2 was studied in different reactors under various experimental conditions by tunable diode laser absorption spectroscopy (Becker et al. 1994; Wiesen et al. 1994). Figure 1 shows a typical concentration-time profile for NO_2, HONO and N_2O which was measured in a 64 *l*-quartz glass reactor which was equipped with a White mirror system. In all experiments the formation of HONO was observed reaching a maximum concentration after ~ 40 min followed by a slow decay whereas N_2O was formed continuously.

The concentration-time profiles of HONO and N_2O shown in figure 1 indicate a consecutive reaction,

$$NO_2 \longrightarrow HONO \longrightarrow N_2O \quad (4).$$

The HONO time behaviour appears to be the difference of two exponentials,

$$[HONO]_t = \frac{k_a[NO_2]_{t=0}}{k_b - k_a} \times [\exp(-k_a t) - \exp(-k_b t)] \quad (I),$$

where t represents reaction time and k_a and k_b rate constants for HONO formation and destruction, respectively. This equation was fitted to the HONO concentrations for each set of conditions using non-linear least squares. In agreement with previous findings (Calvert et al. 1994 and references therein) the HONO formation rate was directly proportional to $[NO_2]$, whereas a square dependence of $[NO_2]$ would be expected if one assumes the termolecular gas-phase reaction (3). This result indicates that the HONO formation occurs heterogeneously with adsorbed water on the reactor surface.

In order to show whether the acidity of the surface of the quartz-glass reactor was responsible for the chemical transformation of NO_2, acid aerosols were generated in a 11-*l* Pyrex glass reactor. Using sulphuric acid and NH_4Cl aerosols a behaviour similar to that in the quartz-glass reactor was observed. Starting with an initial NO_2 concentration of 13 ppm a final N_2O concentration of ~1.3 ppm was obtained after a reaction time of one day. This behaviour could be quantitatively explained by the formal disproportionation reaction,

$$8\ NO_2 + 3\ H_2O \longrightarrow N_2O + 6\ HNO_3 \quad (5),$$

from which a maximum N_2O concentration of 1.6 ppm would be expected.

Since it was found to be very difficult to control the surface activity of the different reactors or the aerosols the experimental set-up was modified. Instead of the White cell, a long-path absorption Herriott cell was used for monitoring the different species. In addition, the gas mixtures in the 11-*l* Pyrex glass reactor were pumped continuously through a bubbler

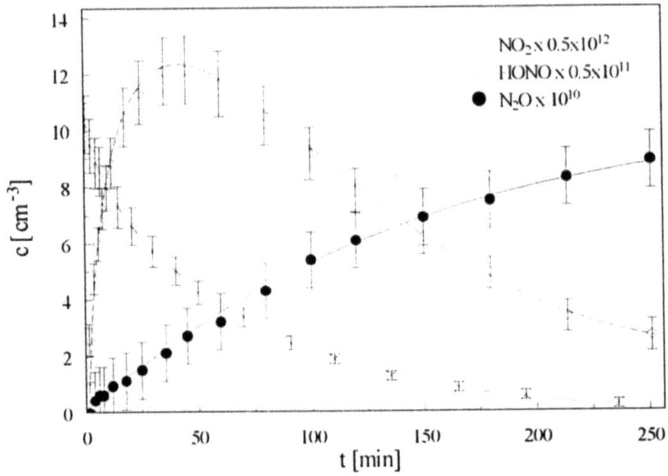

Fig.1. NO_2 decay, N_2O and HONO formation in the 64 *l*-White cell in the dark; $[NO_2]_0$=5.1Y10^{12} cm^{-3}, p_{tot}=6.5 Torr synthetic air.

containing different solutions with well defined parameters such as concentration, temperature, volume, flow velocity and bubble diameter. It was observed that N_2O formation took place in acid solutions such as H_2SO_4, HCl and H_3PO_4 whereas for water or NaCl solutions no N_2O formation was observed.

In another set of experiments the dependence of the maximum HONO concentration and the N_2O formation rate on the concentration of sulphuric acid was studied. HONO was observed in the gas phase only for H_2SO_4 concentrations <55%. The N_2O formation rate was found to have a maximum at ~55% H_2SO_4.

From the experimental results obtained in the present study the following reaction mechanism is proposed:

$$8\,NO_2 + 4\,H_2O \rightleftharpoons 4\,HONO + 4\,HNO_3 \qquad (6)$$

$$2\,HONO + 2\,H_2SO_4 \rightleftharpoons 2\,H_2O + 2\,NO^+HSO_4^- \qquad (7)$$

$$2\,NO^+HSO_4^- + 2\,HONO + 2\,H_2O \longrightarrow (HNO)_2 + 2\,HNO_3 + 2\,H_2SO_4 \qquad (8)$$

$$(HNO)_2 \longrightarrow N_2O + H_2O \qquad (9)$$

$$8\,NO_2 + 3\,H_2O \longrightarrow N_2O + 6\,HNO_3 \qquad (10)$$

In a first step, reaction (6), NO_2 is converted on the surface into HONO and HNO_3. In the reaction system the NO_2 decay constant was found to be directly proportional to the NO_2 concentration and the surface area but was found to be independent of the volume of the sulphuric acid solution. Accordingly, the observed NO_2 decay is mainly caused by an uptake of NO_2 in the sulphuric acid solution.

It should be pointed out that HNO_3 was not detected in the gas phase in any of the experiments. This is probably due to the high solubility of HNO_3 in sulphuric acid. However, HONO was observed in the gas phase with a maximum concentration which was found to be proportional to the inverse volume of the sulphuric acid solution.

In a second step, reaction (7), dissolved HONO reacts with sulphuric acid forming $NO^+HSO_4^-$. It is proposed that the nitrosyl cation (NO^+) which is a strong oxidiser oxidises HONO to form HNO_3. In this oxidation NO^+ is reduced to $(HNO)_2$ which can easily decompose into N_2O and water, reaction (9).

For H_2SO_4 concentrations >55% HONO is converted almost completely into $NO^+HSO_4^-$. Accordingly, no HONO is observed in the gas phase and reactions (8) and (9) are suppressed and the N_2O yield is small. For low H_2SO_4 concentrations the N_2O yield is also small, this is probably caused by low formation yields of $NO^+HSO_4^-$, reaction (7), since it is known that $NO^+HSO_4^-$ is not stable in an excess of water.

In summary, the experimental results from the present study show that NO_2 is heterogeneously converted into HONO as well as N_2O. This conversion seems to take place only on acid surfaces. The kinetic behaviour of both HONO and N_2O indicate that N_2O is formed via HONO as an intermediate.

The findings from the present study imply that HONO formation in the reaction system occurs almost exclusively by the reaction of NO_2 with water and does not require NO. This is in agreement with previous laboratory studies and the most recent findings from field measurements from which the involvement of NO in the surface formation of HONO was excluded. A reaction mechanism explaining the experimental results is proposed. The nitrosyl ion probably plays the major role in converting HONO into N_2O and HNO_3.

b) Formation of alkyl and hydroxyalky hydroperoxides in the ozonolysis of alkenes

Hydrogen peroxide and organic peroxides in the atmosphere are important both as oxidants themselves and as precursors of such radicals as OH, HO_2 and RO_2. Methyl, hydroxymethyl and 1-hydroxyethyl hydroperoxides have been identified in the atmosphere in quantities comparable with those of H_2O_2 (in the low ppbv range). It is not yet known to what extent the two types of organic hydroperoxides as found in the field measurements are formed via OH initiated oxidation of VOC or the ozonolysis of alkenes. Further it is also not known to what extent the processes take place in the gas-phase or in water droplets. OH radical reactions with VOC lead to the formation of RO_2. Under low NO_x conditions the self and cross reactions of RO_2 result in the formation of H_2O_2 and alkyl hydroperoxides. The ozonolysis of alkenes has been shown to produce alkyl hydroperoxides and 1-hydroxyalkyl hydroperoxides (Gäb et al., 1985). It has also been shown that the ozonolysis of alkenes directly forms H_2O_2 independent of HO_2 production (Becker et al. 1990).

Table 1 shows the yields of hydroperoxides found in recent laboratory studies on the ozonolysis of five alkenes in both the gas- and aqueous-phase. The ozonolyses of the different alkenes lead to significantly different hydroperoxide distributions in the gas- and aqueous-phase experiments (Gäb et al. 1995). The total yields of peroxides in the aqueous phase are found to be much higher than in the gas-phase. Ozonolysis of alkenes in water yields hydroxyalkyl hydroperoxides; formation of alkyl hydroperoxides such as methyl hydroperoxide (MHP) was not observed. Only a trace of hydroxymethyl hydroperoxide (HMHP) was observed in the ozonolysis of β-pinene in water. An increase in the yield of H_2O_2 was observed for isoprene, α-pinene and β-pinene on neutralisation of the aqueous samples. This points to the formation of 1-hydroxyalkyl hydroperoxides in these reaction systems which could not be detected directly by the HPLC analytical method used.

Alkene	Phase	H_2O_2	HMHP	1-HEHP	2-HEHP	MHP	Other* 1-HAHP´s	Total
Ethene	gas-phase	0.4	0.2	-	< 0.1	-	n.a.	0.6
	aqueous-phase	-	27.9	-	-	-	-	27.9
Propene	gas-phase	0.3	0.2	-	-	4.1	n.a.	4.6
	aqueous-phase	-	2.7	11.1	-	-	-	13.8
Isoprene	gas-phase	0.7	0.5	-	-	5.0	n.a.	6.2
	aqueous-phase	< 0.1	11.1	-	-	-	13.1	24.2
α-Pinene	gas-phase	0.3	0.1	-	-	0.8	n.a.	1.2
	aqueous-phase	< 0.1	-	-	-	-	11.3	11.3
β-Pinene	gas-phase	0.6	0.3	-	-	-	n.a.	0.9
	aqueous-phase	< 0.1	0.2	-	-	-	16.0	16.2

Table 1: Yields of hydroperoxides in the ozonolysis of alkenes. Legend: As per cent of the limiting reagent, alkene or ozone; (-) not detected; (n.a.) not analysed; (*) determined from the H_2O_2 yield on neutralization

Methyl hydroperoxide is the major hydroperoxide formed in the gas-phase ozonolysis of methyl-substituted compounds, propene, isoprene and α-pinene. The trace of 2-hydroxyethyl hydroperoxide (2-HEHP), observed in the ozonolysis of ethene, is attributed to the reaction of ethene with OH radicals (Gäb et al., 1990; Atkinson et al. 1992).

A closer investigation of the ozonolysis of ethene in the gas-phase in the presence of water vapour has shown (Brockmann et al. 1995) that the HMHP yields increase to about 3% with increasing partial pressure of water. At water partial pressures of more than 4 Torr, however, the HMHP concentration decreases significantly. The reason for this is still unclear. In contrast to another study (Horie et al., 1994) no methyl hydroperoxide was observed.

The formation of all peroxidic products observed in the ozonolysis study could be explained by the Criegee mechanism.Figure 2 shows a simplified reaction mechanism for the ozonolysis of alkenes using propene as an example. The formation of the hydroxyalkyl hydroperoxides is presently interpreted as the result of the reaction of water with the different Criegee intermediates. The fact, that only traces of HMHP were observed in the aqueous phase ozonolysis of β-pinene suggest that the cleavage of the postulated primary ozonide to form the more highly substituted Criegee intermediates is favoured.

The MHP formation observed in the gas-phase ozonolysis is attributed to the reaction of methyl radicals, arising from thermal decomposition of Criegee intermediates with methyl substituents (Atkinson, 1994).

In-situ measurements of the 1-hydroxyalkyl hydroperoxides and a systematic study of the effects of water vapour will be necessary to clarify the observation of HMHP in the gas phase.

c) Photooxidation of Aromatic Hydrocarbons

For detailed accounts on the present status of the photooxidation of aromatics the publications of Atkinson (1990, 91), Becker (1994), Bierbach et al. (1994), Grosjean (1991), LACTOZ Reports 1988-1993 and Wayne et al. (1990) should be consulted. In the troposphere the photooxidation of benzene and the alkylated benzenes, toluene and the xylene isomers is mainly initiated by reaction with OH radicals. The primary OH-kinetics of aromatics are quite well established and the elementary steps can be modelled. A simplified mechanism for the OH initiated oxidation of toluene and p-xylene showing the reaction pathways currently thought important is shown in figure 3. The major pathway involves initial OH radical addition to the aromatic ring to form a hydroxycyclohexadienyl type radical. The minor pathway (~10%) involves H atom abstraction from the substituent alkyl groups to form benzyl type radicals. The subsequent fate of these radicals is analogous to that of the primary alkyl radicals formed in the photooxidation of alkanes.

The reactions of the hydroxycyclohexadienyl and alkyl substituted hydroxycyclohexadienyl radicals formed from the initial addition of the OH-radical to the ring under tropospheric conditions are not well understood. Kinetic studies have been performed on the reactions of OH-aromatic adducts with O_2, NO and NO_2. The kinetic data show that the hydroxycyclohexadienyl radicals react rapidly with NO_2, slowly with O_2 and the reaction with NO

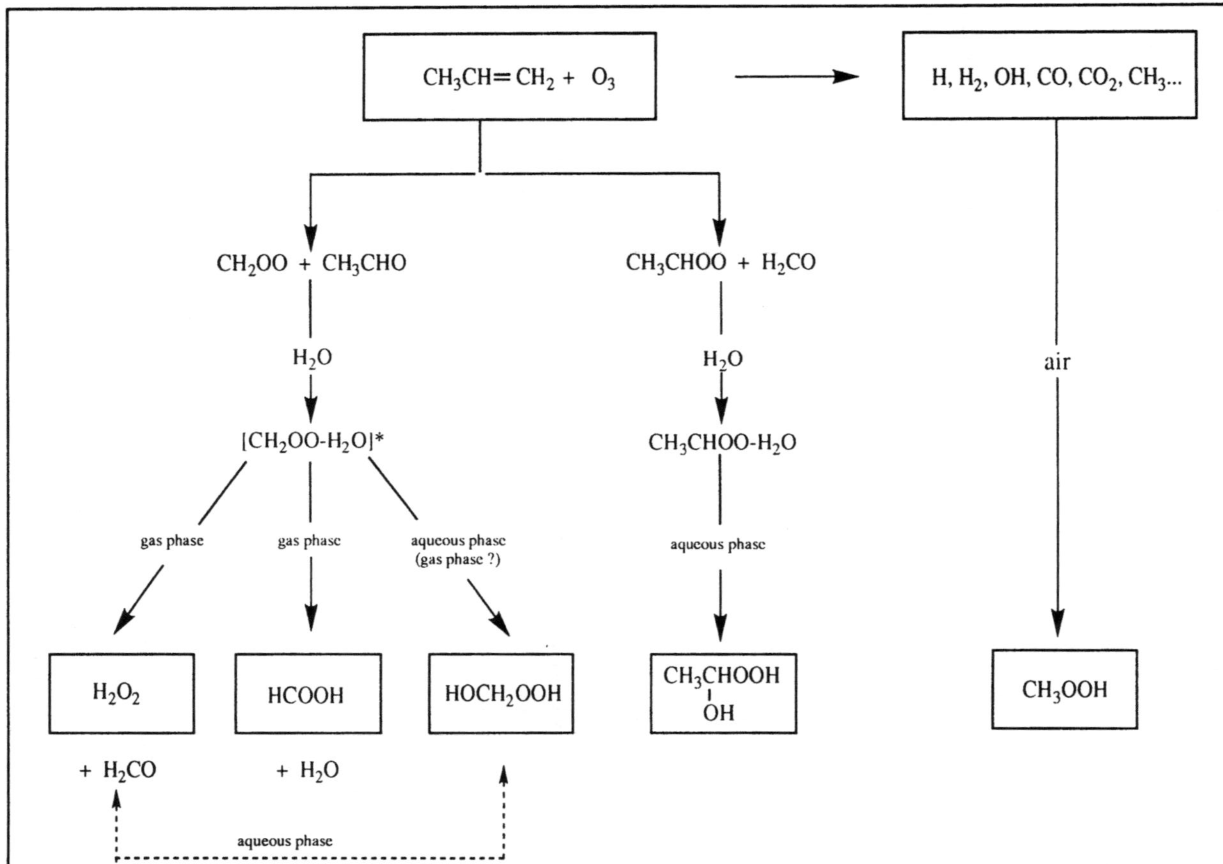

Fig. 2: Simplified reaction scheme for the ozonolysis of propene

is negligible. Recent product studies on the OH radical initiated oxidation of benzene and toluene are consistent with these kinetic data. In the presence of NO_2 concentrations of about 3 ppm the hydroxycyclohexadienyl radicals will react predominately with NO_2 and not with O_2. However, the NO_x concentration in the atmosphere is generally low and the reaction with O_2 is fast enough to suppress the reaction with NO_2 under almost all tropospheric conditions (with the possible exception of concentrated car exhaust and flue gases). This finding has consequences for the interpretation of smog chamber data on aromatic oxidation mechanisms in which relatively high NO_x concentrations are used and implies that the chemical mechanisms currently in use need reassessment.

It is known that the OH-radical addition pathway not only leads to the formation of ring-retaining products such as phenols and nitroaromatics but also to the formation of ring-cleavage products which include a- and g-dicarbonyls. Table 2 shows an example of the types of products which have been identified in the OH initiated oxidation of toluene and p-xylene under NO_x free conditions. It has been found that the presence of NO_x in the reaction system does not lead to significant changes in the product distribution. For both toluene and p-xylene the identified products account for approximately 50 % C. Determination of the aerosol concentration in aromatic reaction systems show that 10-20% of the reacted carbon is retained in particles in, as yet, unidentified form.

The reaction steps of the hydroxycyclohexadienyl and alkyl substituted hydroxycyclohexadienyl radicals leading to ring-opening products are still unclear. The laboratory studies support, however, that the ring-opening process does not involve NO_x. No peroxy type radicals containing the aromatic ring, apart from those formed after H-atom abstraction, have been observed, or can be inferred from the formation of PAN-type compounds.

Muconaldehydes (di-unsaturated 1,6-dicarbonyls) have been postulated as primary ring-opening products of the aromatic oxidation and there is indirect experimental evidence for their formation Becker 1994; Bierbach et al. 1994). Investigations on the atmospheric chemistry of these compounds show that they form

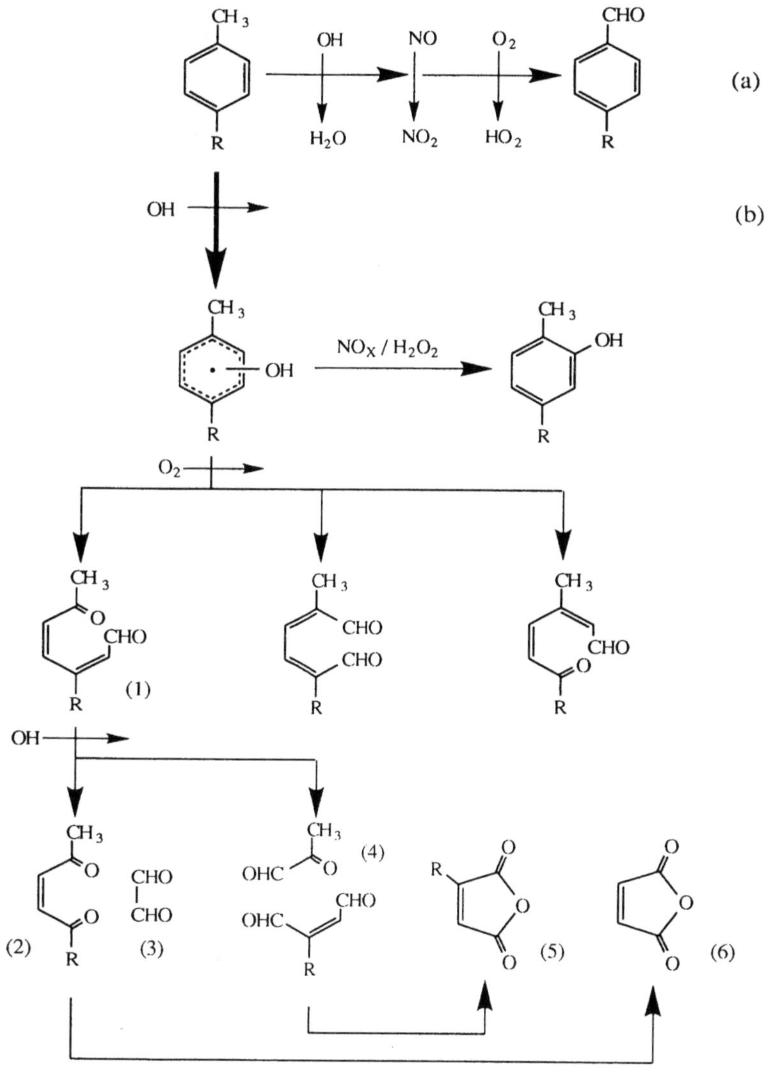

(a) Abstraction Channel; (b) OH-Addition-Channel

(1) Muconaldehyde: 6-Oxo-2,4-heptadienal (R=H), 3-Methyl-6-oxo-2,4-heptadienal (R=CH_3)

(2) 4-Oxo-2-pentenal (R=H), 3-Hexene-2,5-dione (R=CH_3)

(3) Glyoxal ; (4) Methylglyoxal

(5) Methyl maleic anhydride (R=CH_3); (6) Maleic anhydride (R=H)

Fig. 3: Mechanism of Atmospheric Toluene (R=H) and p-Xylene (R=CH_3) Oxidation

Table 2: Product distribution in %C for the reaction of toluene and p-xylene with OH-radicals in 1000 mbar air at 296 ± 2 K in the absence of NO_x. A "+" denotes positive identification without quantification.

Products	**Toluene**	**p-Xylene**
CO[1]	2.7 ± 1.4	8.0 ± 1.4
CO_2[1]	3.5 ± 1.8	4.5 ± 0.4
Formaldehyde[1]	2.3 ± 0.6	1.2 ± 0.2
Formic acid[1]	12.9 ± 3.2	5.1 ± 1.0
Methanol[1]	1.4 ± 0.4	+
Methylhydroperoxide[1]	1.2 ± 0.3	+
Ketene[1]	0.5 ± 0.1	0.3 ± 0.1
Ethyne[1]	0.7 ± 0.2	–
Acetic acid[1,2,4]	5.2 ± 1.3	3.8 ± 0.3
Glyoxal[1]	3.7 ± 0.9	4.0 ± 1.7
Methylglyoxal[1]	4.4 ± 1.1	2.2 ± 0.5
Maleic anhydride[1,3]	4.2 ± 1.1	–
Methylmaleic anhydride[1]	–	+
4-Oxo-2-pentenal[1,3]	≈ 2[A]	+
3-Hexen-2,5-dione[1,4]	–	8.3 ± 1.9
Benzaldehyde[1,2,3]	7.1 ± 1.8	–
p-Tolylaldehyde[1,4]	–	6.4 ± 1.5
2-Hydroxytolylaldehyde[4]	–	+
ortho-Cresol[2,3]	+	–
meta-Cresol[2,3]	+	–
para-Cresol[2,3]	+	–
Phenol[2,3]	+	–
2,5-Dimethylphenol[2,4]	–	8.0
Benzylalcohol[2,3]	+	–
4-Methylbenzylalcohol[4]	–	+
Benzoic acid[2,3]	+	+
4-Methylbenzoic acid[4]	–	+
Methylmaleic anhydride[3]	+	–
Methylhexadiendial isomers[3]	+	–
Hydroxymethylhexadiendial isomers[3]	+	–
Butendial[5]	+	–
2-Methylbutendial[5]	–	+
Total amount	**51.8**	**51.8**

A: Estimated value; 1: Identified by FT-IR in a 1.08 m^3 reaction chamber; 2: Identified by GC-FID after solvent trapping of gas from a 1.08 m^3 reaction chamber; 3: Identified by MS in a flow reactor; 4: Identified by GC-MS from a stirred reactor; 5: Investigations have shown that butendial (2-methylbutendial) is the precursor for maleic anhydride (methylmaleic anhydride) formation predominantly 1,2-dicarbonyls and mono-unsaturated 1,4-dicarbonyls which are known important secondary ring-opening products of the oxidation of aromatics. The oxidation of mono-unsaturated 1,4-dicarbonyls leads to the formation of 1,2-dicarbonyls, maleic anhydride and lactones observed in aromatic oxidation systems. Recent product studies have shown that hydrated vicinal polyketones are also important products of the oxidation of mono-unsaturated 1,4-dicarbonyls. From these studies it has been inferred that polyketones are also important products of the photooxidation of aromatics. Estimates show that they can account for 25-30 % of the missing carbon in aromatic oxidation systems which would increase the carbon balances in aromatic oxidation systems to approximately 80 % C. Maleic anhydride appears to be unique to the oxidation of aromatic hydrocarbons and may possibly serve as a "marker" of the extent of the transformation of aromatics in tropospheric air parcels.

The absolute yields of the ring-retaining products phenol and creosol isomers under atmospheric conditions is controversial with different methods giving conflicting yields; determination of the absolute yields of these compounds is important because of the associated "prompt" formation of HO_2 radicals and also their influence on the final product distribution. Very recent work, however, supports that the yield of phenolic-type compounds is negligible under atmospheric conditions and that they are not direct products of the reaction of the aromatic OH-adduct with oxygen.

Methods to study the cycling of radicals (OH, RO_2, NO_x) in the primary oxidation steps in aromatic oxidation systems are currently being developed. Direct studies of the recycling of OH radicals in the oxidation of several of the aromatics in the present of NO infer that relatively large yields of HO_2 are produced. It is presently not clear whether or not short-lived hydrocyclohexadienylperoxy are involved in the cycling of OH radicals.

Smog chambers have been used to investigate ozone formation in aromatic/NO_x/air oxidation systems. In irradiations of individual hydrocarbons in the presence of NO_x, significant differences have been found between the aromatic hydrocarbons and other classes of hydrocarbons with respect to the formation of ozone and also some aspects of the oxidation mechanisms related to ozone production. For example, a high potential with respect to the ozone formation has been found for toluene in hydrocarbon mixtures. The experimental findings support the view that the large contribution of aromatics not only to the formation of O_3 and but also to other photo-oxidants is due the high photolytic activity of secondary ring cleavage products such as carbonyl or dicarbonyl compounds which results in increased levels of radicals for the conversion of NO to NO_2.

The detailed mechanisms by which the aromatics contribute to photo-oxidant formation represents one of the most poorly characterised aspects of the chemical modules in models which are used in planning of photo-oxidant control strategies. Using the information obtained from kinetic and product studies a condensed chemical reaction mechanism has recently been formulated which adequately describes the concentration-time behaviours of the aromatics, ozone and NO_x in the photo-oxidation of aromatic/ hydrocarbon mixtures under simulated atmospheric conditions in a large volume photoreactor. Two important assumptions are necessary for satisfactory fits. Firstly, a highly efficient radical producing step not involving NO_x is required. Secondly, secondary reactions which scavenge NO_x need to be incorporated.

Although much progress have been made in elucidation of aromatic oxidation processes, particularly recently, it will take many more years of intensive research before the oxidation mechanism under atmospheric conditions is completely understood.

Figure 4 summarises the systems described above and shows their importance with respect to the influence of NO_x on radical propagation. In the case of alkane/alkene reactions with OH/NO_3 radicals NO is necessary to propagate the radical chain, however, in the case of aromatics NO does not appear to be important. NO is also not important for ozonolysis where a significant yield of radicals has been observed among the products. The new heterogeneous transformation of NO_x to HONO is a direct radical source which will accelerate the oxidation processes, particularly in the early morning.

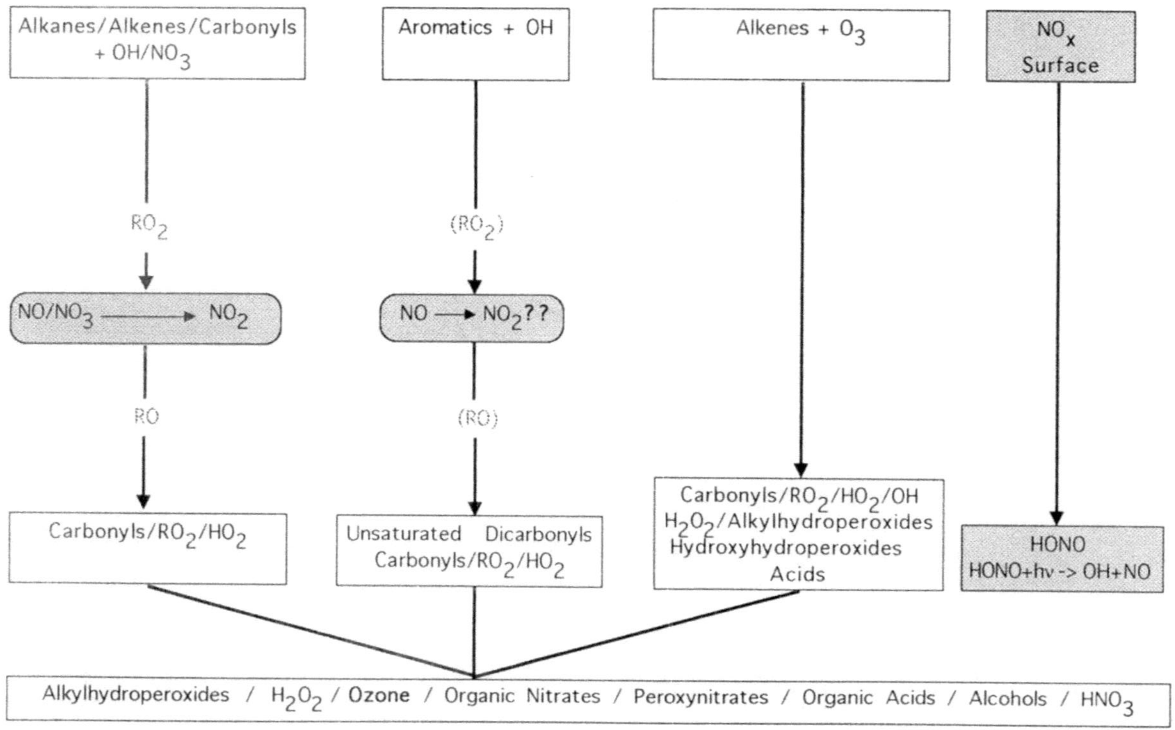

Fig.4: Simplified VOC Oxidation Scheme in the Presence of NO_x

Acknowledgement: Valuable contributions from the work of Prof. S. Gäb, Dr. K. Brockmann and Dr. P. Wiesen, all of the University of Wuppertal, to the contents of the manuscript are gratefully acknowledged. Financial support from the Commission of the European Communities and BMBF is greatly acknowledged.

Literature

Atkinson, R. (1990), Gas-phase tropospheric chemistry of organic compounds: A review.
Atmos. Environ. 24A: 1-41.

Atkinson, R. (1991), Kinetics and mechanisms of the gas-phase reactivity of the NO_3 radical with organic compounds. J. Phys. Chem. Ref. Data 20: 459-507.

Atkinson, R., Aschmann S.M., Arey, J. and Shorees, B. (1992) Formation of OH radicals in the gas phase reactions of O_3 with a series of terpenes. J. Geophys. Res. 97: 6065- 6073

Atkinson, R. Gas-phase tropospheric chemistry of organic compounds. (1994) J. Phys. Chem. Ref. Data, Monograph 2.

Becker, K.H. (1994) The atmospheric oxidation of aromatic hydrocarbons and its impact on photooxidant chemistry. In "Transport and Transformation of Pollutions in the Troposphere", proceeding of EUROTRAC Symposium ´94, Garmisch-Partenkirchen, April 1994, P.M. Borell, P. Borrell, T. Cvitas and W. Seiler (eds.), SPB Academic Publ. bv, Den Haag, pp. 67-74.

Becker, K.H., Brockmann, K.J. and Bechara, J. (1990) Production of hydrogen peroxide in forest air by reaction of ozone with terpenes. Nature 346:256-258.

Becker, K.H., Kleffmann, J., Kurtenbach, R. and Wiesen, P. (1995) Mechanistic study of the heterogeneous conversion of NO_2 into HONO and N_2O on acid surfaces. Faraday Discuss. Chem. Soc., submitted for publication.

Bierbach, A., Barnes, I., K.H. Becker , K.H. and Wiesen, E. (1994) Atmospheric Chemistry of Unsaturated Carbonyls: Butendial, 4-Oxo-2-Pentenal, 3-Hexene-2,5-Dione, Maleic Anhydride, 3H-Furan-2-one and 5-Methyl-3H-Furan-2-One. Environ. Sci. Technol. 28:715-729.

Brockmann, K.J., Gäb, S., Turner, W.V., Wolff, S. and Mönninhoff, S. (1995) Formation of alkyl and hydroxyalkyl hydroperoxides by the ozonolysis of alkenes. In Tropospheric Oxidation Mechanisms; Air Pollution Research Report, K.H. Becker (ed.), Commission of the European Communities, Brussels, in press.

Calvert, J.G., Yarwood, G. and Dunker, A.M. (1994) An evaluation of the mechanism of nitrous acid formation in the urban atmosphere. Res. Chem. Intermed. 20:463-502.

Gäb, S., Hellpointner, E., Turner, W.V. and Korte, F. (1985) Hydroxymethyl hydroperoxide and bis(hydroxymethyl)peroxide from gas phase ozonolysis of naturally occurring alkenes. Nature 316: 535-536.

Gäb, S., Turner, W.V. and Kurth, H.-H. (1990) Formation of hydrogen peroxides and organic hydroperoxides in the gas-phase ozonolysis of biogenic and anthropogenic olefins. In Atmospheric Oxidation Processes; Air Pollution Research Report 33, K.H. Becker (ed.), Commission of the European Communities, Brussels, pp. 41-44.

Gäb, S., Turner, W.V., Wolff, S., Becker, K.H. Ruppert, L. and Brockmann, K.J. (1995) Formation of alkyl and hydroxyalkyl hydroperoxides on ozonolysis in water and in air. Atmos. Environ., submitted for publication.

Grosjean, D. (1991), Atmospheric fate of toxic aromatic compounds. The Science of the Total Environment 100: 367 - 414.

Harrison, R.M. and Kitto, A.N. (1994) Evidence for a surface source of atmospheric nitrous acid. Atmos. Environ. 28:1089-1094

Horie, O., Neeb, P., Limbach, S. and Moortgat, G.K. (1994) Formation of formic acid and organic peroxides in the ozonolysis of ethene with added water vapour. Geophys. Res. Letters 21:1523-1526.

EUROTRAC Annual Reports, Part 8 LACTOZ, International Scientific Secretariat, Garmisch-Partenkirchen, 1988 - 1993.

Wayne, R.P., Barnes, I., Biggs, P., Burrows, J.P., Canosa-Mas, C.E., Hjorth, J., Le Bras, G., Moortgat, G.K., Perner, D., Poulet, G., Restelli, G. and Sidebottom, H. (1991) The nitrate radical: Physics, chemistry and the atmosphere. Atmos. Environ. 25A(1): 1-206.

Wiesen, P., Kleffmann, J., Kurtenbach, R. and Becker, K.H. (1994) Surface catalysed conversion of NO_2 - A new source of atmospheric N_2O? In: Non-CO_2 Greenhouse Gases [Proc.], J. van Ham, L.H.J.M. Jansen and R.J. Swart (eds.), Kluwer Academic Publishers, Dordrecht, pp. 311-316.

Laboratory studies for understanding atmospheric chemical processes

Jens Hjorth,
European Commission,
Environment Institute,
JRC-Ispra,
I-21020 Ispra (VA)

Introduction

Noxious pollutants in the urban environment may be primary, i.e. emitted directly from combustion and other sources (e.g. CO, benzene, polycyclic aromatic hydrocarbons and SO_2) but may also be secondary, i.e. the result of chemical reactions (photochemical and non-photochemical) in the atmosphere. Laboratory studies serve to understand the chemical processes leading to the formation of noxious pollutants, to set up models allowing to simulate these processes and thus also to help formulating efficient control strategies. Issues of particular relevance to the urban environment, requiring laboratory studies, are

- *'Photo oxidants'.* Noxious species such as ozone, peroxyacetylnitrate (PAN) and H_2O_2 (commonly referred to by the ambiguous term 'photo oxidants') are formed by the photochemical transformation of primary pollutants.
- *Atmospheric acidity.* The strong mineral acids H_2SO_4 and HNO_3 as well as carboxylic acids (e.g. formic, acetic and oxalic acid) are formed by chemical transformations in the atmosphere.
- *Reduced visibility.* NO_2 and aerosols, causing urban haze phenomena, are formed by chemical processes.

NATO ASI Series, Partnership Sub-Series, 2. Environment – Vol. 8
Urban Air Pollution
Edited by Ivo Allegrini and Franco De Santis

- *Toxic chemicals*. The rate of removal of noxious chemical compounds emitted from anthropogenic sources is often determined by their reactivity towards atmospheric reactants (mainly the OH and NO_3 radicals and ozone).

All of these issues require detailed laboratory investigations to be dealt with properly. In the following some examples of applications of laboratory studies in understanding aspects of atmospheric chemistry will be discussed with main emphasis on studies of atmospheric night-time chemistry (NO_3 chemistry) and its consequences for the daytime ozone-formation.

Experimental techniques

A variety of different experimental techniques are used in laboratory work on atmospheric chemistry and it will not be attempted here to make an exhaustive list. The most commonly applied techniques can be divided in two categories:

- The static chambers are typically applied in studies aiming at identifying end products of reactions with relevance to atmospheric chemistry and at studying the kinetics of their formation under conditions similar to those found in the lower atmosphere, often using long path IR or UV spectroscopy or gas chromatography as analytical techniques (Fig. 1).
- Flow tubes are typically applied in experiments aiming at studying the kinetics of fast elementary reactions (e.g. radical-radical reactions of atmospheric relevance), through direct measurement of short-lived species. Most often these experiments must be performed under conditions much different from those of the lower atmosphere (e.g. low pressure, bath gases different from air); frequently applied analytical techniques are mass spectrometry, electron spin resonance and laser induced fluorescence.

Flash photolysis techniques (particularly modulated photolysis techniques) are frequently applied in flow systems as well as static systems.

Each of these techniques have got their advantages as well as their disadvantages. An obvious advantage of the static chamber systems is that of being able to work under conditions (pressure, temperature, chemical environment) simulating those of the real atmosphere. A clear disadvantage is that of having most often to deal with relatively complex chemical systems meaning that experimental observations can only be interpreted by applying an intricate scheme of elementary chemical reactions. Also the fact that often the most reactive radical species (such as OH) cannot be measured directly due to their low concentrations contributes to the difficulties associated with the interpretation of experimental data from such systems.

Contrary to the static chambers, flow tube techniques (and flash photolysis) allow often to do direct measurements of relevant radical species. Another

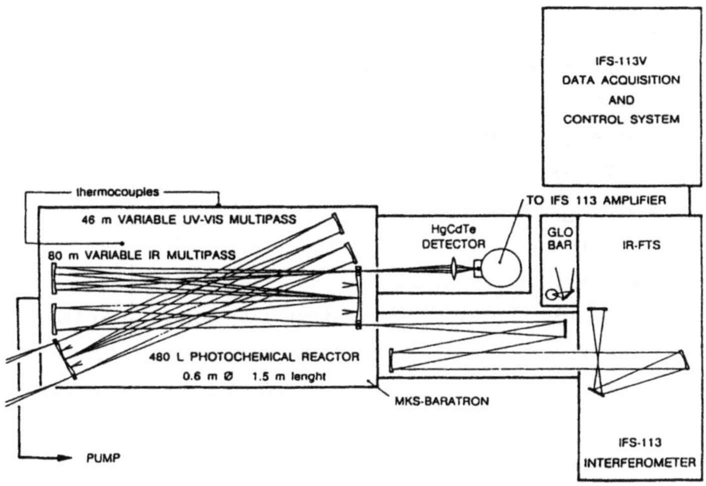

Fig. 1. Photochemical reactor at JRC-Ispra (static chamber) with on-line IR (FTS and TDL) and UV/VIS spectroscopic detection.

advantage of flow tube techniques is that it often is possible to isolate the initial reaction steps from the subsequent ones because only the initial fractions of a second of the reaction is observed; this may greatly simplify the analysis of the kinetics. The obvious disadvantage of applying such techniques is that most often the experimental conditions are very different from ambient, tropospheric conditions and thus some of the results (in particular those regarding the reaction products) may not be directly applicable to real tropospheric conditions.

Tropospheric ozone and the role of NO_x

Episodes of high ambient tropospheric ozone concentrations in industrialised regions are related to emissions of volatile organic compounds and NO_x (a major source of both is car traffic) and to meteorological conditions (the occurrence of high pressure systems favour the build-up of ozone). Although not all aspects of the detailed chemistry involved in the in-situ formation of ozone in the troposphere is known (important gaps remain e.g. in our understanding of the atmospheric oxidation of aromatic compounds), the fundamental mechanism is well-established:

(1) $VOC+OH+O_2 \rightarrow ROO\bullet$(+other products)

(2) $ROO\bullet+NO \rightarrow RO\bullet+NO_2$

(3) $NO_2+h\nu+O_2 \rightarrow NO+O_3$.

NO destroys ozone by the reaction

(4) $NO+O_3 \rightarrow NO_2+O_2$;

thus a so-called 'photo stationary equilibrium' will be established by reaction 3 and 4 while reaction 1 and 2 will tend to shift this equilibrium towards higher concentrations of O_3.

Ozone is also destroyed by other photochemical reactions than direct photolysis, in particular the reaction

(5) $HOO\bullet+O_3 \rightarrow HO\bullet+2O_2$.

Thus the concentration of NO is of pivotal importance in the chemistry related to tropospheric ozone levels: at low [NO], typical of remote marine areas, photochemistry represents a sink for O_3 while in other areas a net photochemical formation of O_3 in the troposphere is observed.

Oxidation of VOC

Understanding the rates and mechanisms involved in the oxidation of volatile organic compounds is a key issue in atmospheric chemistry both because of the interest in the products formed by these processes and because of the need to know how fast contaminants are removed from the atmosphere.

Most often, reactions with OH appear to be the main sinks of VOCs; however, it is not unusual to find that other chemical reactions are giving relevant contributions to the degradation of VOCs; most important seem to be

- reaction with NO_3 radicals
- reaction with O_3
- direct photolysis
- reactions with other radicals (e.g. BrO and HO_2).

NO_3 is formed by the reaction between O_3 and NO_2, during daytime it is rapidly photolysed or destroyed by the reaction with photolytically formed NO but at night-time NO_3 builds up to reach concentrations at the ppt-level; some tens of ppt appear to be typical of a moderately polluted area, and that makes it a potentially important sink for several organic compounds (Wayne et al., 1991). The most rapid reactions of the NO_3 radical with organic compounds are those taking place with alkenes and dialkenes (e.g. isoprene), with reduced sulphur compounds (such as dimethylsulphide) and with phenols (Atkinson, 1991, Wayne et al., 1991). Reactions with NO_3 radicals at night can be considerably faster than the corresponding reactions of OH at day in polluted atmospheres; e.g. in the case of the most important biogenically emitted alkenes (isoprene and some terpenes) the reaction with NO_3 is expected to be at least twice as important as the corresponding reaction with OH. Thus studies not only of the rates but also of the products of these reactions appear to be relevant and in this regard laboratory work has given much useful information. In the following, the issue of tropospheric NO_3 chemistry will be applied as an example of how laboratory studies can interact with field studies in an effort to clarify potentially important aspects of the complex chemistry of the troposphere.

Atmospheric NO_3 chemistry

The basic features of atmospheric NO_3 chemistry can be summarised by the reaction scheme below:

(6) $NO_2+O_3 \rightarrow NO_3+O_2$

(7) $NO_3+NO_2+M \rightleftharpoons N_2O_5+M$

(8) $NO_3+NO \rightarrow 2NO_2$

(9) $N_2O_5+H_2O_{liq} \rightarrow 2HNO_3$

(10) NO_3+substrate $\rightarrow$ products.

Reaction (10) represents all of the many reactions between NO_3 and organic substrates that lead to various classes of products.

For the understanding of the importance of these reactions in the overall chemistry of the troposphere it is relevant to consider their impact on

- formation of HNO_3
- long range transport of NO_x by formation of relatively stable compounds capable of releasing NO_x during their degradation (organic nitrates)
- oxidation of organic compounds
- formation of toxic products.

In the following some laboratory investigations of NO_3 radical reactions of relevance to atmospheric chemistry will be discussed.

Formation of HNO_3

Whenever the NO_3 radical reacts by hydrogen-abstraction the product will be nitric acid, thus representing a sink for NO_x in the atmosphere. A great variety of NO_3-initiated H-abstraction reactions will take place, but generally such reactions are slow compared to reactions where the NO_3 radical adds to unsaturated moieties (with the exception of reactions with reduced sulphur compounds and with HO_2 radicals as discussed later on). However, though the homogeneous gas phase reaction between N_2O_5 and water is relatively slow (Tuazon et al. 1983, Hjorth et al. 1987) there is much evidence for a fast heterogeneous reaction of N_2O_5 with liquid water in the troposphere and maybe also a fast reaction of NO_3 radicals on wet surfaces, which would have a very significant impact on deposition of nitrogen as well as on the global distribution of NO_x, O_3 and OH as discussed by Dentener and Crutzen (1993).

<u>NO_3+alkenes:</u>

NO_3 reactions with alkenes have been studied by several laboratories and discussed in reviews by Wayne et al. (1991) and Atkinson (1991). Information about rate constants have been obtained by flow tube experiments (e.g. Ravishankara and Mauldin 1985) and by relative rate constant studies (e.g. Atkinson et al. 1988). In the latter type of experiments the decay of an alkene caused by reaction with NO_3 radicals is compared to the decay of a reference compound reacting with NO_3 with a known rate constant. The value of the unknown rate constant for the alkene reaction can then be determined by the simple relationship

$\ln([A]_0/[A]) = k_A/k_B \ln([B]_0/[B])$,

where $[A]_0$ and $[B]_0$ denominate the initial concentrations of A and B, [A] and [B] are the concentrations after some elapsed time and k_A and k_B are the rate constants for the reactions between NO_3 and compound A and B respectively.

The reaction mechanisms have mainly been clarified by application of FTIR and GC techniques in experiments in reaction chambers (e.g. Barnes et al. 1990, Hjorth et al. 1990 and Skov et al. 1992). These studies, performed in air using the thermal dissociation of N_2O_5 as source of NO_3 radicals, clearly indicate that the reaction proceeds via addition of NO_3 to the double bond followed by addition of O_2 and formation of peroxynitrate intermediates, easily observable due to their strong spectral features in the IR region:

(11) $R_1R_2C{=}CR_3R_4 + NO_3 + M \rightarrow R_1R_2C(NO_3)C\bullet(R_3R_4) + M$

(12) $R_1R_2C(NO_3)C\bullet(R_3R_4) + O_2 + M \rightarrow R_1R_2C(NO_3)C(OO\bullet)(R_3R_4) + M$

(13) $R_1R_2C(NO_3)C(OO\bullet)(R_3R_4) + NO_2 + M \rightleftharpoons R_1R_2C(NO_3)C(OONO_2)(R_3R_4) + M$.

The subsequent reaction steps lead to formation of nitroxy-carbonyl compounds and -alcohols, unsubstituted aldehydes or ketones and some dinitrate. The main products were identified (in the studies of Barnes et al. 1990 and Hjorth et al. 1990) by comparison with IR spectra of synthesised pure compounds. The dinitrate product, formed by the reaction between alkoxy radicals and NO_2, can be important at the high NO_x levels applied under laboratory conditions but it is unlikely to be of relevance under atmospheric conditions. It is seen that there is a competition between pathways leading to formation of organic nitrates that may act as reservoir compounds or sinks for NO_x in the troposphere and those leading to reduction of NO_3 to NO_2 and oxidation of the alkene to unsubstituted carbonyl species. The laboratory studies show that the more branched alkenes tend to give the lowest yields of nitrate and thus higher yields of unsubstituted carbonyl compounds, e.g. 2,3-dimethyl-2-butene was found to give a yield of 52±13 % of acetone while cis- and trans-2-butene gave only yields of 17±6 % of acetaldehyde. Reactions of NO_3 with branched alkenes are faster than those with unbranched, e.g. the reaction of NO_3 with trans-2-butene is approximately 100 times faster than that of 2,3-dimethyl-2-butene but an evaluation of the overall impact of these reactions must of course be based on a knowledge of ambient concentrations of hydrocarbon species in the atmosphere.

The reaction of NO_3 with terpenes are generally very fast, having rate constants in the range of 10^{-12} to 10^{-10} cm^3 $molecule^{-1}s^{-1}$. They are likely to give the same type of products as the reactions between NO_3 and simpler alkenes but little experimental information is available. The reaction with 1-methyl-cyclohexene, which has a structural similarity to alpha-pinene, has been studied (Wängberg et al., 1993) and found to follow a mechanism similar to that of the aliphatic alkenes, however yielding mainly non-nitrate products.

NO_3+dialkenes:

The reaction of NO_3 with isoprene is of particular environmental importance because of the large natural emissions of this species, in fact, its global emission rate has been estimated as 450×10^{12} g yr^{-1} (Rasmussen and Khalil 1987) against a total anthropogenic emission of non-methane hydrocarbons of 90×10^{12} g yr^{-1}. The rate constant of the reaction between isoprene and NO_3 has been determined as $(8.1 \pm 1) \times 10^{-13}$ $cm^3 molec^{-1} s^{-1}$ (Wille et al. 1991), thus the reaction is fast enough to be a potentially important sink for isoprene, particularly in areas influenced by urban emissions of NO_x (the lifetime of isoprene with respect to this reaction is 1.4 hour for an NO_3 concentration of 10 ppt, i.e. shorter than the typical lifetime with respect to the OH reaction). The products of this reaction have been studied by several laboratories (Jay and Stieglitz 1990, Barnes et al. 1990, Skov et al. 1992). Based on these studies it appears that the reaction mainly yields organic nitrates, the predominant product being 3-methyl-4-nitroxy-2-butenal.

The overall impact of this reaction (and of NO_3 radical reactions with alkenes in general) on tropospheric chemistry will obviously depend on the fate of the organic nitrate products formed. Laboratory studies of rate constants for the reactions with OH and of photolysis rates indicate that the main fate in the atmosphere of carbonyl-nitrates is photolysis (lifetime of less than one day), while the hydroxy-nitrates may be removed mainly by wet deposition (Roberts and Fajer 1989, Zhu et al. 1991, Barnes et al. 1993). Organic nitrates are photolized to form NO_2 in the atmosphere and can thus serve as temporary reservoirs of NO_x (Barnes et al. 1993) similar to what PAN does, though the lifetimes of the bifunctional nitrates in the troposphere appear to be significantly shorter than that of PAN.

Formation of oxiranes

In addition to the products discussed previously, also oxiranes have been suggested as major products of the reaction between NO_3 radicals and alkenes and dialkenes, presumably formed in reaction (11) followed by

(14) $R_1R_2C(NO_3)C{\bullet}(R_3R_4) \rightarrow R_1R_2C(O{\bullet})C(R_3R_4) + NO_2$.

(15) Flow tube studies performed at pressures of few millibar showed high yields of oxirane-products e.g. in the case of isoprene. However, oxiranes were not found in chamber studies at atmospheric pressure. An explanation for this apparent discrepancy was eventually given by Skov et al.(1994) and Benter et al. (1994): The NO_3-alkene adduct formed in reaction (14) will form oxiranes at high yields at low pressures and in the absence of oxygen while increasing pressure as well as increasing oxygen concentrations favour the formation of alternative products. In fact, at ground level atmospheric pressure the yield of oxirane-products from the reaction of isoprene was found to be negligible as in the case of cis- and trans-butene while the highly branched 2,3-dimethyl-2-butene still gave a significant yield of tetramethyloxirane (20%).

<u>NO_3 + aromatic hydrocarbons</u>

Though the OH-initiated oxidation of aromatics is generally more important than the reaction with NO_3 radicals the latter type of reaction becomes more important with increasing alkyl-substitution and predominates for phenolic and heterocyclic aromatics (Zetsch, 1994). The predominant pathway for the reactions between methyl-substituted benzenes and NO_3 appear to be hydrogen-abstraction from the alkyl-substituent, leading to products such as benzaldehyde, benzylalcohol and benzyl-nitrate, according to laboratory studies performed using FTIR and GC-MS analysis for measurements of reactants and products (Chiodini et al., 1993). Under the conditions of ambient air influenced by anthropogenic NO_x-emissions the main product is likely to be benzaldehyde, formed by the following sequence of reactions:

(15) $PhCH_3 + OH \rightarrow PhCH_2\bullet + H_2O$

(16) $PhCH_2\bullet + O_2 + M \rightarrow PhCH_2O_2\bullet + M$

(17) $PhCH_2O_2\bullet + NO \rightarrow PhCH_2O\bullet + NO_2$

(18) $PhCH_2O\bullet + O_2 \rightarrow PhCHO + HO_2$,

where Ph indicates a phenyl-moiety.

However, in addition to the hydrogen-abstraction reactions also addition of NO_3 to the ring appears to take place; an mechanism of addition of NO_3 followed by addition of NO_2 and finally elimination of HNO_3 has been proposed as explanation of the observed formation of minor yields of aromatic nitro-derivatives by the reaction of NO_3 with methyl-substituted benzenes (Chiodini et al., 1993) and for the more substantial yields of nitrophenols obtained by the NO_3-initiated oxidation of phenols (Atkinson et al.,1992) and of naphtalene (Kwok et al., 1994).

OH-generation at night?

OH radicals are normally associated with photochemistry, but laboratory experiments showed that HO_2 radicals could react with NO_3 to form OH. HO_2 radicals may be formed by several other reactions of the NO_3 radical so this observation lead Platt and coworkers (1990) to propose a nighttime mechanism for the generation of OH (Fig. 2).

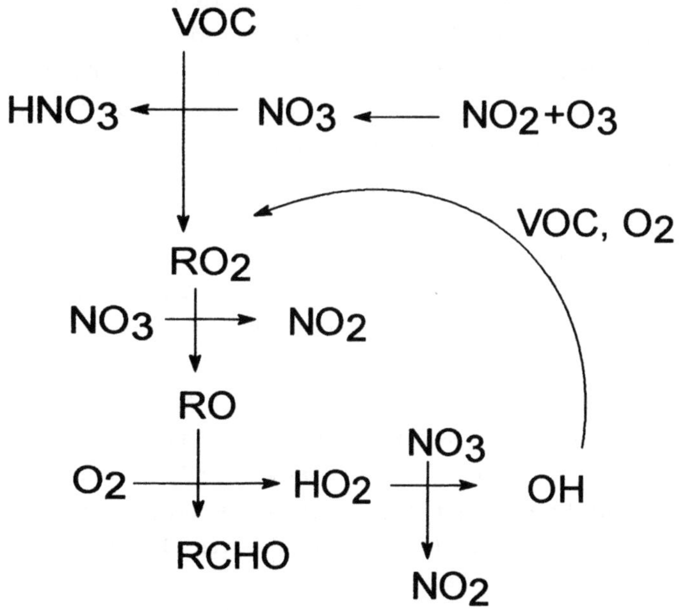

Fig. 2. Mechanism for generation of OH at night, as suggested by Platt et al. (1990).

The experiments where OH could actually be observed (Mellouki et al. 1988, Becker et al. 1992, Mellouki et al. 1993) were performed at low total pressures (2 Torr or less). In Ispra it was attempted to look at the OH-generating reaction at higher pressures (50 or 750 Torr) (Hjorth et al. 1992) in a static reaction chamber where HO_2 was generated by the thermal dissociation of HO_2NO_2

(19) $HO2NO2+M \rightleftharpoons HO2+NO2+M$

while NO_3 was formed by the thermal dissociation of N_2O_5. OH radicals were measured by observing the rate by which isotopically labelled carbon monoxide, $C^{13}O^{18}$, was oxidised to $O^{16}C^{13}O^{18}$. While previous estimates of the rate constant of the OH-forming channel of the reaction were in the range 1-3 x $10^{-12}cm^3molec^{-1}s^{-1}$, this experiment pointed to a rate constant below 0.8 x 10^{-12} $cm^3molec^{-1}s^{-1}$. This lower limit was determined applying a model with a relatively complex chemistry scheme and the reason for this discrepancy may be incorrect values of rate constants applied in this model. However, it seems worthwhile to investigate further if there is a pressure-dependence of the rate constant for this OH-forming reaction which makes it less important at atmospheric pressures. Thus a smog-chamber type of experiment to confirm the hypothesis presented by Platt et al. (1990) seems relevant.

A new frontier in laboratory investigations of atmospheric chemistry: multiphase systems

The study of atmospheric gas phase chemistry, though still leaving many important problems open, has become a rather well-established scientific discipline that has developed a number of standard experimental techniques as well as a basic understanding of the reaction mechanisms of relevance to atmospheric conditions. The role of aerosols and clouds in atmospheric chemistry has been less extensively studied in the laboratory as well as in the field, however, this is a field that appears to gain increasing interest among atmospheric chemists.

There are (at least) two reasons for the present interest in the aerosol and cloud chemistry. One is the awareness of a potentially important impact of clouds on tropospheric photochemistry, particularly the modelling study by Lelieveld and Crutzen (1990) has demonstrated that cloud chemistry may reduce ozone levels significantly. This interaction between gas phase and liquid phase chemistry is caused by the incorporation of very water-soluble and reactive species such as OH, HO_2 and aldehydes into droplets.

The second reason for paying interest to aerosol and cloud chemistry is the issue of their role in climate change: Several authors (e.g. Charlson et al., 1986, Kiehl and Biegreb 1993) have estimated the direct climatic effect of aerosols (due to scattering or absorption of radiation) as well as the indirect effect (due to the influence of aerosols on the physical properties of clouds). and found that increasing aerosol concentrations may have had a significant cooling effect on the earth. This may to some extent compensate for the warming effect of greenhouse gases, in polluted areas even a net cooling effect has been calculated.

The processes controlling the formation and growth of aerosol particles represent a challenge to atmospheric chemists as well as physicists. Also in this area laboratory studies are required to understand processes forming condensable products as well as reactions of gaseous species on surfaces or dissolved in deliquescent particles or droplets.

REFERENCES:

Atkinson R. (1991), 'Kinetics and mechanisms of the gas phase reactions of the NO_3 radical with organic compounds', J. Phys. Chem. Ref. Data 20 (3), 459-507.

Atkinson R., Aschmann S.M. and Arey J. (1992), 'Reactions of OH and NO_3 radicals with phenol, cresols, and 2-nitrophenol at 296 K', Environ. Sci. Technol. 26, 1397-1403.

Atkinson R., Aschmann S.M. and Pitts J.N. Jr. (1988), 'Rate constants for the gas-phase reactions of NO_3 radicals with a series of organic compounds', J. Phys. Chem. 92, 3454 - 3457.

Barnes I., Bastian V., Becker K.H. and Tong Z. (1990), ' Kinetics and products of the reactions of NO_3 with with monoalkenes, dialkenes and monoterpenes', J. Phys. Chem. 94, 2413-2419.

Barnes I., Becker K.H. and Zhu T. (1993), 'Near UV absorption spectra and photolysis products of difunctional organic nitrates: possible importance as NO_x reservoirs', J. Atmos. Chem. 17, 353-373, 1993.

Becker E., Rahman M.M. and Schindler R.N. (1992), 'Determination of the rate constants for the gas phase reactions of NO_3 with HO_2 and OH', Int. J. Chem. Kinet. 25, 795-803.

Charlson R.J., Lovelock J.E., Andreae M.O. and Warren S.G. (1987), 'Oceanic phytoplankton, atmospheric sulphur, cloud albedo and climate', Nature 326, 655-661.

Chiodini G., Rindone B., Cariati F., Polesello S., Restelli G. and Hjorth J. (1993), 'Comparison between the gas-phase and the solution reaction of the nitrate radical and methylarenes', Environ. Sci. Technol. 27 (8), 1659-1664.

Dentener F.J. and Crutzen P.J. (1993), 'Reaction of N_2O_5 on tropospheric aerosols: impact on the global distribution of NO_x, O_3 and OH', J. Geophys Res. 98, 7149-7163.

Dlugokencky F.J. and Howard C.J. (1989), 'Studies of NO_3 radical reactions with some atmospheric organic compounds at low pressures', J. Phys. Chem. 93, 1091-1096.

Hjorth J., Cappellani F. and Restelli G. (1992), 'A TDL and FTIR study of the reaction $HO_2+NO_3 \rightarrow OH+NO_2+O_2$', Joint CEC/Eurotrac Workshop, Lactoz-Halipp Working Group, Leuven, Belgium, September 23-25, CEC Air Pollution Research Report 45, 125-130.

Hjorth J., Lohse C., Nielsen C.J., Skov H. and Restelli G. (1990), 'Products and mechanisms of the gas-phase reactions between NO_3 and a series of alkenes', J. Phys. Chem. 91, 1565-1568.

Hjorth J., Ottobrini G., Cappellani F. and Restelli G. (1987), 'A Fourier transform infrared study of the rate constant of the homogeneous gas-phase reaction N_2O_5 + H_2O and determination of absolute infrared band intensities of N_2O_5 and HNO_3', J. Phys Chem. 91, 1565-1568.

Jay K. and Stieglitz L. (1990), 'The gas phase addition of NO_3 to olefins', Chemosphere 1939-1950.

Kiehl J.T. and Briegleb B.P. (1993), 'The relative role of sulphate aerosols and greenhouse gases in climate forcing', Science 260, 311-314.

Kwok E.S.C., Atkinson R. and Arey J. (1994), 'Kinetics and mechanisms of the gas-phase reactions of the NO_3 radical with aromatic compounds', Int. J. Chem. Kinet. 26, 511-525.

Lelieveld J. and Crutzen P.J. (1990), 'Influence of cloud photochemical processes on tropospheric ozone', Nature 343, 227-233.

Mellouki A., Le Bras G. and Poulet G. (1994), 'Kinetics of the reactions of NO_3 with OH and HO_2', J. Phys. Chem. 92, 2229-2234.

Platt U., Le Bras G., Poulet G., Burrows J.P. and Moortgat G.K. (1990), 'Peroxi radicals from night-time reaction of NO_3 with organic compounds', Nature 348, 147-149.

Rasmussen R.A. and Khalil M.A.K. (1988), 'Isoprene over the Amazon basin', J. Geophys. Res. 93, 1417-1421.

Ravishankara A.R. and Mauldin R.L. III (1985), 'Absolute rate coefficient for the reaction of NO_3 with trans-2-butene', J. Phys. Chem. 89, 3144-3147.

Roberts J.M. and Fajer R.W. (1989), 'UV absorption cross sections of organic nitrates of potential atmospheric importance and estimation of atmospheric lifetimes', Environ. Sci. Technol. 23, 945-951.

Skov H., Benter Th., Schindler R.N., Hjorth J. and Restelli G. (1994), 'Epoxide formation in the reactions of the nitrate radical with 2,3-dimethyl-2-butene, cis- and trans-2-butene and isoprene', Atmospheric Environment 28 (9), 1583-1592.

Skov H., Hjorth J., Lohse C., Jensen N.R. and Restelli G. (1992), 'Products and mechanisms of the reactions of the nitrate radical (NO_3) with isoprene, 1,3-butadiene and 2,3-dimethyl-1,3-butadiene in air', Atmos. Environ. 26A (15), 2771-2783.

Sverdrup G.M., Spicer C.W. and Ward G.F. (1987), 'Investigation of the gas phase reaction of dinitrogen pentoxide with water vapor', Int. J. Chem. Kinet. 19, 191-205.

Tuazon E.C., Atkinson R., Plum C.N., Winer A.M., Pitts J.N., Jr. (1983), 'The reaction of gas phase N_2O_5 with water vapor', Geophys Res. Lett. 10, 953 - 956

Wängberg I. (1993), 'Mechanisms and products of the reaction of NO_3 with cycloalkenes', J. Atmos. Chem. 17, 229-247.

Wayne R., Barnes I., Biggs P., Burrows J.P., Canosa-Mas C.E., Hjorth J., LeBras G., Moortgat G.K., Perner D., Poulet

G., Restelli G. and Sidebottom H. (1991), 'The nitrate radical: physics, chemistry and the atmosphere', Atmospheric Environment 25A, 1-206.

Wille U., Becker E., Schindler R.N., Lançar I.T., Poulet G. and LeBras G. (1991), 'A discharge flow mass spectrometric study of the reaction between the NO_3 radical and isoprene', J. Atmos. Chem. 13, 183-193.

Zetzsch C. (1993), 'Atmospheric oxidation processes of aromatics studied within LACTOZ and STEP', im proceedings of Sixth European Symposium on Physico-Chemical Behaviour of Atmospheric Pollutants, Varese 18-22 October 1993, p. 118-128.

Studies of photochemical ozone formation in toluene/NO_x/air systems by empirical and numerical simulations

Montserrat Martín-Reviejo[1], Klaus Wirtz[1] and Karl H. Becker
Physicalische Chemie FB9
Bergische Universität-GH Wuppertal
Gaußstr. 20
42097 Wuppertal
Germany

Introduction

Benzene and the alkyl-substituted benzenes, toluene and xylenes, are the most abundant aromatics observed in urban atmospheres (see Table 1). The aromatic fraction reaches 20 to 40 % of the total non-methane hydrocarbons with the traffic emission being the major source. Their contribution to the photochemical smog and the formation of photooxidants, especially ozone, have been studied in smog chamber experiments and by simulations using atmospheric chemical models (Atkinson et al., 1980; Killus and Whitten, 1982; Leone et al., 1985). As a result of these studies, the aromatic hydrocarbons are considered to contribute significantly to the tropospheric ozone formation (Carter and Atkinson, 1987). However, the oxidation mechanisms of aromatic hydrocarbons in the atmosphere are up to now not well understood. Recent product studies from the OH-radical initiated oxidation of toluene under NO_x-free conditions and in the presence of NO_x (Atkinson and Aschmann, 1994; Bierbach et al., 1994) gave a poor carbon balance. Therefore, a detailed oxidation mechanism for the aromatic hydrocarbons is at present not possible.

In order to analyze the ozone formation for the toluene/NO_x/air system, smog chamber experiments with toluene alone and with mixtures of toluene/cyclohexane were carried out and simplified chemical models were tested to describe the observed concentration-time profiles of NO, NO_2 and O_3.

[1]Present address: CEAM, Plaza del Carmen 4, 46003 Valencia, Spain

NATO ASI Series, Partnership Sub-Series, 2. Environment – Vol. 8
Urban Air Pollution
Edited by Ivo Allegrini and Franco De Santis

Experimental set-up

Two different photo-reactors have been used as smog chamber systems. The first was a glass reactor of 500 l volume and the second a quartz reactor of 1080 l volume. The reactants were injected under vacuum using microliter syringes and the reactor was then filled with 1000 mbar of synthetic air. The irradiations were carried out at room temperature over a time periods of six or more hours. In both reactors, fluorescence lamps, Philips TL/05 40 W, were used as light sources.

Urban Atmosphere	Benzene	Toluene	Xylenes
Budapest[1]	1.2-9.2	1.2-17.8	0.3-29.5
Hamburg[2]	4.8	9.8	8.1
15 urban USA cities[3]	1-9	1-17	0.3-10
Mexico-city[4]	4-14	8-59	10-34
Los Angeles, 1980's[5]	12-29	20-68	15-58
Los Angeles, 1990's [6]	8.2	19	11.6

Table 1: Range of concentrations in ppbv (1 ppbv=10^{-9} by volume) of the most abundant aromatics, benzene, toluene and xylenes, in different urban atmospheres.
[1] Haszpa et al., 1991. Mean value 6:00-9:00 a.m.
[2] Bruckmann et al., 1988. Annual means.
[3] Singh et al., 1985. Average concentrations.
[4] From a measurement campaign performed in Mexico-city, March 1992.
[5] Grosjean and Fung, 1984. Average concentration of 23 days.
[6] Zielinska and Fujita, 1994. Mean value 10:00-12:00 a.m.

The photolysis frequency of NO_2 was used to establish the light intensity. Concentration-time profiles of NO_2 and the primary hydrocarbon were obtained by means of long-path absorption spectroscopy FTIR (Nicolet 520 in the 500 l reactor and Bruker IFS 88 in the 1080 l reactor, with optical path lengths of 97.2 m and 492 m, respectively).

The spectra were collected using between 4 to 10 min integration time with a spectral resolution of 1 cm^{-1}. NO concentrations were monitored using a Thermo Electron chemiluminiscence $NO-NO_2-NO_x$ analyzer Model 14 D/E. Ozone concentrations were monitored using a Bendix Model 8002 ozone analyzer. Data from both monitors, temperature and pressure values inside the reactor were collected by a PC computer (see Figure 1).

Description of the oxidation mechanism

The used photochemical mechanism is reported in Wirtz et al., 1994. This mechanism is divided into three parts for the inorganic reactions, the cyclohexane reactions and the toluene reactions. The inorganic part consists of 33 reactions similar to those used in other current atmospheric photochemical mechanisms. A simplified mechanism for the cyclohexane oxidation is used which includes the primary steps to describe the photochemical ozone formation in this system. The oxidation of cyclohexane proceeds via the primary attack of the OH-radical forming the alkyl peroxy radical under atmospheric conditions. The further reaction steps of these peroxy radicals include, in addition to the RO_2 + NO reaction, the recently determined rate coefficients for the RO_2 + HO_2 and RO_2 + RO_2 reactions to account for the radical budget, but further reactions of the products are not considered. All tested oxidation mechanisms for toluene are detailed by the primary steps of the OH-abstraction channel but simplified for the unknown and, therefore, speculative reactions and reaction products following the OH-addition. The further reactions of the formed peroxy radicals from the abstraction channel are similar to those from n-alkane oxidation (see Figure 2).

Results and discussion

The concentration-time profiles for ozone, which were obtained by irradiations of toluene/NO_x/air and cyclohexane/NO_x/air systems as well as from mixtures of these two hydrocarbons in the presence of NO_x in the 1080 l are shown in Figure 3. The ozone formation rate in the cyclohexane system is lower than in the toluene system but the maximum ozone concentration is higher. The addition of small amounts of toluene to the cyclohexane system was sufficient to cause an increase in the ozone formation rate but did not influence the concentration of the ozone maximum with respect to the cyclohexane/NO_x/air system. Single hydrocarbon irradiations have been carried out in the 500 l chamber in order to compare the behaviour of toluene and other hydrocarbons with respect to the concentration-time profiles of NO_x, during the runs (Becker et al., 1994). The results are given in Figure 4. Toluene shows higher NO_x loss rates before and after the NO_2 maximum than the tested hydrocarbons.

These results can be explained within the oxidation mechanism of toluene, by an additional radical forming step which increases the O_3 formation rate in the first stages of the oxidation and by a NO_x-loss step which may be the responsable for the fact that the toluene system becomes rapidly NO_x-limited. Numerical simulations were performed with different simplified mechanisms to support these assumptions. The subsequent reactions of the OH-addition channel which account for about 90% of the toluene degradation were changed in such a manner, that the peroxy radical formation step, the NO_x-loss step and the additional radical source are located at different positions within the degradation processes (see Figure 5).

In *mechanism 1,* the reaction of the adduct with O_2 forms a peroxy radical which converts NO to NO_2. The formed alkoxy radical reacts further with O_2 forming HO_2 and the ring-opening product OPEN. This product can remove NO_x from the system or generate additional radicals by a photolytic step. As it can be seen from Figure 6, the concentration time profiles for ozone, NO_2, NO and toluene could be fitted to the experimental values by adjusting the rate parameters for the radical forming and NO_x-loss step. However, the toluene decay and the maximum ozone concentration are slightly overestimated.

In *mechanism 2,* the reaction of the adduct with O_2 forms, in contrast to mechanism 1, directly a ring-opening product and HO_2. This ring-opening product removes NO_x, from the system or reacts further in a photolytic step to generate peroxy radicals and additional HO_2. Also for this mechanism the rate parameters were adjusted to get the best fit to the experimental data (see Figure 7). The NO, NO_2 and ozone concentration-time profiles could be represented by the simulation but the decay of the toluene in this system is higher which indicates an unbalanced radical budget.

In *mechanism 3,* the first reaction of the adduct with O_2 forms a stable dicarbonyl compound and HO_2 for the radical propagation. The further oxidation of this dicarbonyl is initiated by OH radicals forming peroxy radicals. These peroxy radicals convert NO to NO_2 and generate, in a subsequent reaction with O_2, HO_2 radicals and photolytic active products which remove NO_x and serve as additional radical sources. In comparison to mechanism 1, the peroxy radical formation step is limited by the OH initiated oxidation of the dicarbonyl. By adjusting the rate parameters for the NO_x-loss step and the radical formation step, it was not possible to represent the experimental results (see Figure 8). The toluene decay is described by the simulation but the ozone concentration is overestimated by a factor of two and the NO to NO_2 conversion starts too late compared to the experiments.

In summary, for all the simulations it was necessary to introduce a radical propagation step within the first reactions via HO_2 and a peroxy radical formation to account for the radical budget. The additional radical source from the ring-opening product and the NO_x-loss step were used to describe the concentration-time profiles in comparison to the cyclohexane system.

At the present state of this work, it is not possible to apply one of these mechanisms in a photochemical model to describe the degradation of toluene under atmospheric conditions. More experimental work under different conditions as well as some improvements in the simplified mechanisms, especially regarding the PAN formation, are necessary.

Conclusions

In this work, toluene has exhibited high reactivity in single hydrocarbon irradiations and in mixture with another hydrocarbon which has been attributed to an additional radical source in the chemical mechanism of its oxidation. On the other hand, the O_3 formation was strongly NO_x limited because NO_x was rapidly removed from the system reducing the capacity of toluene to form O_3.

Using a simplified chemical mechanism for the oxidation of toluene, agreement between the experimental and the simulation data could be obtained when the following steps are included in the oxidation mechanism of toluene: (i) the formation and decomposition of the OH-adduct from the addition channel, (ii) radical propagation via HO_2 within the first reaction of the OH-adduct, (iii) peroxy radical formation either from the reaction of the OH-adduct with O_2 or from peroxy radicals generated after the ring-opening, (iv) additional radical source from the ring-opening products, and (v) additional NO_x loss by the products after the ring-opening.

References

Atkinson R., Carter W.P.L., Darnall K.R., Winer A.M. and Pitts J.N. (1980). A smog chamber and modeling study of the gas phase NO_x-Air photooxidation of toluene and the cresols. Int. J. Chem. Kin., 12:779-836.

Atkinson R. and Aschmann S.M. (1994). Products of the gas-phase reactions of aromatic hydrocarbons: effect of NO2 concentration. Int. J. Chem. Kinet., 26, 929-944.

Becker K.H., Wirtz K., Martin-Reviejo M. and Millan M. Millan (1994). Photochemical ozone production in hydrocarbon-$N0_x$-air systems. In Proc. 6th Eur. Symp.Physico-chemical Behaviour of Atmospheric Pollutants, G. Angeletti and G. Restelli (eds.), EU-Report EUR 15609/1 EN, Brussels and Luxembourg, pp. 163-168.

Bierbach A., Barnes I., Becker K.H., Klotz B. and Wiesen E. (1994). OH-radical initiated degradation of aromatic hydrocarbons. In Proc. 6th Eur. Symp. Physico-chemical Behaviour of Atmospheric Pollutants, G. Angeletti and G. Restelli (eds.), EU-Report EUR 15609/1 EN, Brussels and Luxembourg, pp. 129-136.

Bruckmann P., Kersten W., Funcke W., Balfanz E., König J., Theisen J., Ball M., and Päpke O. (1988). The occurrence of chlorinated and other organic trace compounds in urban air. Chemosphere, 17:2363-2380.

Carter W.P.L. and Atkinson R. (1987). An experimental study of incremental hydrocarbon reactivity. Environ. Sci. Tecnol., 21:670-679.

Grosjean D. and Fung K. (1984). Hydrocarbons and carbonyls in Los Angeles air. JAPCA, 34:537-543.

Haszpra L., Szilàcyl I., Demeter A., Turànyi T., and Bérces T. (1991). Nonmethane hydrocarbon and aldehyde measurements in Budapest, Hungary. Atmos. Environ., 25A:2103-21 10.

Killus J.P., Whitten G.Z. (1982). A mechanism describing the photochemical oxidation of toluene in smog. Atmos. Environ., 16:1973-1988.

Leone J.A., Flagan R.C., Grosjean D. and Seinfeld J.H. (1985). An outdoor smog chamber and modeling study of toluene-NO_x photooxidation. Int. J. Chem. Kinet., 17:177-216.

Singh H.B., Salas L.J., Cantrell B.K. and Redmond R.M. (1985). Distribution of aromatic hydrocarbons in the ambient air. Atmos. Environ., 19:1911-1919.

Wirtz K., Becker K.H., Martin-Reviejo M. and Millan M. Millan (1994). Computer simulations of recent experimental results on the oxidation of aromatic hydrocarbons. EUROTRAC annual Reports, Part 8-LACTOZ. Commission of the European Communities, Brussels, Belgium.

Zielinska B. and Fujita E. (1994). The composition and concentration of hydrocarbons in the range of C2 to C18 in downtown Los Angeles, CA. Res. Chem. Intermed., 20:321-334.

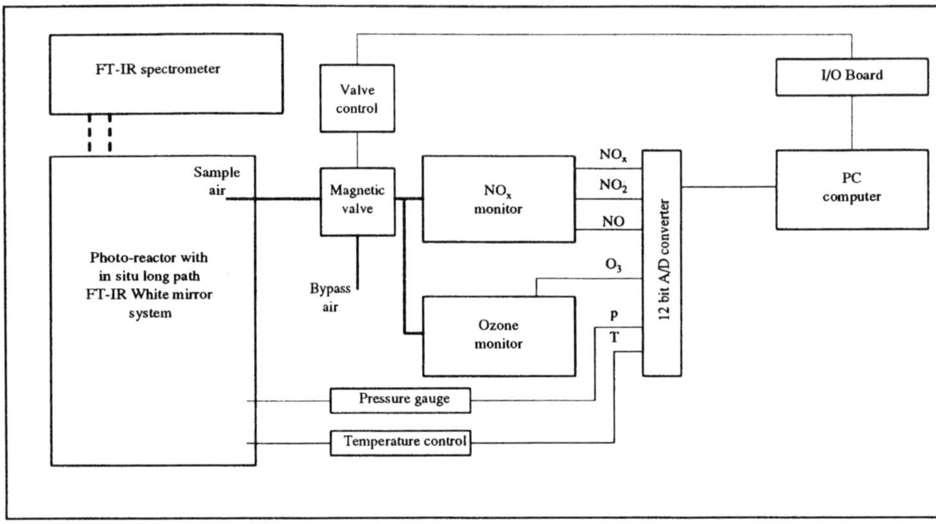

Figure 1: Schematic diagram of the reaction chamber and the monitoring system.

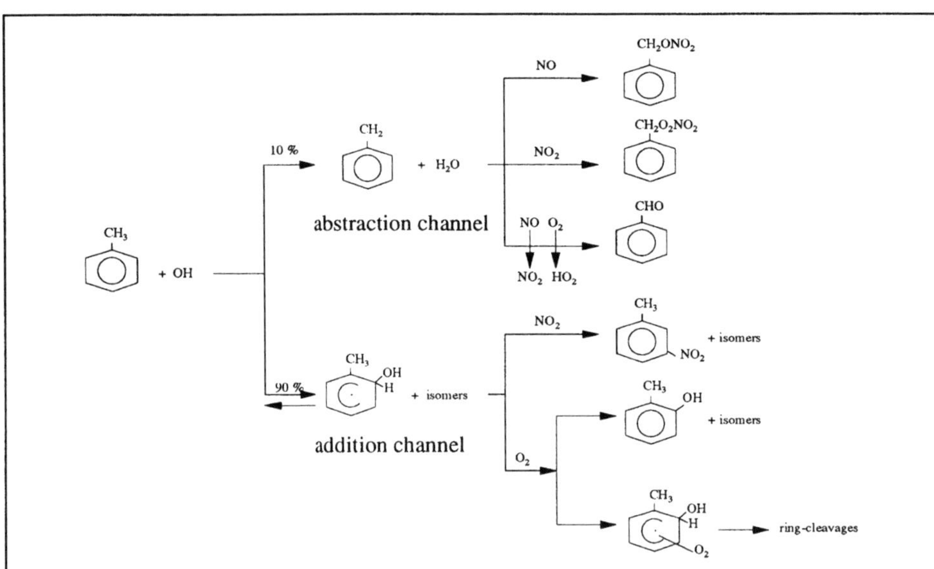

Figure 2: Simplified oxidation mechanism of toluene in the presence of NO_x.

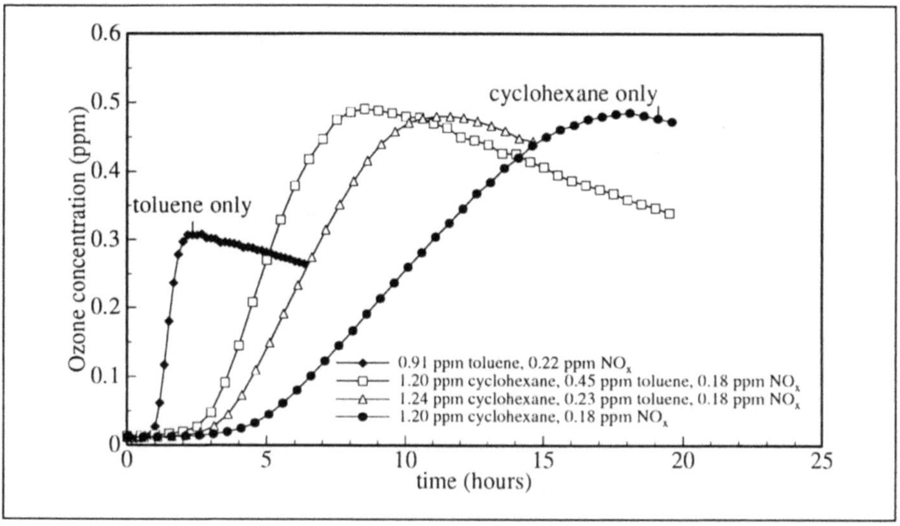

Figure 3: Ozone formation during the irradiation of toluene/cyclohexane/NO_x/air mixtures.

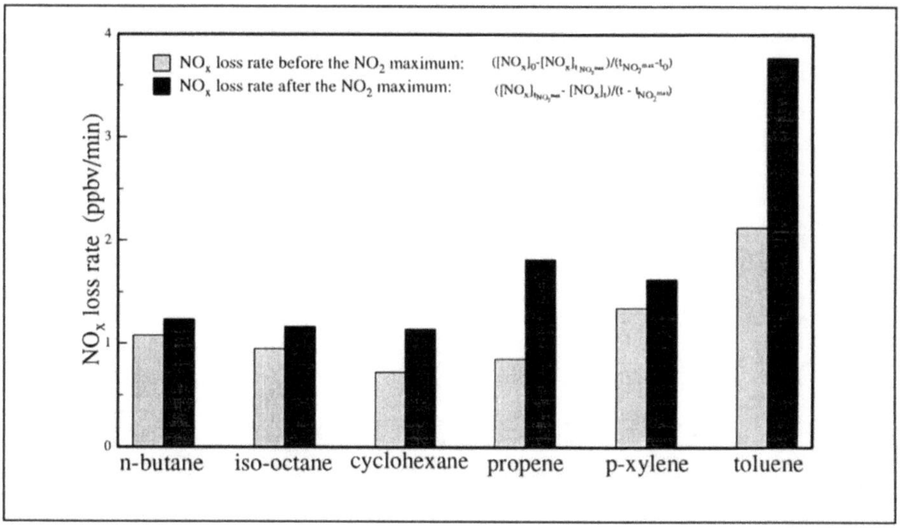

Figure 4: NO_x loss rates before and after the NO_2 maximum derived from the concentration-time profiles of NO and NO_2 during single hydrocarbon irradiations.

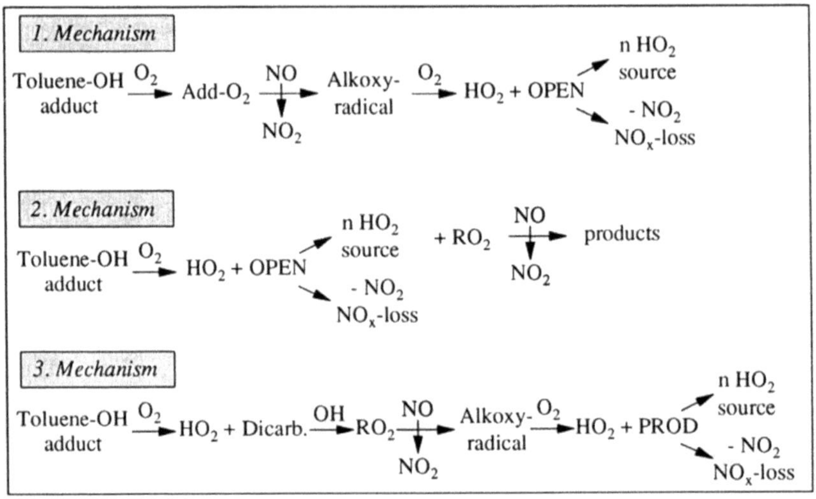

Figure 5: Different chemical mechanisms for the subsequent reaction of the toluene-OH adduct.

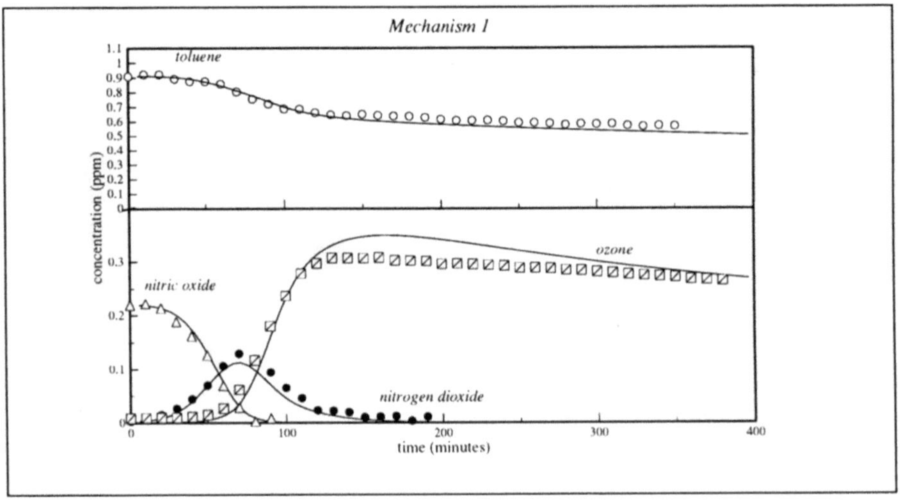

Figure 6: Experimental and simulated concentration-time profiles of NO, NO_2, ozone and toluene. for the toluene-only irradiation (mechanism 1).

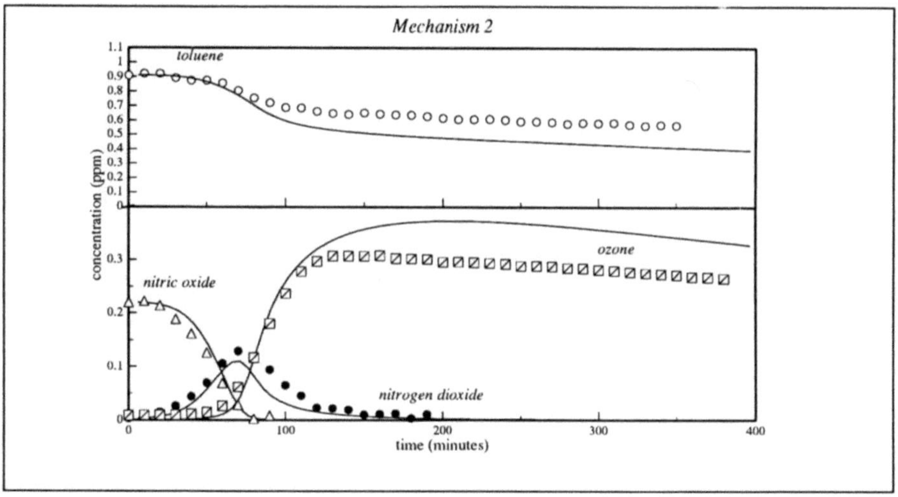

Figure 7: Experimental and simulated concentration-time profiles of NO, NO_2, ozone and toluene for the toluene-only irradiation (mechanism 2).

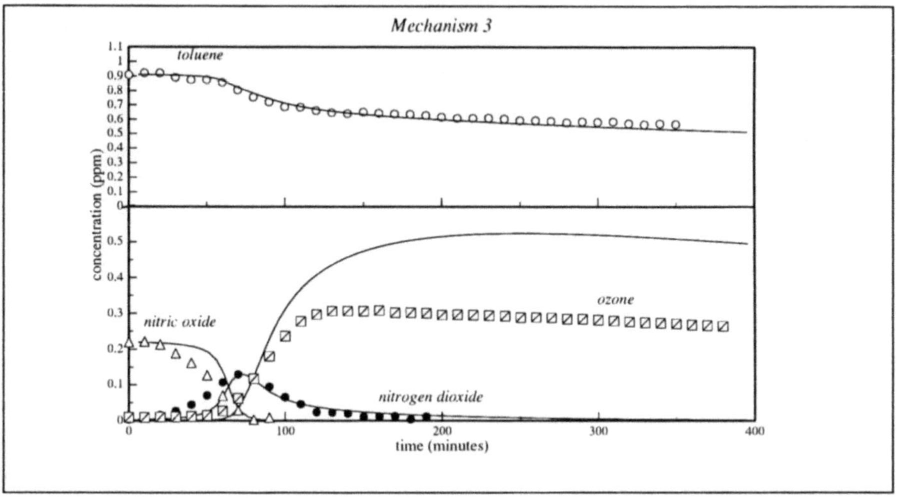

Figure 8: Experimental and simulated concentration-time profiles of NO, NO_2, ozone and toluene for the toluene-only irradiation (mechanism 3).

Laboratory Studies of Elementary Reactions

Christa Fittschen
Laboratoire de Cinétique et Chimie de la Combustion
CNRS URA 876
Bâtiment C11
F-59655 Villeneuve d'Ascq
France

Introduction

About 90 % of the total atmospheric mass resides in the troposphere. Besides the main constituents (N_2, O_2, Ar, CO_2) one can find hundreds of different chemical species. They may have biogenic and / or anthropogenic origin. Tropospheric chemistry mainly consists in the oxidation of these trace gases.

An approach to understand the atmospheric chemistry and to predict the chemical changes of the composition of the earth's atmosphere, is to model the system. Among other things rate constants of elementary reactions are needed as input parameters.

Chemical reactions that occur in the atmosphere are essentially bimolecular and termolecular reactions of radical-molecule or radical-radical type. The two major experimental set-ups which are used to measure absolute kinetic data of atmospheric reactions are the discharge flow reactor and the pulsed photolysis. Even if there are other methods e.g. continuous photolysis, this paper is restricted to the description of these two methods.

Pulsed Photolysis method

The pulsed photolysis consists in photolytic production of radicals by a pulsed light source, either a laser or a flash

NATO ASI Series, Partnership Sub-Series, 2. Environment – Vol. 8
Urban Air Pollution
Edited by Ivo Allegrini and Franco De Santis

lamp. The light pulse has to be short in comparison to the radical loss via chemical reaction. To prepare e.g. OH-radicals by laser photolysis, one has several possibilities:

$$HNO_3 + h\nu\ (248\ nm) \rightarrow OH + NO_2$$

$$HNO_2 + h\nu\ (351\ nm) \rightarrow OH + NO$$

$$H_2O_2 + h\nu\ (248\ nm) \rightarrow 2\ OH$$

$$N_2O + h\nu\ (193\ nm) \rightarrow O\ (^1D) + N_2$$

$$O\ (^1D) + H_2O \rightarrow 2\ OH$$

The kinetics of the radical is then monitored under real time conditions by optical methods mainly fluorescence and absorption. These two methods will be described in some detail

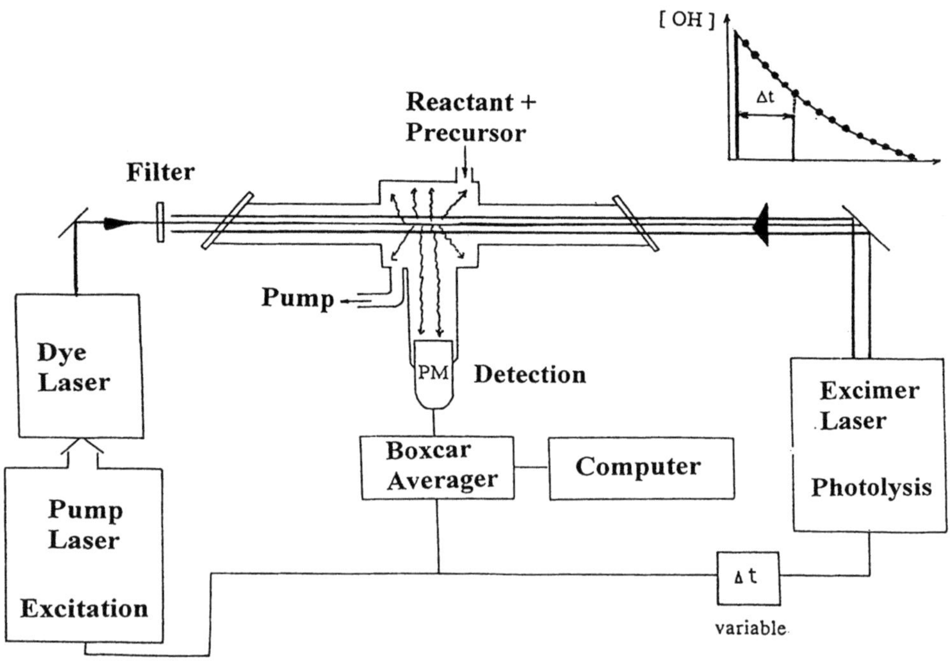

Fig. 1: Typical set-up for Laser-Photolysis / Laser Induced Fluorescence

Because the reaction zone is in the center of the cell there are no complications due to heterogenous wall processes. The low pressure limit is ≈ 10 Torr. Below this value radical loss due to diffusion outside of the observation volume becomes too excessive. The limit for high pressures is given only by mechanical problems, this means that experiments at atmospheric conditions can be easily performed. A photolysis cell can be constructed with heating and / or cooling facilities so rate constants can be measured in a wide temperature range.

Discharge flow method

In a discharge flow tube the radicals are produced by microwave discharge either directly like H-atoms by discharge of H_2/He-mixtures or by a sequence of reactions, e.g. the reaction of F-atoms (generated by microwave discharge) with H_2O to form OH radicals. The radicals are highly diluted in a rare gas flow with high speed (≈ 10 - 50 m/s) in the cylindrical flow tube. The reactant is added via a movable injector somewhere between the radical generation and the detection point.

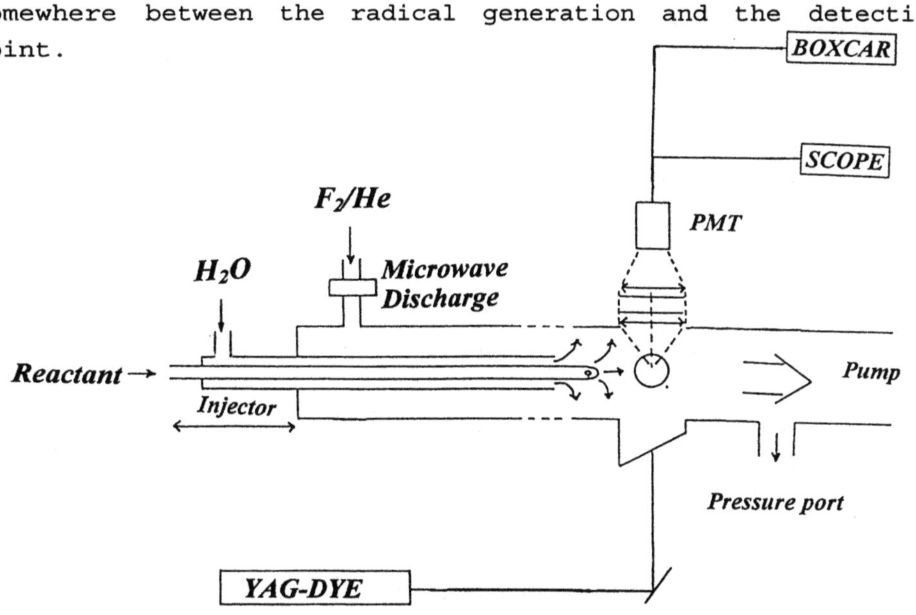

Fig. 2: Discharge flow reactor with LIF-detection

The reaction time is then calculated from the distance and the velocity of the gases. To obtain the kinetics of the radical, its concentration has to be measured with the injector in various positions.

The detection is done at the end of the tube by numerous methods like absorption and fluorescence, but also electron spin resonance or mass spectrometry. Due to the fact that discharge flow experiments are carried out under steady-state conditions, there is no limitation to the detector speed.

A problem in flow tube studies are unavoidable wall reactions. This problem becomes normally more pronounced with decreasing temperature and can thus limit the temperature range. To minimize the correction for radial diffusion, one has to work at pressures below ≈ 10 Torr. Under these conditions diffusion is fast enough to maintain a uniform radial profile.

The advantage of the flow tube compared to the pulsed photolysis method is the wider choice of detection methods due to the steady state conditions. There is also the possibility of radical generation by a sequence of reactions while the pulsed photolysis method requires an appropriate precursor. The disadvantages are the wall reactions and the very limited pressure range. A disadvantage in pulsed photolysis can be the unavoidable photolysis of some reactants, especially if the radical generation process requires short wavelengths.

Both methods work under so-called pseudo-first-order conditions. This means that the kinetic of one reactant (normally the radical) is measured in the presence of an excess of the other reactant. Under these conditions one can assume that the concentration of the second reactant is constant and the kinetic equation for bimolecular reactions is simplified to a first order equation :

$$\ln \frac{[A]_o}{[A]} = k^1 \times t$$

with $k^1 = k \times [B]_0$

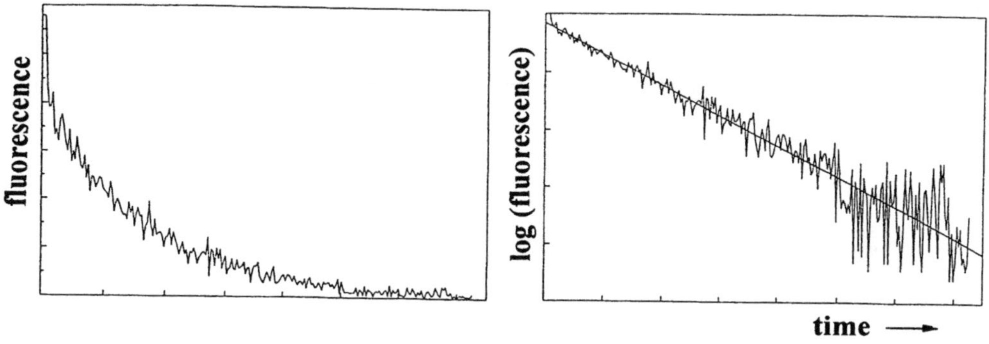

Fig.3: Typical signal of a Laser Photolysis / RF-experiment.

Thus one only needs to know the absolute concentration of the reactant in excess B_0. Measuring the first order rate constant as a function of $[B]_0$ allows us to determine the required rate constant:

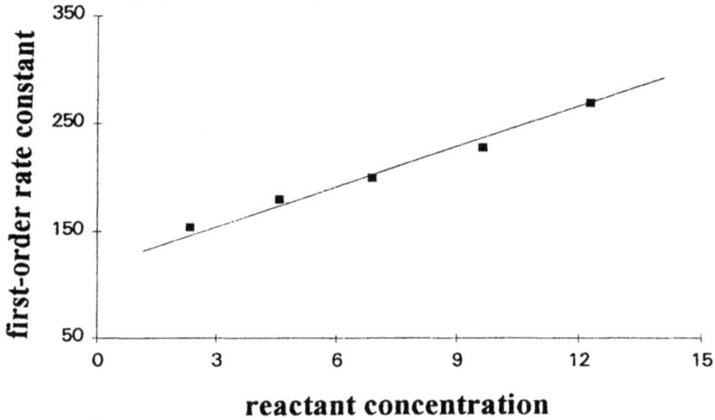

Fig.4: Typical diagram of pseudo first order rate constant versus reactant concentration

The intercept accounts for wall reactions in the case of flow tube studies and for diffusion and reaction of the radical with the precursor in the case of pulsed photolysis.

Fluorescence

The main processes following photochemical excitation are shown in fig. 5:

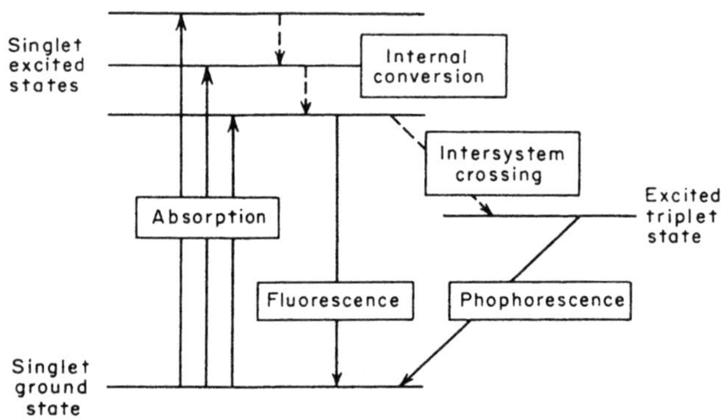

Fig.5: Jablonski diagram

The absorption of a photon leads in most molecules to a transition from the singlet ground state to a singlet excited state. No matter which upper singlet excited state has been reached, there is normally a rapid radiationless transfer of energy to the lowest excited singlet state, called internal conversion. From here there is competition between two processes:

- the first is the radiationless intersystem crossing down to an excited triplet state, followed by emission of a photon, called phosphorescence,
- the second is the direct emission of a photon , called fluorescence.

A third possibility is the deactivation by collision, called quenching. This process becomes more and more important with increasing pressure.

In general the intensity of the fluorescence is proportional to the concentration of the species; nevertheless, there is no simple correlation between intensity and absolute concentration.

In Laboratory studies a distinction is made between Resonance Fluorescence (RF) and Laser Induced Fluorescence (LIF) depending on the source of the excitation photon.

In the case of the RF one produces the radical to observe in an excited state (e.g. OH radicals by microwawe discharge of H_2O/He mixtures at low pressure). The following emission of radiation is then used to excite the radicals in the reaction celll. In the case of LIF the photons are provided by a pulsed laser often a dye laser pumped by a Yag or Excimer laser.

The main advantage of RF, besides its cheap and simple experimental set-up is that the resonance lamp is a continuous light source. So one can observe the entire temporal evolution of a species after one photolysis pulse. To obtain the complete concentration time profile in the case of LIF one has to change the delay between photolysis and excitation pulse, scanning the profile point by point. In return LIF is much more sensitive than RF and not limited to radicals that can be produced by microwave discharge.

Absorption

The fraction of incident light absorbed by a species is directly proportional to its concentration as shown iin the Beer-Lambert law :

$$\lg \frac{I}{I_o} = - \varepsilon \times c \times x$$

with ε being the absorption coefficient, c the concentration and x the path length.

Thus measuring time resolved the intensity of light at a wavelength, where the radical absorbed, allows us to observe the concentration time profile after a photolysis pulse. The light source can be either a lamp (then a monochromator is needed to select the desired wavelength) or a cw-laser.

The advantage of the absorption technique compared to fluorescence is that all species can be detected, while fluorescence is limited to those emitting fluorescence radiation after excitation. The HO_2 radical for example, being a very important species in atmospheric chemistry, can not be detected by fluorescence. In return, absorption is less sensitive than fluorescence, because one has to measure a small difference between two large signals. This might be no longer true at high pressures, due to the increasing importance of quenching processes and with this the decreasing intensity of fluorescence signals.

Design and planning of field campaigns

Finn Palmgren
Department of Emissions and Air Pollution
National Environmental Research Institute (NERI)
Frederiksborgvej 399
DK-4000 Roskilde
Denmark

Introduction

In the past few decades, air pollution in many European cities has changed in character. Improved technology, higher stacks and legislation limiting the sulphur contents in fuels have resulted in a marked reduction in the pollution levels of SO_2. On the other hand, due to the steady growth of road traffic, the pollution from traffic has increased substantially during the same period. In Western European countries the importance of air pollution from sulphur dioxide is diminishing. This is not the case in some Southern and Eastern European cities. Of growing importance in all European cities is the air pollution from traffic. Thus, studies of emissions, dispersion and chemical transformations of traffic air pollution in urban areas are becoming of increasing interest. This is felt both at European and national levels (i.e. COST Action 615 "CITAIR" on "*Database, monitoring networks and modelling of urban air pollution*")

Monitoring and control strategies for urban air pollution can e.g. be based on permanent monitoring programmes and possibly supplemented with field measurement campaigns. To obtain the optimal yield of the measurements, it is necessary to design and plan the measurements in accordance with the objectives and the hypotheses to be tested.

Air quality monitoring has been carried out in many urban areas for many years, and many useful data have been collected. The monitoring programmes have been established in accordance with national legislation or international conventions and agreements. In addition, monitoring programmes have been implemented in order to fulfil local requirements for information and for evaluation

NATO ASI Series, Partnership Sub-Series, 2. Environment – Vol. 8
Urban Air Pollution
Edited by Ivo Allegrini and Franco De Santis

of the impact of the air pollution on health, vegetation and materials.

The permanent monitoring programmes are necessary for obtaining long term series for evaluation of the trends and for control of the measures taken to reduce the air pollution in urban areas. However, it is often too costly or at least not cost-effective to measure many different pollutants by special and advanced measurement methods at the permanent monitoring sites, because these sites may be selected by other criteria. A more effective use of the resource are field measurement campaigns, if they are designed in a proper way. Field campaigns and permanent monitoring programmes can also be supported by application of models.

Air quality models have been developed for all different geographic scales, e.g. global, regional, urban, local and micro scale. For urban, local and micro scales they have been developed for point sources (tall stacks), area sources (domestic heating or small industries) and line sources (road traffic at highways and in street canyons) (*Olesen and Mikkelsen, 1992*). They have been used in many countries and it has been demonstrated that they are very useful e.g. as planning and control tools and for estimates of exposure of humans and vegetation to air pollutants. Results from measuring programmes can also be applied to support further development of models.

This paper will address the use of field campaigns designed especially for studies of the atmospheric dispersion and chemical processes in urban air. The understanding of these processes is a crucial basis for the development of air quality models and also for the establishment of measurement programmes for further evaluation of the impact on the environment.

Objectives of field campaigns

Measurement campaigns, including measurements of all the important pollutants, primary as well as secondary pollutants, and meteorological parameters, are important tools for obtaining information about the physical and chemical processes in the urban atmosphere. Generally, the field campaigns should be linked to the existing permanent monitoring programme in order to estimate how well the short field campaigns represent the air pollution situation of the urban area. The objectives of field campaigns are numerous, but some main objectives are listed in the following:

<u>Chemical and physical processes</u> can be studied by field campaigns. However, it is very important to design the campaign in a way, that the key parameters are measured with the optimal time resolution and at sites, which are representative of the processes to be studied.

<u>Test of air quality models</u> could be a part of the process studies, but for evaluation of models it is necessary to measure under many different conditions in order to test how well the models describe the real world.

<u>Measurements in connection with EU frame directive</u>. The new EU frame directive has promoted investigations of air pollution in an urban area prior to a possible establishment of permanent monitoring stations. If it is documented that the levels are below certain limits, monitoring may not be necessary or may be replaced by field campaigns and/or model calculations.

<u>New measurement methods</u> can be tested under realistic conditions in field campaigns by comparison with established methods or reference methods, which may be too complicated to use on a routine basis.

Urban air pollution

Three main types of air pollution sources exist in urban areas: the large stationary sources with tall stacks (point sources), the small stationary sources distributed in an area (area sources) and the mobile sources, i.e. the traffic (line sources). The concentration of a pollutant at a location in an urban area consists of a mixture of contributions from these sources, secondary formed air pollutants and long range transported pollutants from other urban or industrial areas.

Many monitoring networks are addressed particularly to air pollution from car traffic. With the growing number of cars, it will be even more important to find out, whether the primary air pollutants (e.g. carbon monoxide, nitrogen monoxide, benzene, toluene, diesel soot etc.) or the secondary air pollutants (e.g. ozone, nitrogen dioxide, PAN etc.) are increasing to harmful levels.

The air pollution at a certain location in an urban area is a result of

- <u>the regional background air pollution</u>, which is transported over long distances, e.g. from other urban areas or by recirculation in the region. The air pollutants have had time to react with other pollutants or natural components of the atmosphere,

- <u>the urban background air pollution</u>, which is the regional background plus the air pollution from the urban area. This air pollution is a result of the city as a whole and the chemical reactions, which has reached a certain level and has been influenced i.e. by solar insolation and other components in the air,

- <u>the local air pollution</u>, which is the air pollution emitted close to the measurement/exposure location. Only the fastest chemical reactions are important for the local emitted pollutants. The concentration of these pollutants are mainly determined by the dispersion.

Air pollution at all three levels must be taken into account in order to obtain a total description.

The regional background can be difficult to estimate, depending on climatic conditions. In the northern part of Europe the dispersion of pollutants is mainly determined by the air flow due to the prevailing wind and surface roughness. In the southern parts of Europe the dispersion is highly determined by thermal convection and local wind flow systems. The regional air pollution can be measured at stations outside greater urban areas, typically 25-50 km from the city center or by tracing the air parcels.

On the large scale, especially in the southern part of Europe, the dispersion is more complicated, due to complex flow structure (*Millan et al., 1992*). The long residence time may also lead to more complicated chemical processes, which are not easy to describe. In these cases the models for dispersion and chemical reactions need further development and our present knowledge is mostly based on measurements (*Febo, 1994*).

The urban background can be measured by a few measurement stations located far away from local sources, e.g. above the roof tops. The representativity of such stations for the whole urban area must be documented.

The local air pollution depends very much on local sources, e.g. the traffic in the street, and local dispersion conditions, e.g. the surrounding buildings.

The main properties of the wind flow in an urban area and especially in a street canyon are well-known. When the wind direction is perpendicular to the street direction a vortex is generated in the street canyon, whereby the wind flow at street level is opposite to the flow above the roof level. Such a wind circulation results in a characteristic dependence of pollution in the street on wind direction. Concentrations on the windward side are much lower than the ones on the leeward side.

The simple and very important chemical processes in the local scale involving NO, NO_2 and O_3 are also quite well-known. The residence time of the air in the street canyon can be estimated and by this the transformation to NO_2 of the emitted NO from the traffic in the street. However, many processes are not understood to such a degree that they could be included into models, e.g. the formation of radicals and the transformation of the VOCs.

A description by measurements requires many measurement stations. However, many suitable air pollution models exist for the calculation of dispersion on a local scale and for fast chemical reactions, if all relevant components of the urban background and local meteorological parameters are available (*Berkowicz et al., 1994*).

Design criteria for field campaigns

The design of field campaigns must take into account the above described processes. The design can be supported by model estimates.

The quality assurance and documentation of the quality of measurements is decisive for the result of field campaigns (and all other measurements). The precision and the accuracy must be quite good. The measurement methods must be documented with respect to detection limits, interferences, traceability etc. and all measurements must by synchronized in time.

The field campaign in urban areas should be coordinated with the permanent monitoring network and the quality of these data should also be documented. This could be included in the data analysis.

It is an advantage to measure many components at more than one location in order to be able to compare results. A certain degree of redundance is useful.

Meteorological data are necessary for all field campaigns. They could be obtained from the permanent meteorological measurement networks. In some cases, it is necessary to establish special meteorological stations or apply meteorological models.

Examples on field campaigns in European cities.

In the following, results from three different field campaigns in Milan, Brussels and Copenhagen are used as illustration.

Milan. In the frame of the National Strategic Plan "Aree Metropolitane ed Urbane", the Institute of Atmospheric Pollution of the National Research Council, Rome, organized a measurement campaign in the Greater Milan Area from February 8 to 21, 1993 (*Ciccioli, 1993*). All the national institutes and laboratories involved in research activities in the field of air pollution and in particular meteorology, micro physics, chemical and biological monitoring and modelling, concentrated their efforts in characterizing a typical winter smog situation in Milan.

Joint Research Center (JRC) Ispra participated by the application of a DOAS (Differential Optical Absorption Spectrometry) system (*Platt and Perner, 1980*) for the characterization of urban background by measurements above the roof tops near the center of Milan. The DOAS measurements included SO_2, NO_2, O_3, benzene (C_6H_6), toluene (C_7H_8), P-xylene (C_8H_{10}), formaldehyde (CH_2O) and nitrous acid (HONO). The data from this contribution to the campaign have been published by (*Palmgren et al., 1993*) and (*Palmgren and Sandroni, 1993*). Selected results are used to illustrate the design of field campaigns. The main objectives of JRC's contribution to the campaign was to test the DOAS method and to investigate the urban background.

At the Observatory of Milan (Brera) a permanent monitoring station equipped with traditional monitors has been established. The data from the permanent station can be compared with DOAS results. Figure 1 shows the results for SO_2. The agreement is good. This was also the case for O_3 (figure 2). The NO_2 results showed a less good agreement that appeared to be due to incorrect calibration of the chemiluminescence monitor.

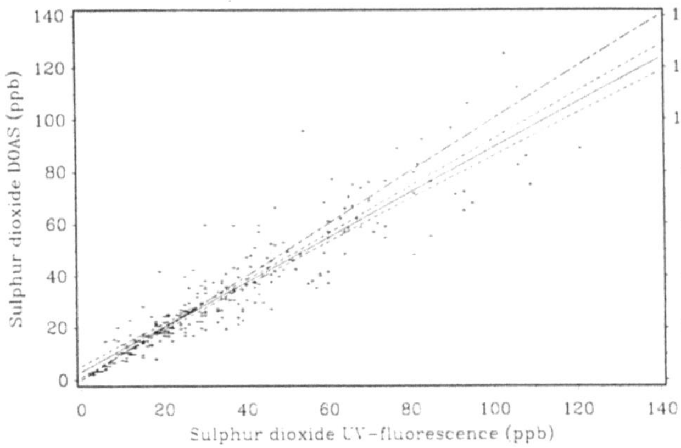

Figure 1. SO_2 measured at the Observatory of Milan by DOAS and by UV-fluorescence.

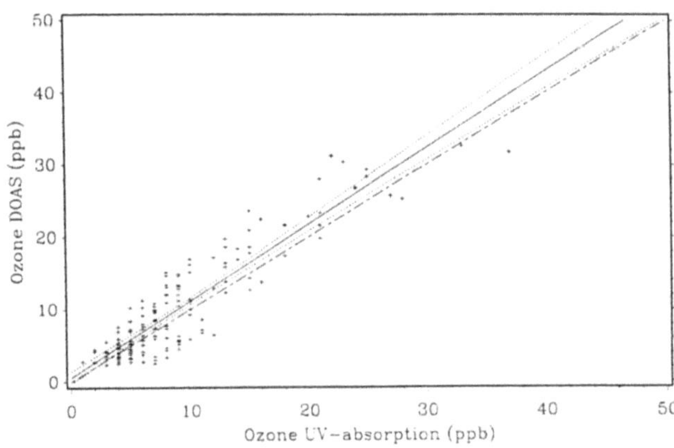

Figure 2. O_3 measured at the Observatory of Milan by DOAS and by UV-absorption.

In the figures 1 to 6 the solid lines are the regression lines, the dotted lines are 95% confidence lines and the dot and dash lines are 1:1.

Measurements of many pollutants were carried out permanently at a site near the Cathedral by DOAS at roof level. These measurements were compared with the DOAS measurements during the campaign at Brera. The distance between the two locations is a few kilometers. NO_2 and O_3 may not vary much over the roof tops near the center of a large urban area, because they are secondary pollutants formed within the area or transported to the urban area. Figure 3 and figure 4 show, that the correlation between the concentrations measured at the two sites is very good and that the levels are the same, indicating correct calibration of the instruments and homogeneous urban background for these components. This was also confirmed for toluene, whereas benzene showed incorrect calibration and the other pollutants were most of the time below the detection limits of DOAS during this winter campaign.

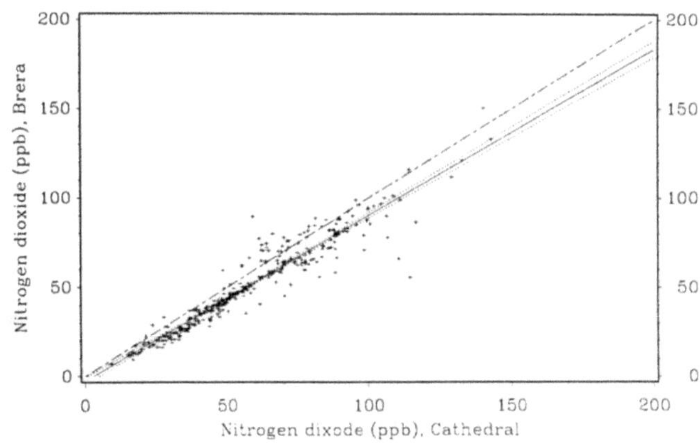

Figure 3. NO_2 measured at the Observatory of Milan and at the Cathedral of Milan.

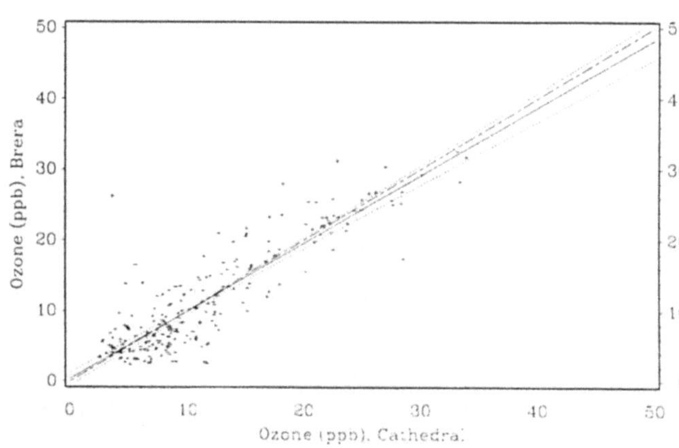

Figure 4. O_3 measured at the Observatory of Milan and at the Cathedral of Milan.

The traffic is the one of the major air pollution sources in Milan. Benzene and toluene are mainly emitted from the petrol engines by evaporation from the fuel system or incomplete combustion. Therefore, a good correlation should be expected for the two pollutants and this was also observed during the campaign (figure 5). The ratio between the concentrations in air was the same as the ratio for petrol.

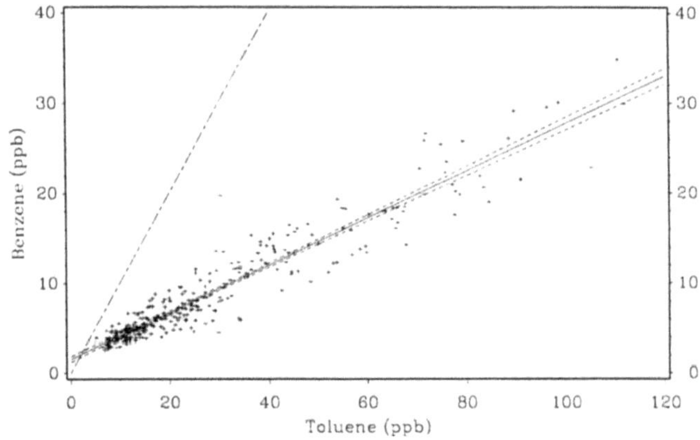

Figure 5. Benzene versus Toluene measured at the Observatory of Milan.

The correlation between CO and toluene, both originating from the traffic, was also good, figure 6. However, the ratio between CO, toluene and NO_x did indicate that the emission of CO from the car park in Milan was relatively high compared with data from Brussels.

A good correlation was also observed between NO_x and toluene. If the concentration of benzene and toluene were assumed to be proportional with the CO and NO_x, annual statistics of benzene and toluene and maybe other components, could be established on the basis of CO and NO_x measurements from the permanent monitoring station and from the monitoring campaigns (ref). However, the monitoring campaign has only been carried out during winter, when the photochemical activity was low. Indication of transformation of toluene during hot summer periods has been observed in Brussels (*Sandroni et al., 1994*) and possibly also

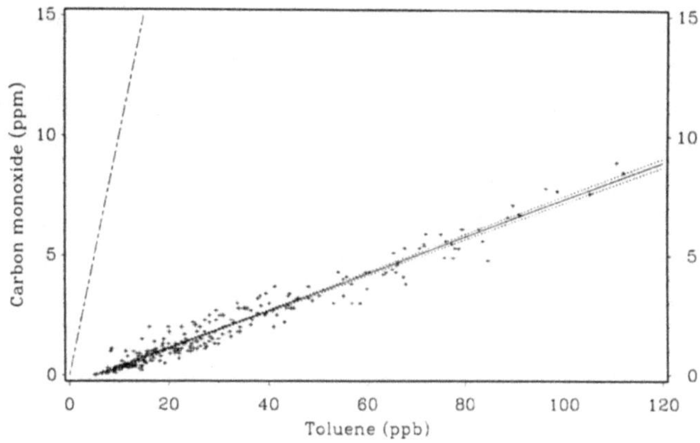

Figure 6. CO versus toluene measured at the Observatory of Milan.

Berlin. This illustrates one of the limitations of the short campaigns. Campaigns during other weather conditions are needed.

Brussels. The field campaigns in Brussels have been carried out by Atmospheric Chemistry Unit of the Environment, JRC under an agreement between EEC DG XI/B3 Urban Pollution Unit, the Brussels Environment Authorities. Two intensive campaigns were planned and carried out in summer 1993 and in winter 1994. Each campaign lasted 2 weeks (*Sandroni et al., 1994*).

The objectives of these campaigns were:

- to gain knowledge of the concentration levels of traditional and new noxious substances at a representative site of the city,
- to demonstrate the use of non-conventional monitors such as DOAS systems in the real-field conditions, and
- to estimate the vertical dispersion of pollution and possible chemical transformations in the street canyon and at the roof tops.

The main emphasis was laid on description of the dispersion and the atmospheric chemistry in a street canyon.

A map of the site is shown in figure 7. Three mobile measurement vans were installed at street level, near the kerb side. Two DOAS systems were installed one in the street and one above the roof tops. Further details are published in *Sandroni et al., (1994)*, in which also an analysis is presented that explains the behavior of the observed concentrations with respect to the actual meteorological conditions, and elucidates the representativeness of the measurements performed during the short campaigns. The differences in pollution levels observed at the different sites is also explained.

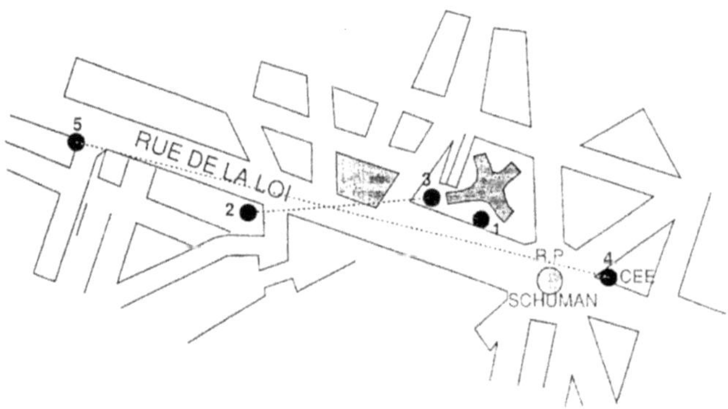

Figure 7. The measurement sites of Brussels.

Rue de la Loi is a one way street with 60,000 cars per day. It is mainly a street canyon with high buildings. However, it has not been possible to locate the monitoring stations (van1, van2 and van3) in the street canyon itself, but at parking lots (with no buildings behind) and at more open parts of the street (near the Berlaymont building). In figure 8 the difference between the measured concentrations of NO_x at van1 and van2 is shown . They were located on the opposite sides of the street. By subtraction of the two sets of data, the background (urban and regional) was eliminated and the contribution from the street could be illustrated.From the previous figure it appears, that even if the vans were located, where the street is not a street canyon, the dispersion was very similar to that street canyon. The concentration levels were higher on the leeward side of the street than on the windward side.

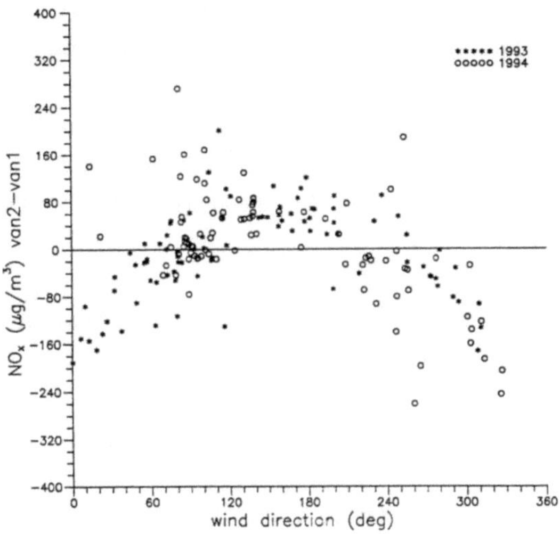

Figure 8. The difference between the NO_x concentrations measured at the two sides of Rue de la Loi in Brussels plotted against the wind direction at Uccle at the outskirts of Brussels.

During the first week of the summer campaign the weather was windy with showers, whereas the weather was very sunny and hot during the last week. This resulted in low concentrations of most

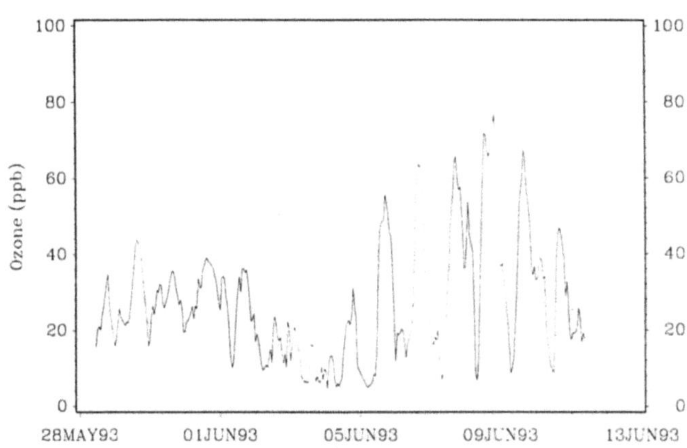

Figure 9. The measured concentration O_3 during the monitoring campaign in Brussels in May-June 1993.

of the pollutants during the first week and higher in the second week and also some photochemical activity with high O_3 levels (figure 9).
Most of the NO_x emitted from vehicles is emitted as NO; only 5-10 % is emitted directly as NO_2, possibly formed in the exhaust system. In the following it is assumed, that 8% NO_2 is emitted directly.

The only important fast chemical transformation of NO to NO_2 in a street canyon is oxidation by O_3:

$$NO + O_3 \rightarrow NO_2 + O_2$$

It means that the concentration of NO_2 in a street canyon is often limited by the supply of O_3 from the surroundings, especially from the air above the roof tops. In figure 10 NO_2 is plotted against NO_x, both measured at van1. The 1:1 line, where all NO is oxidized to NO_2, and theoretical lines for different levels of O_3 are drawn in the figure; the slope of these lines corresponds to 8% direct emission.

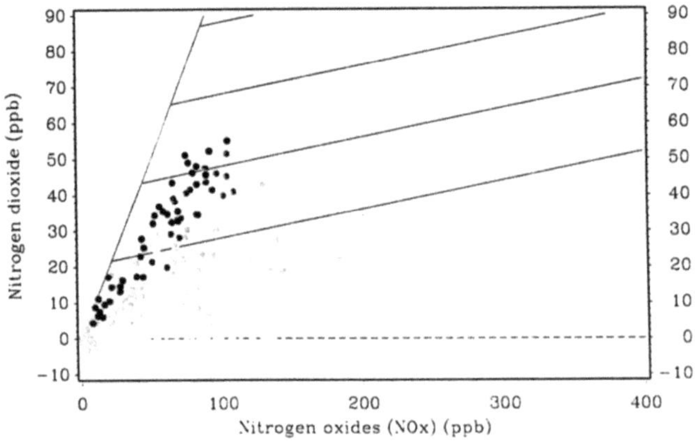

Figure 10. NO_2 versus NO_x at van1 for different levels of O_3 above the roof tops (0-20, 20-40, 40-60 and above 60 ppb O_3 showed by increasing intensity of the dots. The solid lines are the 1:1 line and the lines corresponding to different values of O_3 (20, 40, 60 and 80). The urban background of NO_2 measured at the roof tops were subtracted from the data from the street site.

The plot shows the measured concentrations of NO_2 and NO_x at van1 for different intervals of O_3 measured by DOAS above the roof tops. The measured values were in good agreement with the hypothesis. The scattering and the negative measured values for low O_3 concentrations were due to the normal experimental uncertainty.

The following observations have been made during the hot period of the field campaign in Brussels. The measurements were carried out above the roof tops, which means, that the pollutants have been transported over a certain distance and the chemical processes have had time to occur.

During the hot summer, period peak concentrations of NO_3-radicals were observed. NO_3-radicals are produced by reactions between NO_2 and O_3 during photochemical episodes. The highest concentrations of NO_3-radicals were observed at night because they are rapidly photolyzed during day-time. HONO are formed by reactions between NO_2 and radicals.

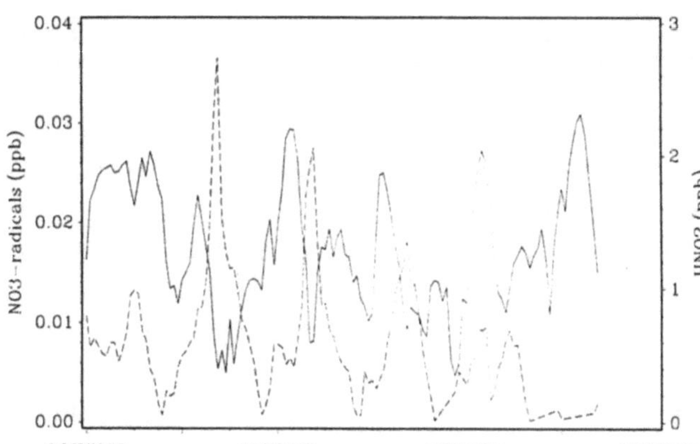

Figure 11. The measured concentration of HONO (solid line) and NO_3-radicals (dotted line) during a hot summer period in June 1993 in Brussels.

The concentration of NO_2 is determined by the emission of NO and the concentration of O_3 as mentioned earlier. The concentration of toluene correlated with the emission. However, during the hot period toluene was decomposed in the afternoons, figure 12.

Many other observations of the chemical processes have been observed during the campaigns and are reported in *Sandroni et al., (1994)*.

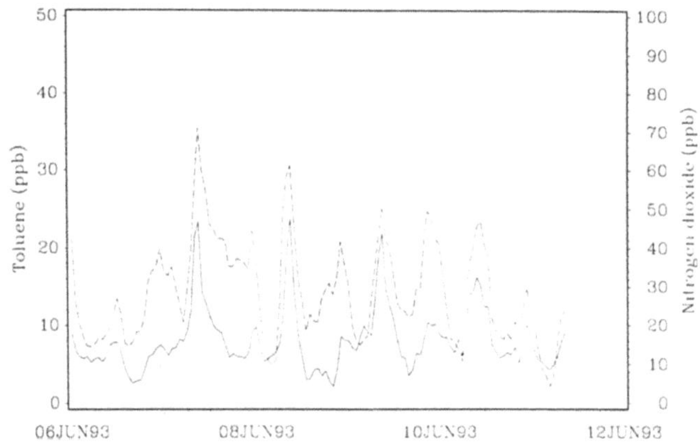

Figure 12. The measured concentration of NO_2 (solid line) and toluene (dotted line) during a hot summer period in June 1993 in Brussels.

Copenhagen. A project devoted to studies of air pollution from traffic in urban areas is being carried out within the framework of the new governmental Danish Environmental Research Programme, which was initiated in 1992/1993 (*Berkowicz et al., 1994*). Several Danish institutions are participating in this project. It includes continuous measurements of air pollution, traffic and meteorology at Jagtvej, a street in Copenhagen. Measurements of background air pollution are available from a monitoring station placed on the roof of a nearby building. Including measurements using DOAS.

The dependence of wind direction is shown at figure 14. NO showed a very clear dependence of the wind direction corresponding to the dispersion in the street canyon, whereas O_3 did not show a significant dependence of wind direction. SO_2 was highest, when the wind blew from south and southeast, where some industries and big coal fired power plant are located.

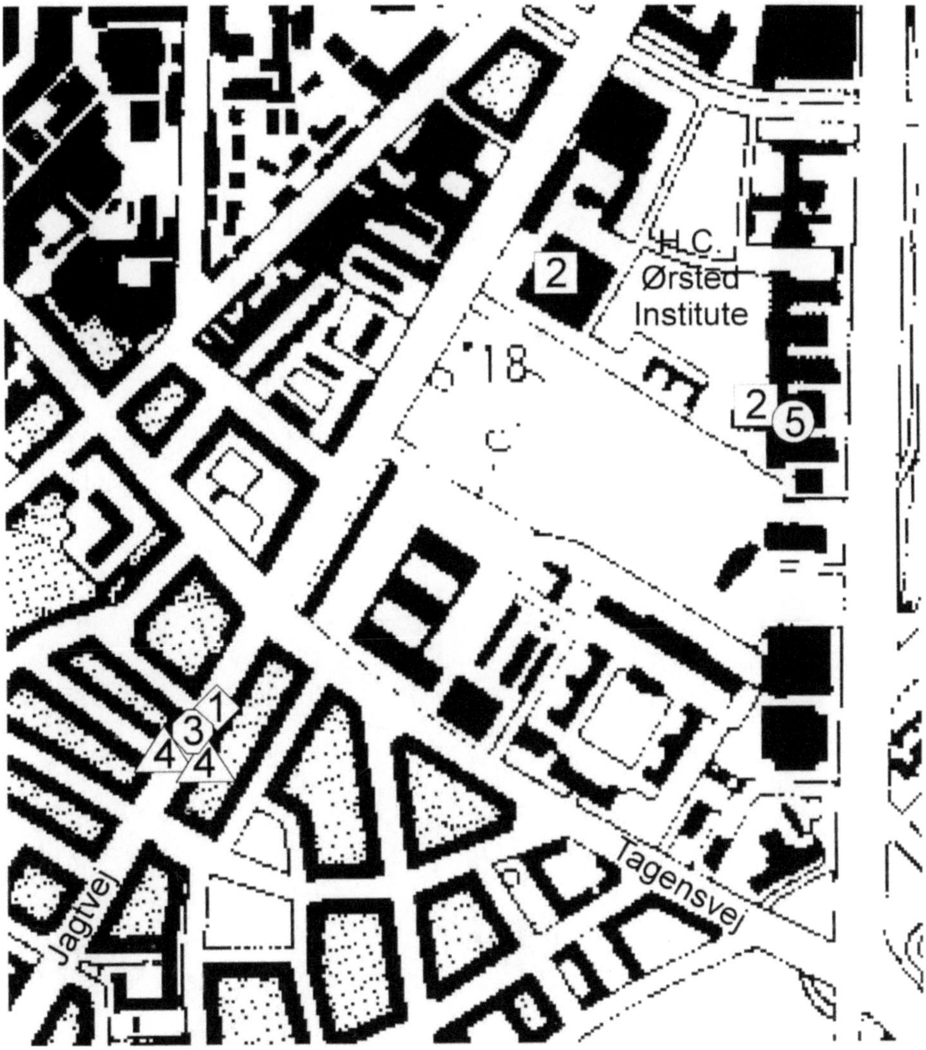

Figure 13. The air pollution measurements are carried out in connection with the Danish Air Quality Monitoring Programme. 1) the permanent monitoring station, 2) Urban background station (DOAS and traditional monitors), 3) Automatic traffic counts, 4) Special meteorological masts and 5) standard meteorological mast.

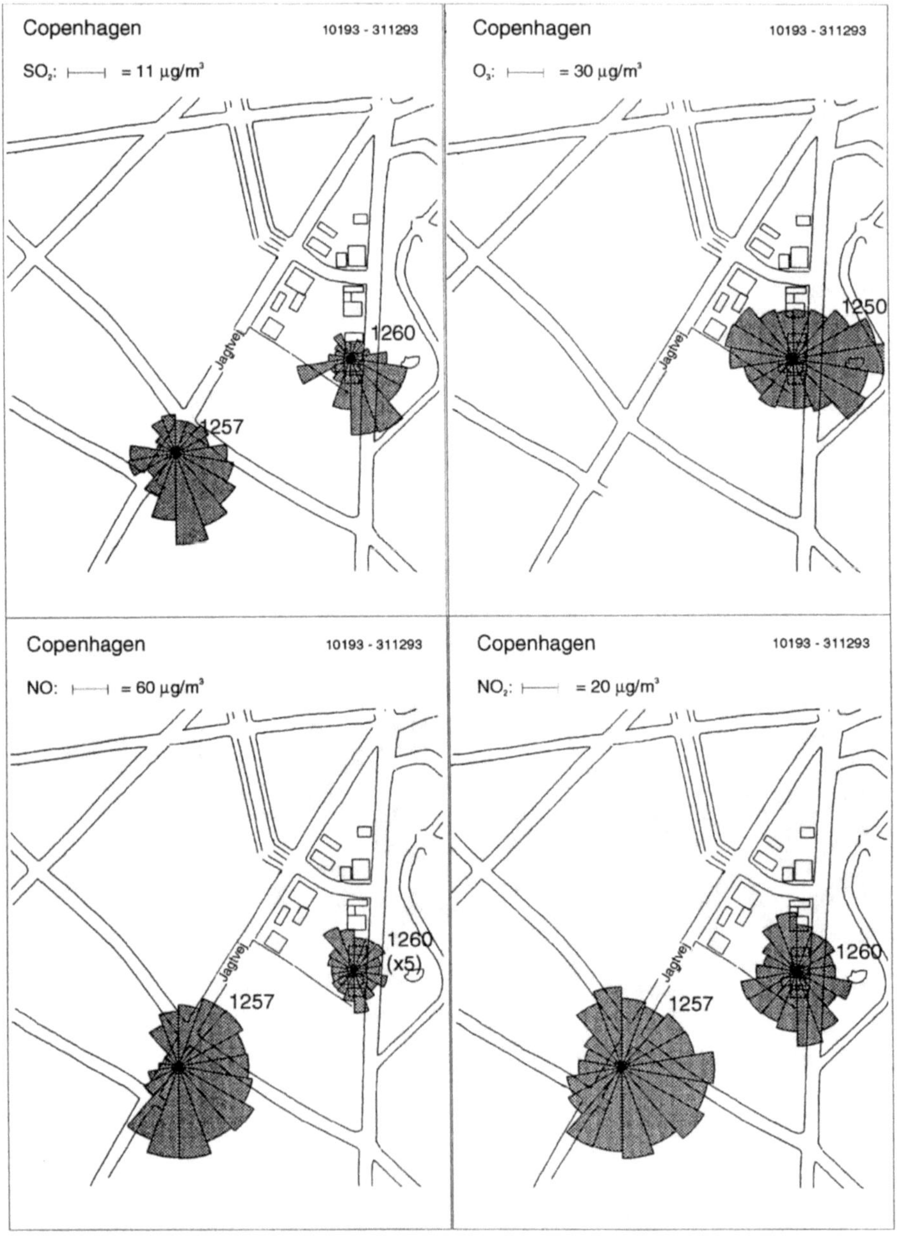

Figure 14. The concentration of SO_2, O_3, NO and NO_2 for different wind directions at the street station (Jagtvej) and the urban background station in Copenhagen.

An important part of the project is the study of the influence of traffic pollution on the formation of ozone and other photochemical oxidants. The transformation rate for oxidation of NO to the more noxious NO_2, as well as the formation rates of photo-oxidants such as ozone and PAN, are strongly dependent on the concentrations of volatile organic compounds (VOC) in the air. Among the main VOC sources, road traffic appears to be the most difficult to quantify accurately; measurements at street level combined with dispersion modelling will make it possible to estimate the contribution from traffic of various VOC species, and their distribution. Presently, few VOC species (benzene, toluene, p-xylene and formaldehyde) are measured continuously on the roof station by DOAS. Several measuring campaigns both in the city background area and in the street are performed in cooperation with the European Joint Research Centre ISPRA.

The concentration of toluene measured in Copenhagen correlated well with the concentrations of NO_x. These results could be compared with similar data from Milan. The levels were much

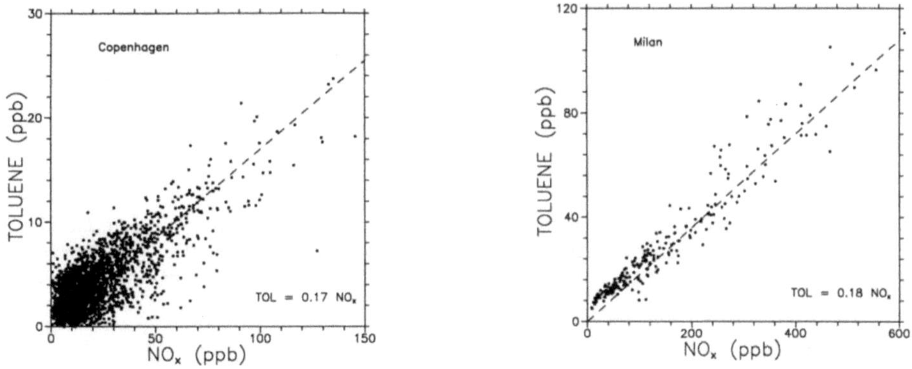

Figure 15. Toluene versus NO_x measured over roof tops in Copenhagen and in Milan. The dotted lines are the linear regression lines.

higher in Milan than in Copenhagen, but the ratio between toluene and NO_x was about the same, figure 15. In both cities the road traffic was the most important source of toluene and NO_x.
The data from the project are used for testing and developing the Operational Street Pollution Model (OSPM) (*Berkowicz et al., 1994*). In figure 16 the measured and calculated dependence of the NO_x concentrations on the wind direction is shown. The model explains the observed behavior very well. When the wind direction is such that the measuring point is in the windward side (the shaded area), the concentrations are much smaller than for the leeward case. The chemical reaction between O_3 and NO that is responsible for the formation of NO_2 is also included in the model and good results have been obtained (*Hertel and Berkowicz, 1989b*).

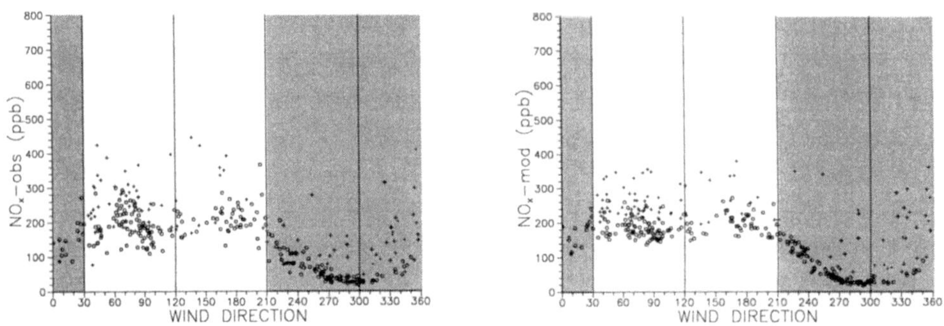

Figure 16. The concentration of NO_x measured and modelled for different wind directions at the street station (Jagtvej) in Copenhagen. + = wind speed: 1-3 m/s; o = wind speed: 4-6 m/s.

Conclusions

Field campaigns provide very useful information about the chemical and physical processes in urban air, but the yield of the campaigns is substantially improved by coordination with permanent monitoring network and support by air quality models for planning the campaigns and the interpretation of the data.

The following recommendations for urban field campaigns can be given:

- Measurement campaigns must as far as possible be coordinated with the existing permanent monitoring networks.

- Data should be collected for regional background, urban background and local scale. The data can be based on permanent measurements, measurement campaigns or model calculations.

- Siting criteria should be based on considerations about the chemical processes and dispersion processes to be studied, possibly based on complex meteorological and chemical models.

- Parameters to be measured should be based on considerations about the chemical processes and dispersion processes to be studied. The time resolution must be in accordance with the variations of emission and meteorological parameters. Redundance can be useful.

- The campaign should be as long as possible in order not to be too dependent on weather conditions and problems of instrumentation.

- Meteorological parameters and emission data (e.g. traffic counts) must be available.

- Quality assurance is very important, i.e. calibration and synchronization of measurements.

Acknowledgement
I wish to gratefully acknowledge Dr. G. Restelli, Dr. S. Sandroni and their colleagues for the cooperation during my stay at JRC Ispra. I am also grateful to Dr. R. Berkowicz and Ms. E. Vignati for the fruitful discussions and the analysis of the data from Copenhagen and Brussels.

References

Berkowicz, R., Hertel, O., Sørensen, N.N. and Michelsen, J.A. (1994). Modelling air pollution from traffic in urban areas. *IMA Conference Proceedings on Flow and Dispersion Through Groups of Obstacles,* University of Cambridge, Cambridge, England, 28-30 March 1994.

Ciccioli, P. (ed.) 1993. Proceedings of the "1st Italian Symposium on the Strategies and Techniques for the Monitoring of the Atmosphere", September 20-22, 1993. Societá Chimica Italiana, Rome. 758 pp.

Febo (1994). Description of Urban Pollution: Use of Proper Variables. Proceedings of. NATO Advanced Research Workshop. "Monitoring and Control Strategies for Urban Air Pollution", October 10-14, 1994, Erice, Sicily, Italy.

Hertel, O. and Berkowicz, R. (1989a). Modelling pollution from traffic in a street canyon. Evaluation of data and model development. National Environmental Research Institute. DMU Luft A-129. 77p.

Hertel, O. and Berkowicz, R. (1989b). Modelling NO_2 in a street canyon. National Environmental Research Institute. DMU Luft A-131. 31 pp.

Kemp, K., Palmgren, F. and Manscher, O.H. (1994). The Danish Air Quality Monitoring Programme. Annual Report for 1993. National Environmental Research Institute, Technical Report No. 113. 86 pp.

Millan, M.M. (ed.) (1992). Mesometeorological Cycles of Air Pollution in the Iberian Peninsula. Contract EV4V-0097-E. Report No 44 in the Air Pollution Research Report Series of the Environmental Research Programme of the Commission of the European Communities, DGXII. EUR No 14834. 220 pp.

Olesen, H.R. and Mikkelsen, T (eds.) (1992): Proceedings of the workshop "Objectives for Next Generation of Practical Short-Range Atmospheric Dispersion Models", May 6-8, 1992, Risø, Denmark. Danish Centre for Atmospheric research, P.O. Box 358, DK-4000, Denmark. 262 pp.

Palmgren, F., Cerutti, C.; Noriega, A., Melandrone, G., and Sandroni, S. (1993). Air Quality Measurements at Milan in February 1993 by a DOAS technique, *EUR 15392 EN*.

Palmgren, F. and Sandroni, S. (1993). Differential Optical Absorption Spectroscopy (DOAS). Principles and Applications for Air Quality Monitoring. Proceedings of the "1st Italian Symposium on the Strategies and Techniques for the Monitoring of the Atmosphere", September 20-22, 1993. Edited by P. Ciccioli. Societá Chimica Italiana, Rome.

Platt, U. and Perner, D. (1980). Direct Measurements of Atmospheric CH_2O, HNO_2, O_3, NO_2 and SO_2 by Differential Absorption in the Near UV. J. Geoph. Res., 85, C12 7453-7458. In: Killinger, D.K. and Mooredian, A. (eds.), 1993.

Sandroni, S., Cerutti, C.; Noriega, A., and Palmgren, F. (1993). Air Quality Measurements in Brussels (1993-1994). To be published as *EUR-report*. 40 pp.

Diffusion Denuders: tools for simultaneous sampling of NO_2 and PAN and for the determination of reactivity of marble surfaces

F. De Santis, I. Allegrini, P. Di Filippo, D. Pasella
CNR - Istituto Inquinamento Atmosferico
A. d. R. Roma, Via Salaria Km 29,300, CP 10
00016 Monterotondo Scalo, Roma
Italy

Introduction

The variety of both criteria and non-criteria pollutants which are simultaneously present in an urban atmosphere and the presence of different physical forms, present a delicate analytical problem and impose high demands on the sampling techniques especially with regard to selectivity. It is well known that an accurate assessment of gas/particle distribution by means of traditional sampling techniques is hampered by gas-phase adsorption on filter media and volatilization from collected atmospheric particles. The determination of nitric acid in the presence of nitrate or of ammonia in the presence of ammonium can be cited as common examples.
The goal of the research on diffusion techniques was to develop a sampling system which was not subject to the sampling artifacts mentioned above. Diffusion denuders were originally used to scrub out undesired gases from aerosols. Beginning in the early 80's, denuder diffusion techniques have developed into a well established analytical technique which is now widely used for general air sampling.
The rationale behind denuder technique is simple: denuder tubes preconcentrate an air sample by removing a reactive gas from a stream through selective adsorption on a specific coating. A diffusion denuder in its simplest form, consists of a glass tube coated on the inside with a reagent specific for the gaseous analyte of interest (Ferm, 1982). When a sample stream is passed through a denuder device, the analyte accumulates on

NATO ASI Series, Partnership Sub-Series, 2. Environment – Vol. 8
Urban Air Pollution
Edited by Ivo Allegrini and Franco De Santis

the coating. After sampling the collected species can be removed by using an aqueous solution and then analyzed. This simple concept has been further extended into the annular denuder (Possanzini et al., 1983) in order to improve temporal resolution.

The collection efficiency of a diffusion denuder can be calculated by the Winiwarter equation (Winiwarter, 1989) a solution of Fick's diffusion equation which holds at any geometry of the denuder.

This equation describes the diffusion limited accomodation of the analyte on the walls of the device with laminar flow in the assumption that the mixing ratio of the analyte is zero at the denuder wall i.e. when the walls can be considered as a "perfect sink" for the analyte. Theoretical predictions made according to the Winiwarter equation simply establish an upper limit in the calculated penetration of a given pollutant. On the basis of the fact that this theoretical value can be reached or not as a function of the reactivity of the coating, we have developed a technique for the determination of the reactivity of a surface. In this regard, it is well-known that a potential drawback of the majority of the studies reported in the literature on this subject is that they overlook the fact that it is important to study the intrinsic reactivity of a surface independent of mass-transfer limitations. In fact, since the diffusion of pollutants is usually slow at atmospheric pressure, most of the reported corrosion rates for e.g. metals and lithoid materials, both in the field and in the laboratory, are flawed by mass transfer limitations. One of the aims of this paper is to describe a simple method based on the use of a diffusion denuder, capable of measuring parameters related to heterogeneous reactivity of pollutants on surfaces in the laboratory at ambient pressure and realistic concentration levels of pollutants and without mass-transfer limitations. As an example of application of this technique we report here on an investigation intended to provide a better understanding of the processes related to dry deposition of SO_2 alone and in combination with NO_2 on three types of marble surfaces.

As an example of the versatility of the diffusion denuder technique we report also on a very recent use of this device as a sampler, i.e the simultaneous determination of NO_2 and PAN on a carbon denuder. It was found that on this coating nitrogen dioxide yields nitrite anion whereas PAN yields nitrate. These species are extracted after sampling with water and analyzed by ion chromatography. It was found that the use of an alkali in the coating carbon mixture makes negligible the reduction of NO_2 to NO, which occurs on a simple surface of active carbon. Laboratory and field tests concerning sorption efficiency, capacity and effect of humidity have been carried out. The pre-exponential and esponent for the first term of the Winiwarter equation were estimated. It was found that for PAN the agreement between the experimental and calculated values is quite good and compatible with a "perfect sink" behaviour of carbon, whereas, in the case of NO_2 the experimental collection efficiency is lower than theory but still suitable for ambient sampling.
The response for NO_2 has been compared during a field campaign in Milan with the corresponding integrated response of a chemiluminescence analyzer, while data for PAN have been compared with the response of a GC-ECD technique. The results obtained are reported and discussed.

Experimental

Diffusion denuders for studying the reactivity of marble surfaces.

A flat diffusion denuder has been built. It consists of an enclosed polyethylene box containing two parallel briquettes of the material under investigation (12 cm long x 5 cm high x 0.8 cm thick). The marble samples, after being polished with 400-grit silicon-carbide powder were washed with distilled water in ultrasonic bath, and conditioned in a sealed container for > 24 h at the same relative humidity to be used in the test. The briquettes are placed orizontally face to face and spaced 0.3 cm apart. Air is drawn into the box and flows parallel to the plates at variable rate in order to vary the residence time of the pollutant inside the box. The admission of air is carried out through a circular opening that gradually widen in order to insure a uniform distribution of flow at the inlet and outlet. This permits full development of laminar flow (Reynolds number less than 50) before heterogeneous wall removal occurred.

The flat denuder is contained in a temperature controlled chamber. The experiments were carried out as follows: first a flow of air at controlled relative humidity was admitted to the flat denuder for conditioning the surfaces at the chosen r.h. After two hours SO_2 (or SO_2 and NO_2 in selected experiments) was added to the air flow and its outlet concentration monitored by using a SO_2 analyzer.

Permeation tubes were used to generate SO_2 and NO_2 which were diluted at the inlet with clean air at controlled relative humidity (Humicon D.A.S. Palombara Sabina Italy). Since the source of NO_2 produces also small concentration levels of HONO and HNO_3, a downstream Na_2CO_3 denuder (coated with a 2.5% solution of aqueous methanol 50/50 v/v) was used as a scrubber. NO_2 and SO_2 were monitored by using a chemiluminescence (Environment Mod. AC 30) and a fluorescence (Monitor Labs Mod. 8850) continuous analyzer, respectively.

Diffusion denuders as a sampler for NO_2 and PAN.

The use of diffusion denuders as samplers has been described fully in previous publications from this laboratory (Possanzini et al., 1983; Allegrini et al., 1987). Pyrex annular denuders, of various dimensions in relation to sampling flowrate, were used for laboratory and field experiments. Typically, laboratory experiments were conducted by using annular denuders 10 cm in length, 1.0 and 1.3 in annulus diameter, operated at 2 l min^{-1}, while annular denuders 22 cm in length, 3.0 and 3.3 in annulus diameter, were used for field experiments at 12 l min^{-1}. The annular walls were sandblasted so as to create a granular texture capable of supporting the fine carbon powder. The adsorbent used in this study was active carbon (BDH Prod. 33204). Prior to use, carbon was extracted overnight in a Soxhlet with water, washed twice with boiling 0.5 M NaOH, twice with boiling water and then dried at 105 °C. The coating solution was prepared by adding 1.5 g of carbon to 10 ml of a 2% NaOH and 1% glycerine in methanol/water 80/20 v/v solution. The sorbent material was coated on the inside of the annular space by pouring the coating solution into the denuder, discarding the excess suspension and then drying with a gentle flux of compressed clean air while rotating, to ensure an even coating. At the end of drying a short period of air-flow higher than the nominal working flowrate was used for blowing loose carbon particles off denuder walls. After sampling in air for a known period of time, the denuder was extracted by shaking for 1 min with 10 ml of distilled water, the suspension was injected into the ion chromatograph (Dionex, QIC) after filtering through a disposable filter assembly (Gelman Acrodisc) and analyzed for nitrite and nitrate. The concentration of NO_2 and PAN were calculated by referring to the calibration graph constructed by using the peak areas of nitrite and nitrate standards respectively. The concentration of NO_2 was calculated from the nitrite amount whereas for PAN a correction on the nitrate amount was applied to take into account that the uptake of NO_2 yields, also a small constant amount of nitrate (1% on molar basis, at 12 l min^{-1} with denuders of length 22 cm, inner and outer annular diameter, 3.0 and 3.3 cm, respectively).

The method gives for NO_2 and PAN detection limits better than 0.05 ppb and 0.06 ppb, respectively at the flow rate of 12 l min^{-1} and for a 24 h sampling. NO_2 atmospheres were generated with a permeation tube (see above). PAN atmospheres were generated by a flow of zero air passing over a thermostated n-eptane PAN solution prepared according the method of Nielsen et al, (1982).

Results of the experiments with the flat denuder

The specific interaction gas-surface can be represented by using the concept of reaction probability γ which is defined as the ratio of the number of molecules which are sorbed on a unit

surface area per unit time to the total number of molecules which strike per unit surface per unit time.
This coefficient may take values between 0 (no reaction) and 1 (perfect sink surface). The flux of pollutant to the surface (J) is equal to the rate of molecolar collision, as predicted by the molecular theory of gases, multiplied by the reaction probability:

$$J = \gamma \left[\frac{kT}{2\pi m} \right]^{1/2} C \qquad [1]$$

where C is the pollutant concentration in air near the surface (strictly speaking the concentration at one mean free path away from the surface should be used. This semplification could be unjustified only when γ is high), k is the Boltzmann constant, T is the absolute temperature and m is the mass of an individual pollutant molecule.
The use of the diffusion denuder for studying the reactivity of marble results from the basic assumption that the surface reaction is the limiting step for mass transport to the walls, i.e. γ is small (below 10^{-4}). Under this condition uptake rates are independent of diffusion and a surface removal constant, dimensioned as a deposition velocity, can be determined.
The use of the continuity equation gives the following solution derived from kinetic theory

$$\frac{C}{C_0} = \exp(-4Uz) \qquad [2]$$

where $U = \frac{Kd}{D}$ and $z = \frac{LD}{V_m d^2}$

with C = outlet concentration
C_0 = inlet concentration
K = surface removal constant
d = equivalent diameter of the channel
D = diffusion coefficient
L = length of the channel
V_m = mean flow speed

By substitution and in the case of parallel plate denuder (with dimension ***a*** corresponding to half depth and ***b*** corresponding to the long dimension of the channel), the equation reduces to

$$\frac{C}{C_0} = \exp\left(\frac{-2L(2a+b)K}{F}\right) \qquad [3]$$

where ***F*** is the flowrate.

Equation [3] has been used to investigate the reactivity of Thassos, Penteli and Carrara marbles. By measuring the inlet and outlet concentrations of SO_2 at different r.h. conditions and in the presence or not of NO_2, it has been possible to relate the decrease of concentration with the reactivity expressed as a function of the surface removal constant K, and therefore of the deposition velocity.

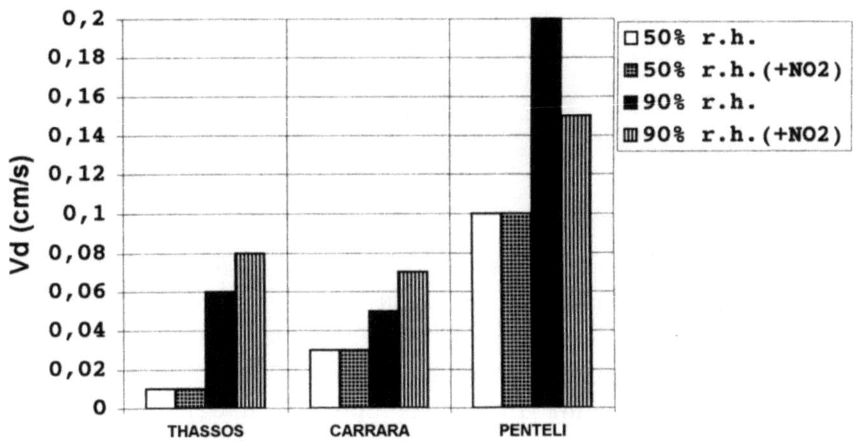

Figure 1. Determination of Vd by flat denuder.

The surfaces tested constantly show a pattern of an initially high rate of deposition which rapidly decreases during the first hour of exposure and reaches values between 0.2 and 0.05 cm/sec at 90% r.h. and between 0.1 and 0.01 at 50% r.h.. The relatively high initial rate of deposition is probably related to SO_2 dissolution in the surface moisture layer of the briquettes. The deposition rate continues to decrease upon prolonged exposure with deposition balancing the reaction with the stone. The addition of NO_2 to the gaseous stream does not cause any effect on the deposition of SO_2 at 50% r.h., whereas at 90% r.h. a sensible increase for Thassos and Carrara marbles can be observed. On Penteli marble the presence of NO_2 surprisingly is able to cause a drop in reactivity.

It is worth noting that mass transport by diffusion has been expressely taken into account in the data analysis. Therefore, the values of deposition deposity reported here would apply to the environment under turbulent conditions and represent the maximum rates of dry deposition.

Sampling of NO_2 and PAN by annular denuders

The collection efficiency *(E)* for NO_2 and PAN of annular denuders coated with the mixture 1.5% carbon + 2% NaOH + 1% glycerine, can be estimated from the first term of the Gormley-Kennedy (1949) equation applied to annular geometry according to Winiwarter (1989)

$$1 - E = 0.91 \exp[-7.52\Delta] \qquad [4]$$

with $\Delta = \dfrac{\pi(d_0 + d_i)DL}{(d_0 - d_i)Q}$

where D is the diffusion coefficient of the gas (0.154 cm2 s^{-1} for NO_2 and 0.093 cm2 s^{-1} for PAN), L is the length of the tube, d_0 and d_i the outer and inner diameters of the annulus and Q is the volumetric flow rate.

Efficiency of the denuder in actual operation can be determined by using sequential pairs of denuders. This was done in laboratory and in some air analyses to verify separation of the analytes. The efficiency was calculated by

$$E = 1 - \frac{b}{a} \qquad [5]$$

where a and b are the amounts of nitrite on the first and second denuder, respectively. From a series of experiments at various flowrates, ranging from 2 to 12 l min^{-1}, it was found that chemisorption efficiencies for NO_2 were close to those expected on the basis of the Gormley - Kennedy equation. The differences found by linear regression analysis in the preexponential term (0.61 instead of 0.91) and in the exponential term (4.15 instead of 7.52) while represent a sensible deviation from the perfect sink assumption, do not have any practical consequence since at working conditions (i.e. at 12 l/min) the efficiency of the denuders used for field sampling is well above 99%.

The determination of efficiency carried out on PAN shows that active carbon behaves as a perfect sink for this species. The preexponential and exponential terms measured experimentally are 0.80 and 6.89, respectively in very good agreement with theory.

Denuder capacity was estimated by measuring the breakthrough of NO_2 at a flow rate of 12 l min^{-1}. The denuder started to deviate from calculated efficiency near 900 ug nitrite/denuder at 50-60% r.h.. This is likely to vary according to the amount of active carbon present on the walls, but is more than adequate for the determination of atmospheric NO_2 over the short sampling periods in a strongly polluted urban environment (4 hrs) described in the present paper. Considering that the atmospheric levels of PAN are at most some tens of ppb, no denuder capacities tests on this species were carried out.

The interference of O_3 was studied by exposing denuders previously loaded with NO_2 to O_3 concentrations, ranging from 50 to 600 ppb. Some experiments were performed by sampling standard atmospheres containing both NO_2 and O_3. The conversion of nitrite to nitrate was constantly below 1%.

HONO and HNO_3 are obvious interferents for the method since this species yield nitrite and nitrate anions on an alkaline substrate as it has been shown by Allegrini et al.(1987). Therefore, HONO and HNO_3 should be removed from the air stream before sampling, by using a Na_2CO_3 denuder. This denuder is able to collect also SO_2 and, as it has been shown by Febo et al. (1993) in higly polluted environments, the layer of sulphite formed on the denuder due to the collection of SO_2 can collect a noticeable amount of NO_2. As a consequence, SO_2 should also be removed. In conclusion, this involves the placement (in line before the carbon denuder) of a TCM denuder and of a Na_2CO_3 denuder for the collection of SO_2, HONO, and

HNO_3. Tests were conducted to ascertain that NO_2 analyses would not be compromised by this configuration. To this end, the proposed arrangement was evaluated in the laboratory in the range of 20 to 90% RH. Losses of NO_2 and PAN on the TCM and carbonate denuders were negligibly low (from 1 to 3 percent), indicating that the sampling line was suitable for the determination of NO_2 and PAN. It is interesting to note that this train can also be employed for collection of HONO on the two carbonate denuders according to the procedure (differential technique) suggested by Febo et al. (1993).

The active carbon denuder has been field tested in urban smog conditions during a measurement campaign from 10th to 21st of February 1993 in Milan. The sampling line, comprised of a TCM, two carbonate and of the carbon denuder arranged in series, was installed about 30 m above ground in the city centre, on the top of the medieval tower of Brera. Side-by-side sampling was conducted for NO_2 with a CL instrument operated by the Institute of General and Applied Physics of the University of Milan and for PAN with a GC-EC instrument (Ciccioli et al., 1992)

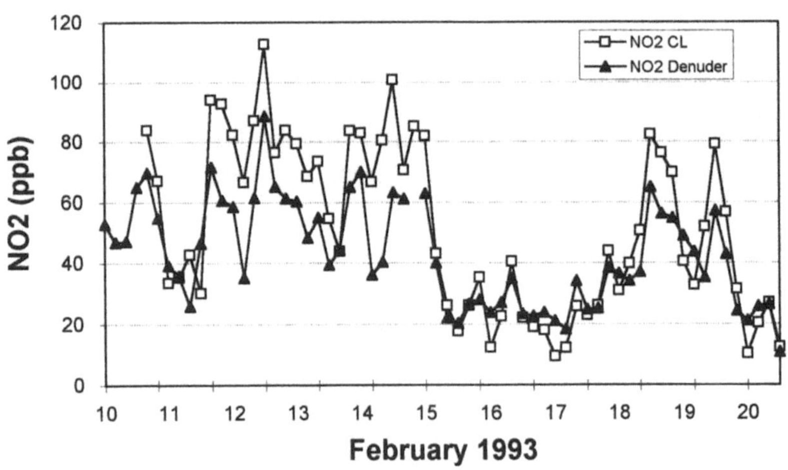

Fig. 2 Time series of NO2 measurements made by CL and Carbon denuder. Milan, Feb. 10-21, 1993.

Fig. 2 shows the time evolution of NO_2 measured by carbon denuder and CL. There was a definite bias towards higher NO_2

values by the CL method as compared to the carbon method which can not be explained on the basis of the inability by the CL instrument to distinguish NO_2 from other secondary nitrogeneous species such as PAN, HNO_3 and HONO.

The CL and denuder measurements agreed with one another on Feb. 11, 15-18 and 20 but showed notable differences, during the days from the 12th to 15th, when the concentration level of other simultaneously monitored pollutants were particularly high (high levels of CO, NO and VOCs).

The regression analysis of the carbon results against the CL results gives a slope of 0.60 and a correlation of $r^2 = 0.87$. This slope becomes 0.83 (with a correlation of $r^2 = 0.84$) when the regression analysis is performed on a subset of data below 60 ppb.

A possible reason that the denuder measurements are lower than the CL measurements, could be a higher production of NO (at the expense of deposited nitrite on the carbon coating in the field) than that determined in laboratory trials. Another possibility is conversion to nitrogenous species that were not detected in laboratory (e.g. N_2O). A poor field collection efficiency or denuder capacity can be ruled out on the basis of a number of samplings (carried out during the first 4 days) with a double carbon denuder; the breakthrough from the first denuder which was monitored by the downstream second denuder showed that the > 99 % collection efficiency was always verified. Consideration was also given to the possibility that the recovery might be only partial and also that nitrite might decompose by some unknown mechanism after the extraction. However, on several occasions we analyzed the extraction solution from field sampling both immediately after sampling and after a week storage at room temperature in the laboratory. No changes either in nitrite (or nitrate) were ever observed.

We feel that another possible explanation for the differences found could be that the CL data are too high because of an incorrect calibration of the two channel instrument at high NO levels. The coincidence of large deviations between results when high levels of NO and other pollutants were measured, might also be indicative of a positve artifact in the CL response.

Figure 3 presents the results of PAN measurement by carbon denuder in comparison to GC-ECD. Since the denuders were exposed for 4 h periods, the GC-ECD values shown are averages of 12 separate determinations. The agreement between the two

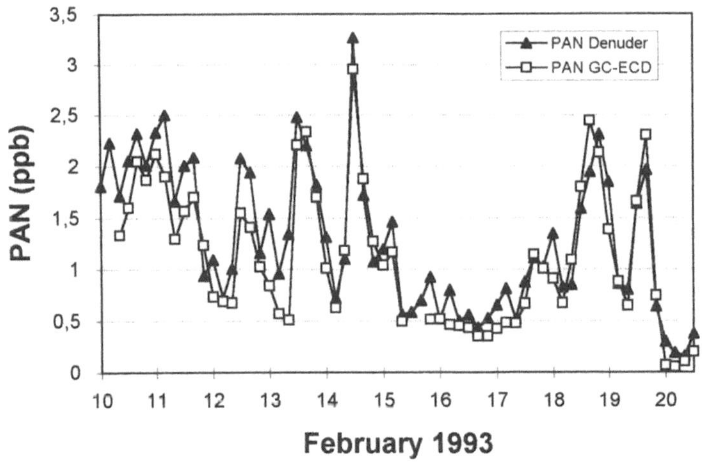

Fig.3 Time series of PAN measurements made by GC-ECD and Carbon denuder. Milan, Fe. 10-21, 1993.

techniques found is quite good. The regression analysis of the carbon results against the GC-ECD gives a slope close to one (0.95) and a correlation of 0.87.

Since other organic nitrates (e.g. alkyl nitrates), peroxypropionyl nitrate (PPN), and peroxybenzoyl nitrate (PBzN) are usually present in the atmosphere at much lower concentration levels than PAN (Fahey et al., 1985), their possible collection should not constitute a problem for the denuder.

Acknowledgement

We wish to acknowledge the technical assistance of S. Brachetti. Resources for this study were provided by the EEC within the framework of the Programme EV5V-CT94-0537 and by CNR within the framework of the PS "Tecnologie Chimiche Innovative" and "Aree Metropolitane e Urbane".

References

Allegrini I., De Santis F., Di Palo V., Febo A., Perrino C., Possanzini M., Liberti A. (1987) Annular denuder method for sampling of atmospheric pollutants. *Sci. Total Envir.* **67**, 1- 16.

Febo A., Perrino C. and Cortiello M. (1993) A denuder technique for the measurement of nitrous acid in urban atmospheres. *Atmos. Environ.* **27A**, 1721-1728.

Ferm M. (1979) Method for determination of atmospheric ammonia. *Atmos. Environ.* **13**, 1385-1393.

Gormley P. G. and Kennedy M. (1949) Diffusion from a stream flowing through a cylindrical tube. *Proc. R. Ir. Acad.* **52**, 163-169.

Nielsen T., Hansen A. M., Thomsen E. L. (1982) A convenient method for preparation of pure standards of peroxy acetyl nitrate for atmospheric analyses. *Atmos. Envir.* **16**, 2447-2450.

Possanzini M., Febo A., Liberti A. (1983) New design of a high performance denuder for the sampling of atmospheric pollutants. *Atmos. Envir.* **17**, 2605-2610.

Winiwarter W. (1989) A calculation procedure for the determination of the collection efficiency in annular denuders. *Atmos. Envir.* **23**, 1997-2002.

A New Approach to Designing of Air Quality Monitoring Networks in Urban Areas

Grazyna Mitosek
Institute of Environmental Protection
Air Protection Division
ul. Kolektorska 4
01-692 Warsaw
Poland

Introduction

At the end of 1980s a general review of air quality measurements carried out in Poland was made. From the review it became clear that the total number of measurement sites (particu- larly for most common pollutants) is much too high and the collected data are of very limited practical value.

As a result of studying numerous monitoring systems operated in different countries, and on the basis of experience gained there during years of monitoring activity, a new concept of air quality monitoring system in urban areas was developed and adopted as a basis for the reorganization of the measurements in towns and cities in Poland. In this concept, environmental conditions typical of the country have been taken into account. This concerns mainly:

- pollutants which still cause a lot of problems i.e. sulphur dioxide, nitrogen oxides and suspended particles originating from burning processes, as well as compounds emitted by vehicles,

- character of urban areas, where residential districts are frequently combined with industrial plants, and where houses are in many cases heated individually by means of coal (especially in old parts of towns).

NATO ASI Series, Partnership Sub-Series, 2. Environment – Vol. 8
Urban Air Pollution
Edited by Ivo Allegrini and Franco De Santis

A similar picture of both environmental and monitoring situation may be seen in many other European countries.

General assumptions

The basic assumption about air quality monitoring (AQM) in urban areas is that the monitoring system should be a source of information to enable appropriate measures aimed at reducing environmental hazards caused by air pollution, and it should assess the effectiveness of measures undertaken. For proper environmental decision making, particularly when very limited resources are available, determination of priorities is necessary. Thus, the monitoring data should be able:

- to identify areas where more severe problems occur - both on local and national levels,
- to assess the importance of a problem indicated by the existing levels of air pollutants.

It is rather impossible to obtain this kind of information from a network of monitoring stations located at randomly selected sites in a town or a city, unless its area is covered with a very dense network of monitoring sites (that would be extremely expensive). In order to obtain the information demanded at a reasonable cost, properly and carefully designed measurement systems are needed.

In the concept presented in this paper it is assumed that a properly organized monitoring system should meet the following requirements:

- location of the monitoring sites and program of the measurements must be closely connected with a purpose of measurements and planned data use,
- the system must assure comparability of results obtained from different stations (both at local and national level) and enable proper data interpretation,
- it should be possible to get maximum information at the lowest cost of measurements, which also means with the fewest monitoring sites.

To achieve these goals, a classification of measurement sites has been proposed, based on two characteristics:

- type of measurement site,
- spatial representativity of the site.

They will be discussed below.

Taking into consideration the need of both long-term and short-term measures aimed at environmental hazard abatement, two kinds of monitoring networks in urban areas have been distinguished:

- general air quality monitoring networks,
- warning air monitoring systems.

The general air quality monitoring networks are considered the basic systems in urban areas. Mainly manual measurements of 24-hour concentrations are carried out in the networks. This type of measurements, with samples taken on a site and then analyzed in a laboratory, still predominates in countries of central and Eastern Europe. Data collected in the system will be used for long-term environmental programs and assessment of effectiveness of long-term policy.

The warning systems - with real time data collected by means of continuous monitors, should be established only in highly polluted areas where air pollution episodes occur and countermeasures are possible. The main task of these networks is to provide data for short-term pollution abatement procedures.

It may happen that both networks exist in a city - in such a case they must form a consistent system of non-overlapping measurements.

Monitoring objectives

A list of possible objectives connected with the considered monitoring systems is long. The general AQM network may be a source of documented data that, among others, may be used:

- to evaluate air quality in relation to national ambient air quality standards,

- to develop and/or update long-term plans for air pollution abatement,
- to evaluate the effectiveness of investment policies and activities aimed at reducing air pollution,
- to determine an impact of industrial plants on air quality and to support enforcement measures,
- for town and country planning,
- to document population exposure to air pollution,
- to provide the public with information on air quality,
- to assess the trends,
- to validate mathematical models.

The main purpose of warning networks is to detect and possibly predict episodes connected with high pollution levels in the atmosphere and to provide data for short-term activity during episodes. The data collected from the warning networks may also serve many other purposes listed above.

Determination of a purpose of planned measurements should be the first step in the designing of any monitoring activity.

Types of measurement sites

The type of a measurement site in the proposed classification shows a character of area from the standpoint of expected pollution levels (i.e. indicates whether the measurements are carried out in highly polluted parts of a city, in areas of average concentrations, etc.) or indicates pollutants origin in the considered area.

Four types of monitoring sites have been distinguished for the general AQM networks:

Type M: sites located in inhabited areas of potentially the highest expected pollutant concentrations (i.e. where the expected long-term concentrations of a particular pollutant are higher than in other areas of the town/city).

Type S: sites representing areas of high population density where concentrations of a particular pollutant may be treated as average for densely populated parts of the city.

Type Z: sites located in areas where the highest influence of specific emission sources (industrial plants) on inhabited areas may be expected.

Type W: sites established for some specific purposes, located in areas selected according to specific objectives (e.g. to be potentially influenced by a newly designed industrial plant; where for health reasons the population exposure to a specific compound should be known; with high exposure of pedestrians to traffic related pollutants, etc.).

Taking into account the differences in concentration distributions of various pollutants, connected with their different origin and dispersion in the atmosphere, it should be stressed that the type of a monitoring site depends on the individual pollutant being examined.

In the general AQM networks, types M and S of the sites are considered the basic ones , recommended to be established as permanent stations. Establishment of sites of type Z or W will depend on local needs and possibilities. In many cases, when the setting up of this type of site is planned, temporary or mobile measurements may be sufficient to collect the required data.

As far as measurement purposes are concerned, it is worth mentioning that the results obtained from each of the proposed types of sites may be used for many purposes. At the same time, a monitoring network comprising the four types of stations should allow the most important objectives of measurements in urban areas to be achieved.

The classification proposed for warning AQM networks is based on analysis of the possible origin of pollutants in areas where high short-term concentrations may be expected. Again, four types of monitoring sites have been distinguished:

Type AM: sites located in densely populated areas with frequent occurrence of high concentrations of pollutants (caused by different emission sources).

Type AZ: sites set up in densely populated areas with frequent occurrence of high concentrations of pollutants emitted from a particular industrial plant.

Type AK: sites representing areas (densely built up) with very high concentrations of pollutants emitted from traffic.

Type AN: sites located in densely inhabited areas exposed to a strong periodical influence of a distant industrial plant.

The classification based on different station types seems important from the standpoint of:

- linking objectives of measurements with site selection procedures,
- proper identification of measurement results,
- proper data interpretation and comparisons.

Spatial representativity of monitoring sites

The second feature adopted in the classification is spatial representativity of a monitoring site.

According to a concept originally developed in US-EPA (Ludwig, Kealoha, 1975) the scale of representativity is assumed to be a qualitative characteristic of dimension of an area for which measurement results may be considered representative (i.e. for which concentrations of a selected pollutant are reasonably similar to those obtained at the monitoring site). Determination of such a feature is indispensable for proper data interpretation, comparisons and use.

The scales proposed for urban measurements are:

Microscale: referring to concentrations in volumes with dimensions of the order of not more than 100 m.

Middle scale: defining concentrations within areas with dimensions of tens to hundreds of meters (usually not more than 500 m).

Neighbourhood scale: representing concentrations over some extended area of the city with relatively uniform land use; dimensions are of the order of kilometres (usually from about 0.5 to 4 km).

Urban scale: characterizing the overall, citywide conditions over areas with dimensions of several to a dozen or so kilometres.

It is very important that both the type of a site and its spatial representativity are attributed to each monitoring site (individually for each pollutant) and inserted into a site code used in data bases.

Priorities

When a new monitoring network is designed or an existing one is reorganized, it is necessary to establish priorities concerning both the compounds to be monitored and the areas where the stations should be located. Distinction of the four types of monitoring sites does not mean that all the types should be set up. The types of sites to be established should depend on a particular situation associated with the actual needs and measurement possibilities.

In the presented concept it is assumed that the priorities should be given to the pollutants creating most problems in a given town and to the areas (inhabited) exposed to potentially the highest pollutant concentrations (particularly if exceedance of ambient air quality standards may be expected). This means that the establishment of at least one site of M type for pollutants of expected high concentrations should be considered as the first priority. More detailed analysis of the issue, together with adequate criteria relevant to the distinguished types of measurement sites, may be found in the literature referred to in the paper (Mitosek et al.,1991).

Remarks on site selection

One of the critical points in the monitoring network designing is proper selection of measurement sites.

Site selection procedure for any monitoring station consists of several steps, such as:

1. Gathering of indispensable documentation and information; making desired calculations by means of mathematical models when possible.

The indispensable information includes: emission inventories, meteorological and climatological data, existing results of measurements in a considered area (carefully and cautiously selected), maps and plans showing existing and prospective land use, traffic distribution maps, etc.

2. Selection of potential location areas (determination of general location of a monitoring site).

To determine the general location for a monitoring site of any type a detailed analysis of many different factors must be made, based on all the information and documentation available. For the reasons mentioned earlier, in the analyses connected with site selection all the preselected compounds should be considered individually.

3. Elimination of areas influenced by local factors and emission sources, selection of potential monitoring sites.

In this phase, the previously chosen general location areas should be reviewed to avoid setting up a station under undesirable local influences (deforming measurement results). The influences may be caused both by local emission sources of pollutants (individual chimneys, dumps, roads, parking lots, etc.) and by other factors, including local topography. To eliminate the risk of carrying out the measurements in sites influenced by local factors, detailed documentation concerning the preselected areas must be collected and analyzed. Besides, an inspection of all the potential areas is necessary. Finally, potential monitoring sites (of the considered type) should be selected. For practical reasons, it is recommended to find a few potential locations for one station.

4. Final selection of location(s) meeting the probe inlet placement criteria (determination of specific location).

The specific location criteria are established to eliminate or minimize distortion of measurement results (connected with very local influences and disturbances) and to ensure comparability of sampling conditions.
The criteria refer, among others, to:

- distances from roads,
- height of the probe inlet above ground,
- distances from horizontal and vertical surfaces,
- distances from obstacles (e.g. buildings),
- distances from trees.

In practice, two sets of criteria are used. The ones adopted for air pollution measurements not related to traffic and the ones valid for measurements aimed at assessment of traffic influence on air quality in the vicinity of roads.

As can be seen from the above outline of the problem, proper site selection is not an easy task and needs a lot of effort. On the other hand, the monitoring stations are established to work for a long time, during which the collected data is used for decisions that need reliable basis.

References

Ludwig FL, Kealoha JH (1975) Selecting sites for carbon monoxide monitoring. US Environmental Protection Agency, Research Triangle Park NC

Mitosek G et al. (1991) Guidelines for designing the air pollution monitoring networks. Part I - Warning networks in urban-industrial areas. Biblioteka Monitoringu Srodowiska PIOS Warsaw

Mitosek G et al. (1991) Guidelines for designing the air pollution monitoring networks. Part II - General air quality monitoring systems in urban areas. Biblioteka Monitoringu Srodowiska PIOS Warsaw

The Monitoring System for Urban Air Pollution in Belarus

Ivan I. Matveenko, Marina A. Podgayskaya
Main Administration of Hydrometeorology of Belarus,
Centre of Radiation Control and Environmental Monitoring
16 Komsomolskaya Street,
220600 Minsk
Belarus

Introduction

The increased pace of industrial and urban development calls for special attention to atmospheric protection. Air control is a necessary condition for the success of atmospheric protection activities. The Main Admistration of Hydrometeorology of Belarus monitors air pollution in 14 cities with more than 65% of urban population and measures gamma radiation dose rates in 53 cities. (Figure 1).

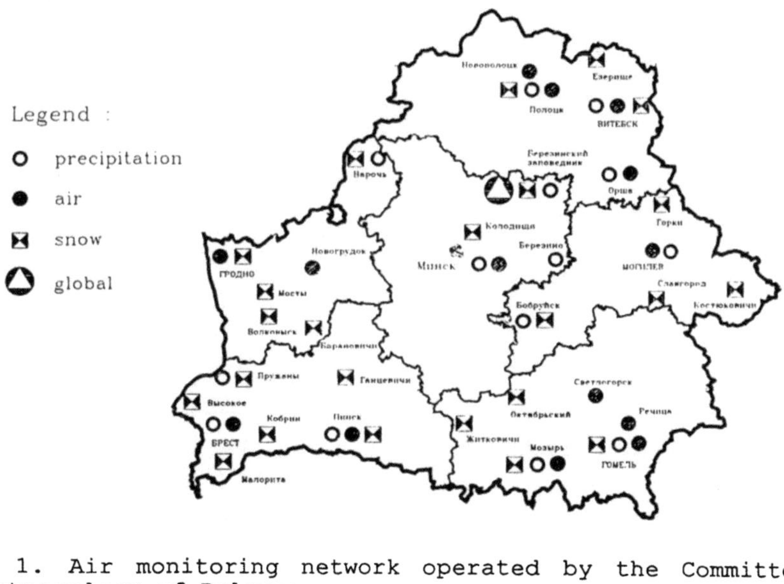

Figure 1. Air monitoring network operated by the Committee for Hydrometeorology of Belarus

The Ministry of Public Health monitors sanitation and working zones of enterprises. The Ministry of Ecology monitors pollution.

NATO ASI Series, Partnership Sub-Series, 2. Environment – Vol. 8
Urban Air Pollution
Edited by Ivo Allegrini and Franco De Santis

Air monitoring provides actual on-line pollution data, preventing adverse anthropogenic impact on the environment and human health.

Monitoring system for urban air pollution in Belarus.

Modern industrial development poses a growing threat to the environment and population in cases of emergency at chemical plants and transport of highly toxic pollutants (HTP) which enter the atmosphere. This necessitates predicting the extent of pollution. Proper control, data analysis, prediction and simulation of atmospheric transport of pollutants are carried out by automatic data processing system operating at the Republican Centre of Radiation Control and Environmental Monitoring. (Figure 2).

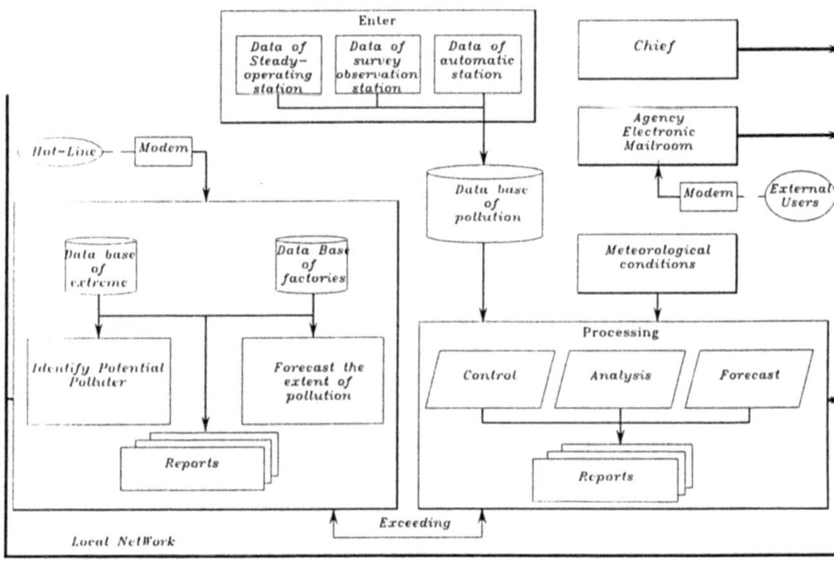

Figure 2. Automatic system for processing the air pollution data.

The system incorporates computerized working stations of the laboratories and departments and utilizes data from automatic, steady-operating and survey observation stations.
Monitoring is conducted at 47 steady-operating urban stations, 3 to 4 times daily. Observational data from steady-operating stations in Minsk arrive at the laboratory of air pollution

control. Results of analyses are logged into air pollution database.Automatic stations utilize telemeters, which take output and discrete control signals, digitize them and via modem transmit data over a telecommunication channel to the air pollution database. Information exchange between laboratories and experts' departments of the Centre occurs through a local computer network. (Figure3).

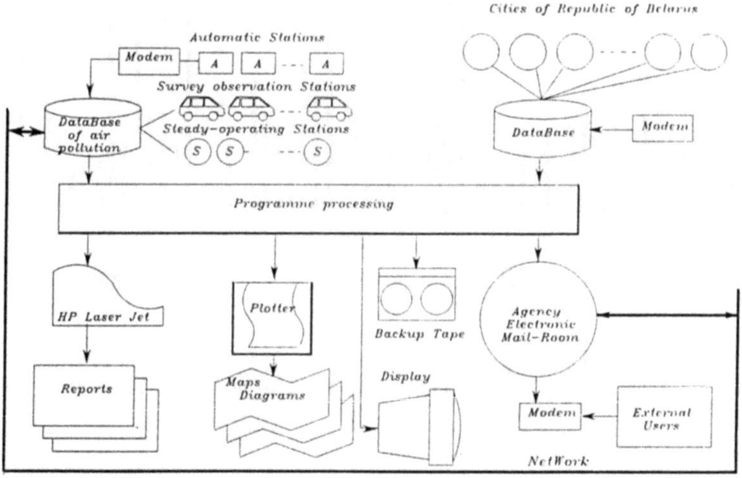

Figure 3. Diagram of automatic data processing.

At first, the data is controlled for excess levels. When maximum permissible concentrations are exceeded more than 10 times, a relevant message is formed and along with excess data is transmitted to the operator on duty for further analysis and response.

The data processing includes computation of various statistical characteristics over the whole range of pollutants (measured both at stations and throughout the city) computation of background concentrations of air pollutants, computation of air pollution parameter and of the air pollution index (API) indicating the excess of total air pollution level over the permissible value. All this data can be displayed graphically or in tabular form and printed out. (Figure 4)

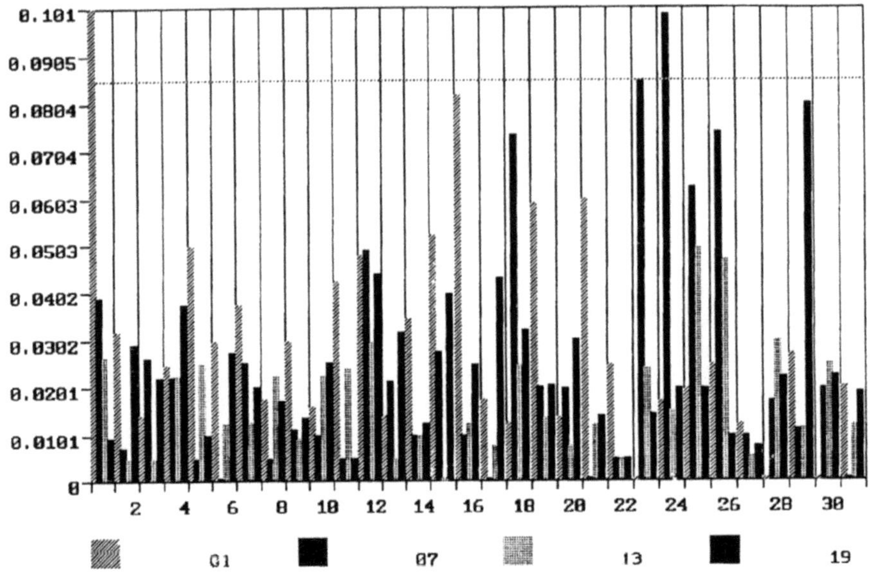

Figure 4. Example of programme: data on NO_2 concentration (mg/m^3) in the air; Misk, 9 stationary post, terms - 01, 07, 13, 19, January 1995.

High levels of air pollution are predicted according to meteorological conditions. The air pollution forecast for the next day is prepared daily. On-line warnings are relayed in order to control emissions. The processed and on-line data are transmitted via local network to the agency's electronic mailroom for informing remote users. Data exchange with remote users, namely, Municipal Commitee for Ecology, Civil Defence Headquarters etc., is carried out via modems (MNP5 V42/42bis). Every user registered in the Centre, may obtain required information on the basis of his class of access. A user may also make requests for specific supplementary data. After receiving data on extreme emissions the operator on duty analyzes them and identifies the potential pollution sources. On-line data is immediately transmitted via modem to associated agencies, e.g. Defence Headquarters. The automatic system allows the forecasting of the extent of pollution

in case of highly toxic emissions and also to simulate emergency processes at chemical- and radiation plants.
Forecasts for chemically- and radiation - hazardous plants are made with due regard to the physical state and parameters under a given meteorological condition. The data transmitted from Belarus meteorological network every 3 hours are processed in the Main Administration of Hydrometeorology and are open to wide applications. This enables us to computate the areas of potential and actual pollution, the extent of pollution zone, the time of pollutant approach to the object, duration of highly toxic pollutant episodes. Daily data on measured rates of gamma radiation , transmitted to the Centre via communication links, makes it possible to control air radiation pollution (Figure 5). A communication link is established with the automatic control system of Ignalina NPP. Efforts are made to institute information exchange with other NPPs.

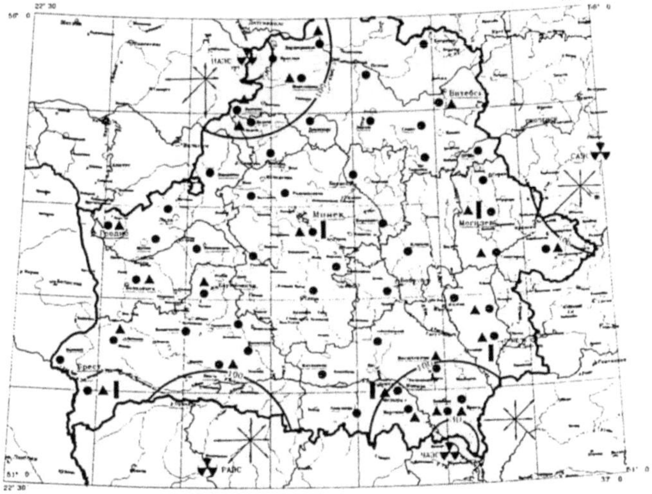

Figure 5. Steady-operating stations of radiation control of the Main Administration of Hydrometeorology of Belarus :●: Site of gamma-radiation measurement, ▲: Site of measurements of natural deposition, ▌: Site of measurements of radioaktive aerosols.

Work is now underway towards software simulation of atmospheric pollutants transport, based on various solutions of turbulence diffusion equation under corresponding limiting conditions, given turbulence factor and meteo elements. Simulation will be made in several directions, using different approaches and solution methods: models of long-range and transboundary transport,regional pollution, and on-line design of local dispersion. These models enable us to design average long-term concentrations of a given pollutant under constant wind direction and velocity, based on meteo statistical data.

Conclusion

The analysis of observational data obtained at the Hydrometereological network shows that air pollution problems in Belarus persist despite the absence of tendency for a pollution increase. Automatic systems for processing air pollution data, which have been set up at the Centre of Radiation Control and Environmental Monitoring, provide timely and real data on the status and trends of environmental changes, solve diagnostic and forecasting problems, enable to make multivariate analyses of radiation and ecological situation and enhance the protection activities in Belarus. The system has module structure and is oriented to widely available hardware and peripherals.

Recent Developments in Atmosphere Pollution Control

G. Elisei
c/o CISE - Tecnologie Innovative[1]
P.O. Box 12081, I-20134 - Milano (I)

Introduction

The air quality control over extended urban and industrial areas, that was starting almost 30 years ago, has clearly proved a worthwile project.
During this period,new performances and improvements on HW/SW tools and methods have been stimulated by research activities, oriented projects and control legislation.
In particular since the Sixties, due to growing interest in the transport and diffusion of pollutants emitted by thermoelectric and nuclear power plants,studies and research activities have been directed to investigate not only the atmospheric ground level concentrations, but also those of upper layers.
During the last few years, pilot experiments have been performed to integrate remote sensing devices into air quality monitoring networks.
An outstanding question is whether we have reached a satisfactory use of advanced HW/SW systems to control air quality.

Remote Sensing Systems[2]

Due to the short space available the description will only

---

1 CISE is a Company that carries out applied research concerning technological innovation and environmental fields

2 Systems described in this paper have been designed and built by ENEL and CISE with the cooperatin of CNR Institutes. However, other remote sensing systems are available on the national and international market.

NATO ASI Series, Partnership Sub-Series, 2. Environment - Vol. 8
Urban Air Pollution
Edited by Ivo Allegrini and Franco De Santis

deal with the measuring principle and with the collected data. All the systems are selfsufficient stations and can be operated by remote control.

Doppler SODAR works by periodically sending a series of acoustic waves into the atmosphere by means of parabolic antennas. The analysis of back scattered echoes allows us to determine the three components of the wind. The instrument acts as a meteorological tower with anemometers placed every 50 m, starting from the 50m level to height of 500-1000 m.
RASS system. The device consists of two radar antennas (transmitter and receiver) and an acoustic antenna.The acoustic antenna periodically sends a vertical sound monitored by a radar that can range up to 1500m.The burst speed detection allows us to estimate local atmospheric density, and so the air temperature profile can be calculated.
The SODAR system acts as a variable height meteorological tower equipped with temperature sensors placed every 10m starting from a height of 50m .

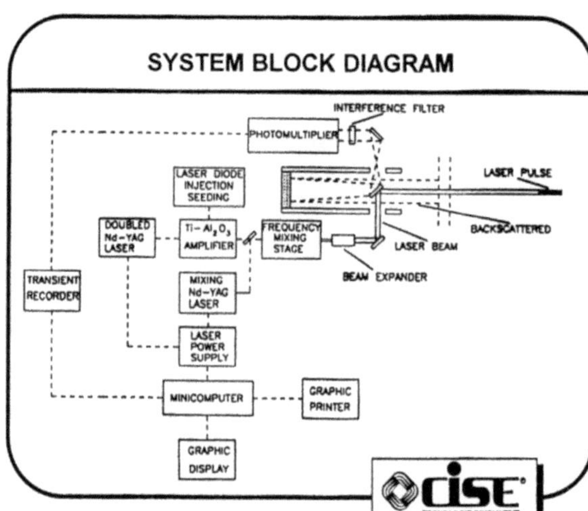

Fig 1

The DIAL system allows remote sensing of gaseous substances in the atmosphere. It can be used as a stationary measuring

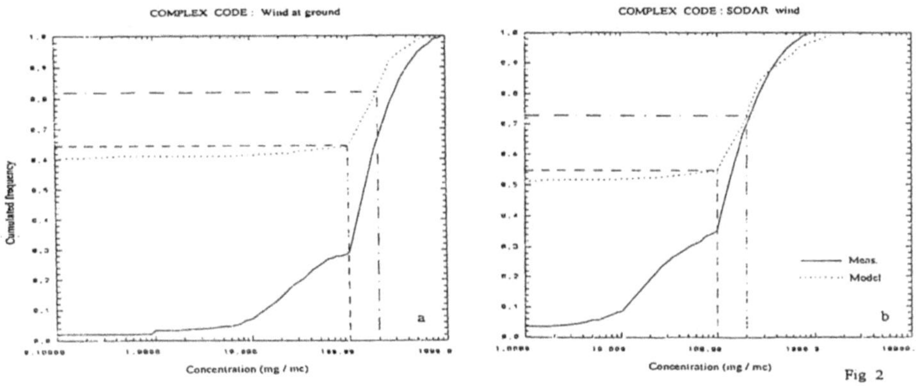

Fig 2

in Fig 2[6]. Concentration values and cumulated frequency distributions (cdf) are shown horizontally and vertically. We can observe that replacing wind at ground level with wind measured at height (215m) cdf tend to overlap for C=100 $\mu g/m^3$. In fact the model changes the forecasted data from :

-35 to 46% with C > 100 $\mu g/m^3$

-18 to 28% with C > 200 $\mu g/m^3$

The two second values (46 and 28%) show quite a good fit with those of the network.

b) a comparison between measured and forecasted data provided by 11 models (10 gaussian models and Lagrangian "particles" diffusion model). The Lagrangian particles model is able to use vertical profile of wind (speed and direction) turbulence parameters (σ_Θ and σ_ω) and standard deviation of wind components. Comparison has been made using data collected during an experimental project organized at the Nuclear Research Center in Karlsruhe (Germany).

In Fig 3 we show, the diagram elaborated by ENEL-CRAM and CNR-ICG -Turin. Cdf are functions of R index (R represents the ratio between measured and model evaluated data). Factor 10 includes 90% of data forecasted by the particles model and 50 to 78% of data forecasted by all other gaussian models.It is evident that the performances of the particles model are better than all the other gaussian ones.

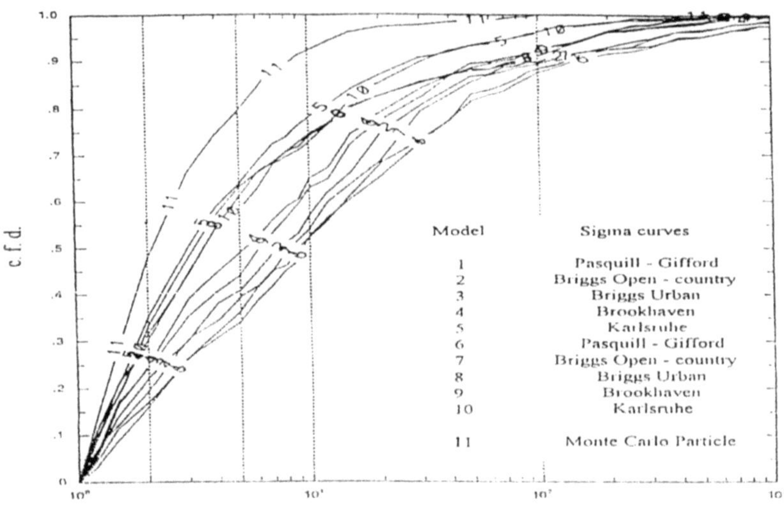

Fig 3

Network Supervisor Center (NSC)

NSC uses data acquired by a Remote Sensing Environmental Station which is equipped with RASS,Doppler SODAR and chemical-meteorological devices at ground. Instruments and models are set up according to the Italian Environment Ministry decree issued on 20 / 05 / 91.The structures have been prepared following basic criteria used in industrial plants: real time data analysis, interaction between intelligent systems of different representativity and importance. The available NSC performances allow air quality data analyses from urban and industrial monitoring networks.

Data is validated before being used for to diffusion models. All functions are available by means of an interactive symbol. Fig 4 shows the Remote Sensing Station: NSC is outlined in its lower part. Two computers have been installed to assure a continuous service and the possibility of using advanced mathematical codes. The less powerful unit (Automatic Unit)is dedicated to standard activity data controls, basic data processes and results visualization. The operator makes essential controls automatically way.

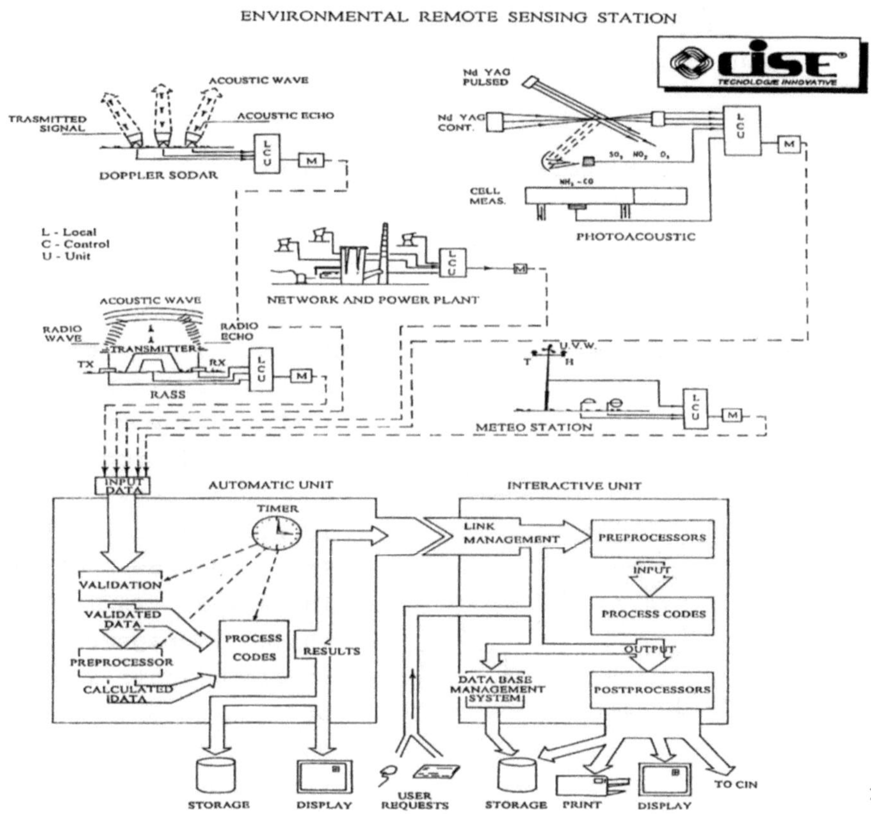

Fig 4

The more powerful unit (Interactive Unit) is managed in a more flexible way and planned to face emergency situations. Specific elaborations can start exclusively at the request of the user, who can be active in computing management. On the whole, the NSC carries out the following activities.

- real time sensor control,
- organization and management of information in acquisition phases (Doppler SODAR, RASS, meteo chemical station, thermoelectric power plant

monitoring network),

- organization and execution activities ready for the user (statistical tests,models,visualisations),
- starting up of man-machine dialogue,
- topography data management,
- connection with other nodes management.

Data Management and Network Supervisor Center

Software architecture

NINAS-NetworkIntegrated Atmospheric Survey.

It implements management activities in the Immediate Availability Data Area (IADA) and controls the programmed processes on the local network (with TCP/IP protocol) Fig 5. It prepares user interface to input phase management of models,graphic analyses of results and displays on monitor data coming from sensors.It guides the use of graphic libraries for scientific visualization making a large quantity of information available. NINAS has been designed by CISE in cooperation with ENEL-CRAM. The applicative part has been written in C language ,while the interface uses UIMS (User Interface Management System) equipped with its own language (D).

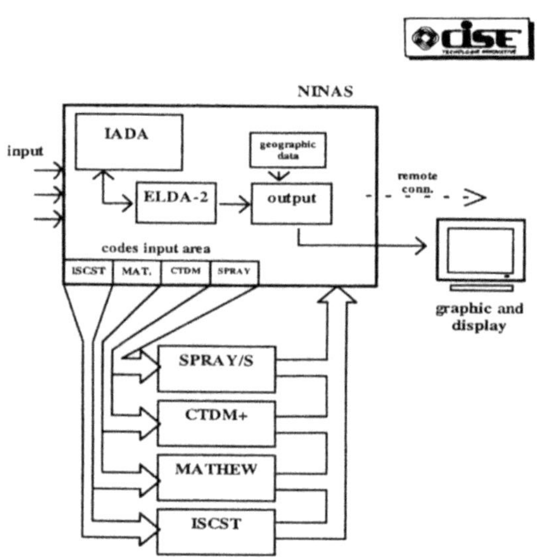

Fig 5

NINAS uses the AIX operative system (UNIX-IBM version) and X11 (or X Window) for graphic processes.

ELDA 2 - (Meteo-climatological Data Process-Version 2) .
It performs controls,preprocessing, organization of incoming data creates the Data Base of collected measurement. It has been designed in order to work in UNIX operative systems and written in Fortran and C languages. SW architecture can organize,store and manage data inside a simple, compact structure. The option of making up many standardization processes and some statistical analyses is available.It produces automatic station reports by tables, graphic syntheses of acquired data and by some daily and monthly analyses.

Mathematical models

ISCST - Industrial Source Control Short Term.
This model is recommended by the United States Environmental Protection Agency (EPA- 1986 / 87) and suggested by the " Working Group on Air Quality Control " (the Italian Institute for Health Service - ISS 1990) . ISCST model is a stationary gaussian type,it uses "Briggs formulae to calculate plume rise ,and Pasquill-Gifford curves or Briggs" urban condition functions to evaluate pollution dispersion. It can also consider wake effects of stacks and buildings near the emission points In the first case and in the presence of straight constant winds, it forecasts the ground impact of plumes having a small dynamic buoyancy; in the second case it considers the distortion flow effect caused by the buildings being of considerable dimensions and the possible impact of plume at ground .
CTDM-PLUS. It is a new hybrid-gaussian model to evaluate the dispersion of pollutants emitted by point sources on complex terrains .In the presence of obstacles,it simulates the flow lines and plume distortion,levelling at different heights from the ground depending on the terrain topography.

It is possible to evaluate the partial or complete overcoming obstacles depending on wind speed, and height, and atmospheric stability conditions. Releases in convective conditions can consider plume splitting phenomena (bi-gaussian pattern) to evaluate the effect of convective bubbles producing a strong vertical mixing of the pollutant substances.

MATHEW . This code was developed by the Lawrence Livermoore Laboratory. It assesses the wind field by means of a procedure including interpolations and smoothing of experimental data (anemometer measurements at ground level and vertical profiles from Doppler SODAR).The model uses cartesian coordinates, with fixed grid steps both in vertical and horizontal directions.

The interpolated wind field is obtained by separately managing wind speed and direction. They are evaluated in all grid points as a weighted average of the values referred to the three nearest points.The resulting wind field is again adjusted to satisfy the mass conservation principle.The last adjustment should be as low as possible.

The vertical component of the wind field is not considered during the interpolation-extrapolation phase, because tridimensional parameters are not normally available ; so vertical wind speed is created by the model during the adjustment phase.

SPRAY/S. It is a "particle" Lagrangian model that considers dispersion in a tridimensional field with complex topography and allowing the simulation of variable meteo-diffusive situations.It is a simplified version (S) of SPRAY code developed by ENEL-CRAM[3],ARIA Technologies [4], and CNR-ICG[5] in order to operate a real time air quality control.

---

3 Italian National Electricity Company - Environmental and Material Research Center.

4 ARIA Technologies - Le Charlebuorg 14-30, rue de Nantes 92700 Colombes - Paris (F)

5 CNR-ICG Consiglio Nazionale delle Ricerche - C.so Fiume, 4 - Torino (I)

It allows the simulation of many different sources (point,areal, linear,volumetric) with continuous or intermittent emissions.It can use tridimensional wind fields and turbulence parameters simulating complex atmospheric conditions such as temperature inversions, calm wind situations,and convective instability.

Possible Uses of Data Management and Network Supervisor Center (NSC)

The NSC is composed of a flexible and modular HW and SW structure, so it is designed to control urban networks and industrial plants with different complexity and power. Computing modules can be equipped in "extended" and expansive configuration (to control extended areas with advanced mathematical models), or "reduced" to control small sources and few network measurement points.

Environmental management can concern :

- power plant emissions,
- urban networks,
- industrial areas (steel plants, cement factories,
  petrochemical plants, waste management plants) with point or diffuse emissions.

Innovative Characteristics

They could be described as:

- the capability to control critical environmental situations by using high-technology devices and advanced tridimensional mathematical models,
- the possibility to exchange real time data with remote high level nodes (regional or national centers) to perform more reliable local forecasts.

Mobile Intelligent Network

The Mobile Intelligent Network (MIN) is a truck-mounted system for data acquisition and processing,making use of Remote Sensing Control (RSC) Software implemented on a DEC MicroVAX computer.The RSC Software is designed to handle and integrate meteo and pollution data coming from different sources,taking full advantage of VAX /VMS operating systems and real-time multiprogramming capabilities.

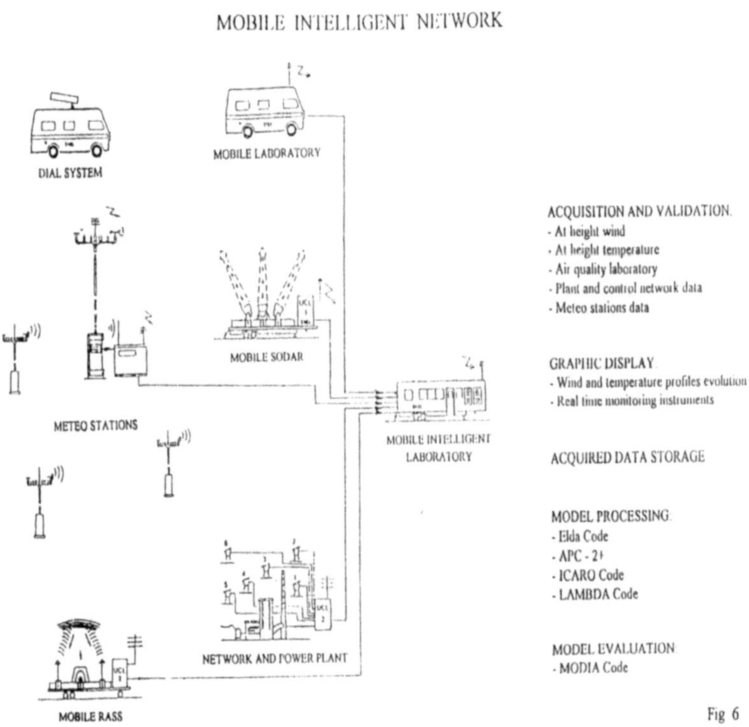

Fig 6

The MIN (Fig 6) can be connected to traditional and/or advanced ("phased-array" type) Doppler SODARs,a mobile RASS system and a radio communication system operating at 44.900 Mhz.The radio communication system is able to connect up to six remote
locations within a 10 Km radius from the MIN,that controls data transmission and diagnostics.

Every 30 minutes the MIN can receive the raw data,store it into an ELDA file format and activate different mathematical models to forecast ground level concentration of pollutants and tridimensional wind field on complex terrain.
The MIN will be connected to remote nodes using the 900 Mhz GCS Communication Network.

Conclusion

The renewal of existing networks,which might be considered first generation, should be made with a wider use of the innovative HW/SW supports recently developed. "Profiler sensors" (for temperature,wind,and pollution concentrations),if used in integrated networks, would allow optimization of the network and cost redution too.

The possibility of acquiring ground level and at height information increases the reliability of mathematical models.It has been verified that a gaussian code implemented with wind at height, improves its own performances and a "particle" model has better performances, than traditional "gaussian "ones.

The best way to reach improved models and network performances could be through a close cooperation between the Research Centers and Industry. Network connection allows the application of a local areas of large scale forecasts, by reaching appropriate decision which can save economic and personnel resources.

At the moment, a Remote Sensing Station, a Data Management and Network Supervisor Center, integrated with classical and advanced mathematical models, are operative at CISE. All this system is classified as a type A station (according to the Italian Environment Ministry law (of 20/05/91).The NSC can operate both as a Local Control Node and as a Regional Control Node.

At present a Mobile Intelligent Network equipped with advanced HW and SW systems is operative, and is able to support Environmental Impact Evaluation studies and to control the representativity of air quality monitoring networks.

References

Bonino G., Elisei G., Marzorati A., Trivero P.: "Results on Planetary Boundary Layer Sounding by RASS", Atmospheric Research, 20 pp. 309-316, 1986.

Brusasca G., Marzorati A., Tinarelli G.: "Software for real-time acoustic remote sensing instrumentation and atmospheric control, int. proc. Conf. Envirosoft 86, 759-769.

Brusasca G., Morselli M.G., Anfossi D.: "Model evaluation criteria on tracer experimental data". Proceedings of the Manno Workshop "Intercomparison of advanced practical short-range atmospheric dispersion models". August 30-September 3 1993, Manno, Switzerland.

G. Tinarelli, D. Anfossi, G. Brusasca, E. Ferrero, U. Giostra, M.G. Morselli, J. Moussafir, F. Tampieri, F. Trombetti. "Lagrangian particle simulation of tracer dispersion in the lee of a schematic two-dimensional hill". Journal of Applied Meteorology - Vol. 33, N. 6, pp. 744-756.

Brusasca G., Elisei G., Finzi G.: " Real time forecast of SO_2 pollution around a power plant by means of a central intelligent unit"Proc. of the WMO Conference on air pollution modelling and its application, Leningrad, 19-24 May 1987.

Brusasca G., Catenacci G., Cavicchioli C., Faggian P., Riva G.M., Sozzi R.: "Validazione dei modelli EPA Complex-1, RTDM, CTDMPLUS sulla base del data set sperimentale relativo alla campagna di Sostanj 1991" CISE report, SAA-93056.

Elisei G., Marzorati A.: "Measurement of wind and temperature proflîles in the lower atmosphere", 2nd Symposium of IMEKO Technical Committee on Metrology Tc8, Helsinki, (Finland), 22-24 August 1988.

Maini M., Zardetto G., Marzorati A., Frego G., Bider M.: "Gestione con calcolatore di una rete di dati ambientali provenienti da apparati di telerilevamento radioacustico", Annul Meeting A.E.I., 2.d.3, pp 1-6, Catania (Italy), 27-30 September 1987

Marzorati A., Mastrantonio G., Fiocco G.: " Criteria for the automatic classification of Micrometeorological Situations by analysis of SODAR intensity", 4th Simposium of the International society of Acoustic Remote Sensing, Camberra (Australia), 16-24 February 1988.

Zanzottera E., Marzorati A.: "Tecniche laser per il telerilevamento dell'atmosfera: stato attuale e prospettive future", F.A.S.T. Seminar, Milan 29th November 1993.

Complex Analysis of the atmosphere contamination in industrial centers of Belarus

A.A.Kovalyov, I.I.Lishtvan

Academy of Science of Belarus
Scientific-technical center "Ecomir"
Republic of Belarus Minsk

A system for the operative control of atmosphere contamination is the most effective means both for detection of harmful releases and identification of trends of the atmosphere quality change at a regional level. Such systems shall be installed in stationary control sites located in potentially dangerous areas or, on the contrary, in background testing sites. To determine the optimum number of control sites and their location, it is necessary to conduct a special investigation of the territory. For the accomplishment of this task and a number of others related to assessment of the atmosphere quality, a sistem of complex analysis of the atmosphere contamination has been developed under the belarussian national program "Remote diagnostics".

The system comprises a complex of ground and airborne equipment and computers intended for processing, storage and analysis of data. Helicopters are used for investigation of territories at a regional level. The system employs a trace UV spectrometer (operating range of wavelengths - 250-455 nm), designed for determination of ozone, SO_2, NO_2 concentration, as well as fine-dispersed aerosol. To define the altitude profile of aerosol the system comprises a laser-location system (LLS).

Specifications of these systems developed in Belarus are given in Table 1.

Table 1. Main specifications of UV spectrometer

Spectral range (nm)	250-455
Spectral resolution (nm)	0,8
Spectrum scanning time (sec)	15
Number of scanning steps	1000

NATO ASI Series, Partnership Sub-Series, 2. Environment – Vol. 8
Urban Air Pollution
Edited by Ivo Allegrini and Franco De Santis

Main specications of LLS

Operating wave lengths of radiator (µm)	0,53; 1,06
Spatial resolution (m)	3
Activity range (km)	3

Mobile laboratories are used for carrying out local tasks as well as for auxiliary ground support to airplane measurements. A mobile ecological laboratory comprises a commercial chromatograph "Chrom 5" produced in the Czech Republic and optical gas analyzers designed in the Academy of Science of Belarus. The former can determine the concentations of a number of substances, even seldom found contaminants. The latter allows the operative determination of the most frequently found substances, for instance CO_2, NO_2, SO_2. Main contaminants and their minimum concentrations are given in Table 2. The mobile lidar station (MLS) which was also developedat the Academy of Science of Belarus allows us to measure altitude profiles of the atmosphere contamination by aerosol particles of different size. The station operates in the visible range of spectrum and its design allows us to build in an additional IR unit for measuring altitude profiles of gaseous contaminant concentrations. Some specifications of MLS are given in Table 3.

Both airborne and mobile instrumentation is equipped with computer systems for data processing. To fix the measurements to locality, the systems of satellite positioning Wild 200 (Leika) has been used. Coordinate fixing allows us to use modern geoinformation technologies which enable the integration of the obtained data with the data of the topographic survey.

To identify ecological trends and support the unit of data processing an archive data base is created and a base of simulation diffusion models. A special dynamic model is developed for simulating contamination diffusion from point, linear and square sources which integrates methods for calculation of impurity diffusion in the atmosphere accepted in the CIS (Central Information System). This is the basis for comparison of calculated data obtained using Gaussian diffusion model which are widely used in developed countries of the West.

Table 2

N	Substance	MPC(mg/m^3)	Sensitivity (mg/m^3)
1	O_3	0.16	0.023
2	$CFCL_3$	100	0.17
3	CCL_4	4	0.5
4	SO_2	0.5	0.03
5	H_2S	0.008	0.016
6	CS_2	0.03	0.6
7	HCHO	0.03	0.1
8	HNO_3	0.4	0.06
9	CL_2	0.2	0.65
10	NO	0.6	0.3
11	NH_3	0.2	0.008
12	NO_2	0.085	0.26

During complex experiments on definition of the industrial influence on the environment in Soligorsk industrial area, investigations were conducted on the atmosphere contamination by salty releases. The data obtained showed that most unfavourable impact on the environment is due to potassium plants, the most powerful contamination source being salty dumps.

Table 3. Main specifications of MLS

N	Operating wavelengths (μm)	Divergence of beam (mrad)	Energy of impulse (mJ)	Frequency of modulation (Hz)	Depth of modulation (%)
1.	1.064	2.5	600	10-20	20
2.	0.532	2.5	250	10-20	20
3.	0.355	2.5	70	10-20	20
4.	0.683	2.5	100	10-20	20
5.	0.416	2.5	28	10-20	20

Results of the experiment conducted using ground lidar measurements are given in Fig. 1.

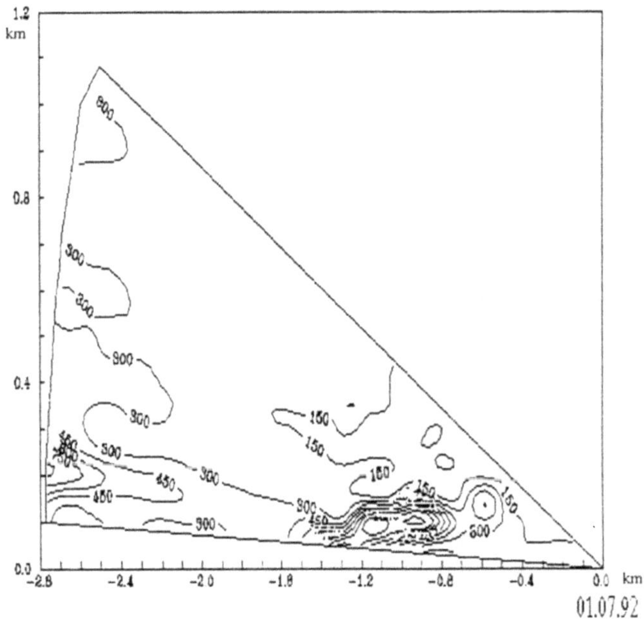

Figure 1. An example of vertical section of aerosol contamination from the salt dump of the first potassium plant with contrary wind, measured by ground lidar

Altitude profile of aerosol contamination with contrary wind is shown there. A model of contamination distribution in this region was developed using experimental data which allows us to follow the dynamic variations of salt particle concentrations as a function of weather conditions. Two figures are given as examples. In the first figure one can see a topographic map with 20 min average isolines of potassium salt concentration in designed weather conditions. In the second figure isolines of daily averaged weather conditions are shown.

The developed system of complex analysis of the atmosphere contamination allows solving the tasks of both planned investigations of territories and operative control of contamination in emergency situations. In future the lidar station will be additionally equipped with IR channel and an extension of data and model base is envisaged.

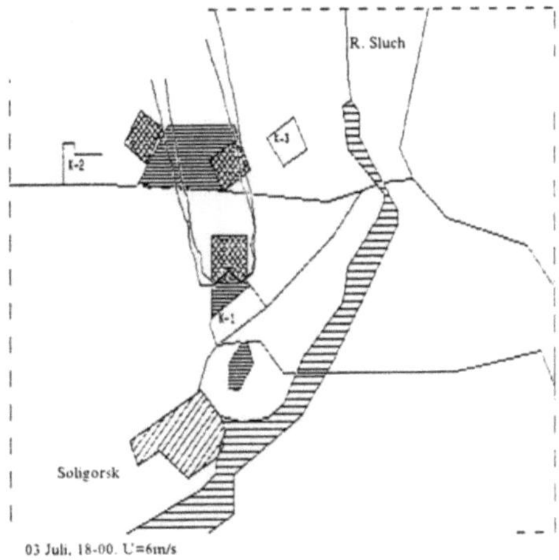

Figure 2. Isolines of daily averaged 20 min KCl conc. from the salt dump of the first potassium plant based on model computation. (C = 0.01 mg/m^3, C = 0.005 mg/m^3).

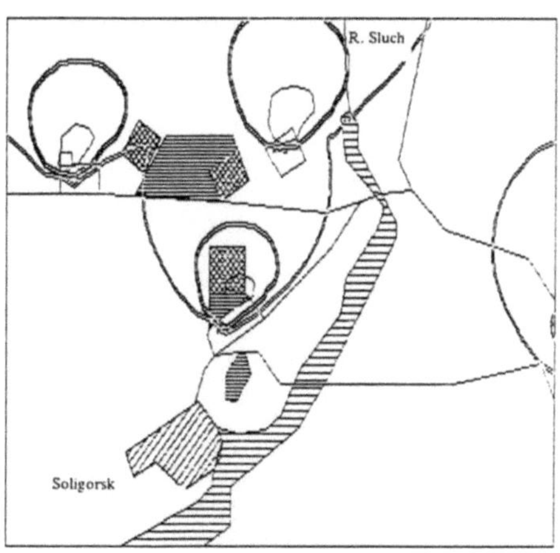

Figure 3. Isolines of daily averaged KCl concentration from salt dumps of four potassium plants based on model computation. (C = 0.04 mg/m^3, C = 0.0015 mg/m^3, C = 0.003 mg/m^3).

Monitoring and Modeling Methods for Developing Air Pollution Control Strategies: A Case Study in the Northwest Czech Republic

R.K. Stevens and J.P. Pinto
U.S. Environmental Protection Agency, Research Triangle Park, NC, USA

R.D. Willis[1], Y. Mamane[2], J.J. Novak,[3] I. Benes[4]

Introduction

Rapid growth in the consumption of lignite in the Czech Republic beginning about 1950 resulted by 1987 in SO_2 and total suspended particulate matter levels exceeding 100 $\mu g/m^3$ (annual average) in Prague and in many urban areas of the northwest Czech Republic (Moldan and Schnoor, 1992). Such elevated levels are thought to contribute to health problems in the Czech Republic (CR) where life expectancy has been estimated (Moldan and Schnoor) to be three to six years shorter than in Western Europe. According to the Czech Hydrometeorological Institute (CHMI), six of the largest electric and home heating power plants in the Czech Republic are located in the northwest region, a basin running southwest to northeast and surrounded by mountains (CHMI, 1994). In addition, many residents in the northwest CR use high-sulfur lignite to fire stoves for heating during the winter. This practice is thought to be a major source of air pollution.

In January 1992 the Czech Ministry of Environment and the Czech Institute of Hygiene, in collaboration with the U.S. Environmental Protection Agency (EPA), began conducting air quality field studies in the northwest CR. These studies were initiated to determine the sources of air pollutants and to assess the impact of lignite combustion and mobile source emissions on air quality in the northern region. Field measurements during the past three years have provided data on

[1]ManTech Environmental Technology, Inc., Research Triangle Park, NC, USA
[2]Technion-Israel Institute of Technology, Haifa, Israel
[3]Czech Hydrometeorological Institute, Prague, CR
[4]Czech Institute of Hygiene, Teplice, CR

NATO ASI Series, Partnership Sub-Series, 2. Environment – Vol. 8
Urban Air Pollution
Edited by Ivo Allegrini and Franco De Santis

ambient concentrations of SO_2, CO, NO_X, polynuclear aromatic hydrocarbons (PAHs), fine and coarse particle mass, elemental composition, and aerosol acidity. The primary monitoring sites were located in the industrial city of Teplice in the northwest CR and the rural city of Prachatice in the southern CR. Emissions from power plants, factories, incinerators, motor vehicles, and residential space heating were also characterized. Ambient and source signature data were input into receptor models to determine the relative contributions of major pollution sources in the northwest region. The results are being used by the Czech government to prioritize pollution emission control strategies based on available control technologies and resources. In addition, data from the study are currently being used in long-term epidemiological and related health studies conducted jointly by Czech and EPA scientists.

Experimental

Ambient Air Sampling and Analysis

The versatile air pollution sampler, or VAPS (Stevens et al., 1993), was the primary sampling apparatus used to collect ambient samples. The VAPS is a modified dichotomous sampler consisting of a PM-10 inlet and virtual impactor that separates the airflow into three channels, two that collect fine particles (<2.5 μm) and another that collects coarse particles (2.5 to 10 μm). An annular denuder (Stevens et al., 1990) followed by a 47-mm Teflon filter pack was added to the first fine particle channel. The denuder was coated with sodium carbonate to remove acidic gases (SO_2, HCl, HNO_3, HF) from the sample stream. The denuder was extracted after sampling and the extract analyzed by ion chromatography (IC). Fine particle mass, elemental concentrations (by energy-dispersive X-ray fluorescence [XRF]), inorganic anions (IC), and acidity data were determined from the Teflon filter (Stevens et al., 1978). The second fine particle channel comprised a 47-mm-diameter quartz filter to collect organic and elemental carbon and a polyurethane foam (PUF) trap downstream of the quartz filter to collect semivolatile organic species. PAHs were obtained by extracting the PUF and the quartz filters and analyzing the extracts by high performance liquid chromatography methods (Method TO-13, 1990). Coarse particles were collected on a 47-mm 0.6-μm pore size polycarbonate filter, which was analyzed for trace metal concentrations and examined by scanning electron microscopy (SEM). The flow rate of the VAPS was 32 L/min; 15 L/min was directed through each of the fine channels and 2 L/min through the coarse channel. At this flow the d_{50} cut point of the inlet was approximately 10 μm.

SEM coupled with energy-dispersive X-ray analysis (SEM/EDX) was performed on selected ambient samples collected in Teplice and on airborne samples collected in the plumes of industrial and power plant sources in the northwest CR. SEM/EDX yielded data on the size, morphology, and chemistry of individual particles, providing input into source apportionment efforts.

Sampling Locations

The main sampling site in this study was located in Teplice, an industrial city 90 km northwest of Prague with a population of about 60,000. Several glass factories and five major power plants providing electricity and steam to area residences are within 50 km of Teplice. In addition, one of the largest open-pit lignite mines is only 30 km from Teplice. The control site was located in Prachatice, a small rural city 130 km south of Prague and exposed to lower air pollution levels than Teplice. Figure 1 shows the location of the sampling sites and major power plants in the Czech Republic.

Source Sampling

Source signatures were obtained from major pollution sources expected to impact air quality in Teplice including the Ledvice, Tusimice, Pocerady, and Prunerov power plants; the Retenice glass factory and a hospital incinerator in Teplice; chimneys of home heating furnaces; and automotive traffic. The VAPS was used to obtain mobile source signatures for trace metals and organic compounds in the Letensky tunnel in Prague. Samples from the major point sources were collected with a

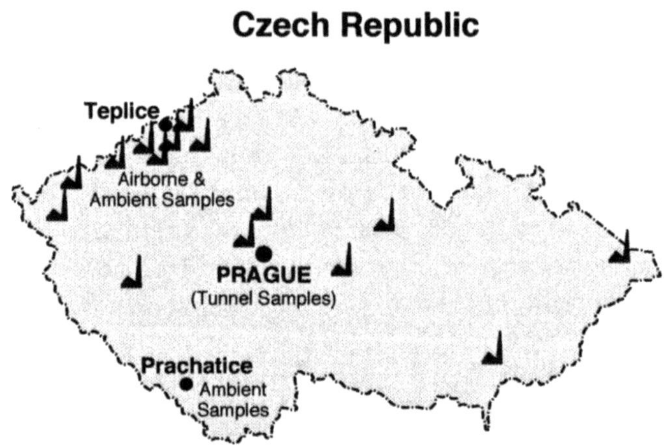

Figure 1. Map of the Czech Republic. The primary focus of the study was the industrial city of Teplice. The small rural city of Prachatice was selected as a background site. Also shown are the locations of the major power plants.

dilution source probe developed jointly by EPA and CHMI. Inorganic (SO_2, metals, and SO_4^{2-}) and organic (semivolatile, elemental and organic carbon, and PAH) pollutants were collected with this sampling system. Air from the stack was first drawn into an isokinetic sampling probe and then allowed to mix with outside clean air in a 0.4-L quartz sampling manifold. The dilution airstream (10:1) was passed through Drierite absorbant to remove water vapor and through Teflon and quartz filters to remove ambient particles. The SO_2 and particles were collected on an annular denuder followed by a Teflon filter pack. A cyclone inlet with d_{50} cut point of 2.5 μm was used as an inlet to the sampling train. Organics were collected by a quartz filter and a PUF trap.

Samples were also collected from chimneys of two residences heated with coal fuel. One chimney was located in Teplice and the other in Sobeslav, about 60 km south of Prague. Sampling in Teplice was performed by placing a 2×2×2-m dilution chamber over an active chimney. Aerosol concentrations were approximately 100 times background levels in this chamber. An annular denuder followed by a Teflon filter pack was used to sample for SO_2, fine particle mass, and elemental composition. A quartz filter and a PUF trap were used to sample for organic species.

Results and Discussion

Concentrations of Aerosol and Gas Phase Pollutants

The mass concentration and elemental composition of the aerosol measured in Teplice during the winter and summer of 1993 are shown in Table 1. Air pollution is most severe during the winter season, when central Europe is under the influence of high pressure, resulting in light winds and stable conditions. These adverse conditions, often accompanied by surface inversions and fogs, may last several days until the passage of a cold front or a low-pressure air mass system (Ireson et al., 1992). The U.S. 24-h ambient air quality standards for SO_2 (0.14 ppm) and particulate matter (150 $\mu g/m^3$) were exceeded on a number of days during the winter of 1993. The mean fine and coarse particle concentrations were 122 and 18.5 $\mu g/m^3$, respectively. Coarse particle mass concentrations typically ranged from 10 to 30% of the fine particle mass. Sulfate and organic carbon concentrations were approximately 42 and 34 $\mu g/m^3$.

Between January 29 and February 6, 1993, an inversion occurred during which the peak fine mass concentration reached 867 $\mu g/m^3$, and the average concentration was 208 $\mu g/m^3$. The SO_4^{2-} and the organic carbon concentrations were 400 and 140 $\mu g/m^3$, respectively. Filter samples collected on February 3 and 4, 1993, had pH values of 3.7 and 4.1, indicating the presence of acidic aerosols.

The February 1993 episode in Teplice is similar in character to the classic London smog episode, December 5-9, 1952, which has been associated with approximately 4000 excess deaths (Logan, 1952; Wilkins, 1954). Both were the result of exposure to high sulfur oxide and particulate concentrations; in both cities, home heating and fuel combustion by power plants were the principal pollution sources; and both occurred during the winter months during strong surface inversions. These similarities suggest that mortality and morbidity records should be examined during such episodes to equate these exposures to health effects. Exceedances of the air quality standards for SO_2 and particulate mass themselves indicate that health problems

Table 1. Mean Aerosol Composition in Teplice, Winter and Summer 1993*

Species	January - March				May - August			
SO_2 (µg/m³)	153	(60)			na			
Particulate Matter (µg/m³)	**Fine**		**Coarse**		**Fine**		**Coarse**	
Total Mass	122	(3.0)	18.5	(4.9)	28.7	(1.2)	10.6	(1.1)
Organic C	33.8	(5.0)	na		na		na	
Elemental C	2.3	(0.3)	na		na		na	
Metal Oxides	6.5	(0.5)	14.0	(2.0)	1.88	(0.14)	7.3	(1.0)
Sulfate†	41.9	(5.7)	nd		10.2	(1.3)	nd	
Sum of PAHs	0.278	(0.012)	na		0.027	(0.001)	na	
Benzo[*a*]pyrene	0.008	(0.0004)	na		0.0005	(0.0004)	na	
Trace Elements (ng/m³)								
Al	830	(130)	2020	(544)	264	(500	850	(210)
Si	940	(180)	3240	(770)	296	(47)	1660	(400)
S	10140	(920)	nd		2460	(190)	nd	
Cl	430	(27)	95.1	(18.3)	31.2	(3.9)	78.3	(6.2)
K	300	(24)	204	(23)	96.8	(7.6)	124	(7.1)
Ca	145	(12)	574	(45)	75.3	(6.3)	435	(33)
Ti	71.3	(7.4)	153	(22)	27.0	(4.4)	73.4	(11.0)
V	11.7	(1.8)	nd		nd		nd	
Cr	5.9	(1.00)	nd		nd		nd	
Mn	18.9	(1.7)	12.3	(1.5)	10.2	(1.0)	10.8	(1.1)
Fe	380	(42)	820	(84)	150	(14)	480	(48)
Cu	14.1	(1.7)	6.6	(1.2)	nd		4.1	(0.7)
Zn	160	(16)	17.6	(4.0)	33.6	(3.6)	10.2	(1.6)
Ga	1.6	(0.2)	nd		nd		nd	
As	44.5	(4.5)	nd		nd		nd	
Se	8.1	(1.0)	nd		nd		nd	
Br	18.5	(2.1)	4.4	(0.7)	5.1	(0.9)	2.0	(0.3)
Pb	108	(11)	6.8	(2.1)	39.8	(4.4)	nd	

*Estimated uncertainty in parentheses; na = measurements were not available; nd = detected at the 3-sigma level in fewer than half the samples.

†Sulfur expressed as ammonium sulfate.

occurring during these episodes may be associated with exposure to high concentrations of SO_2 and particulate matter.

The composition of the fine particles collected during the winter of 1993 was dominated by sulfates, organic carbon, and trace metals. Concentrations of As, Cr, V, Se, and Pb were substantially higher than those observed in urban locations in the United States (Davis et al., 1984). Spring and summer particle concentrations were substantially lower than levels measured during the winter. These findings reflect more favorable meteorological conditions and the absence of home heating emissions.

Polynuclear Aromatic Hydrocarbon Measurements

During 1993, samples were analyzed for nine PAHs. The sampling method, a quartz filter followed by a PUF plug, is expected to provide quantitative measurement of the heavier PAHs - benzo[*a*]anthracene, benzo[*k*]fluoranthene, benzo[*a*]pyrene (B[*a*]P), ideno[*cd*]pyrene, and benzo[*ghi*]perylene. The lighter, so-called semivolatile PAHs would not be collected quantitatively. Total PAH concentrations are several times higher in winter than during the summer in both Teplice and Prachatice; total PAHs measured during the winter of 1993 in Teplice average about 60% higher than those measured in Prachatice over the same time periods; and in the summer season, PAH concentrations are relatively low and are about equal at both sites. Concentrations of B[*a*]P are much higher in the winter months, when home heating emissions are the most prevalent. During the summer, mobile sources are the major contributor to B[*a*]P and the ratio of B[*a*]P to Pb is about 0.01. During the winter, when the ratio is 5 to 15 times higher, emissions from inefficient lignite combustion in home heating systems is the most probable source of the increase in B[*a*]P and PAH concentrations.

Scanning Electron Microscopy Results

Selected samples from the Czech study were analyzed by SEM/EDX, including ambient samples collected in Teplice during the winter of 1992 and the winter and summer of 1993 and source samples obtained from home heating systems, coal-fired power plants, and the Letensky traffic tunnel in Prague.

The ambient coarse fraction showed extremely high concentrations of fly ash, aluminosilicate minerals, and gypsum (some of which may be formed through reaction of calcite with acidic sulfate). Fly ash particles included spherical particles typical of coal-fired power plant emissions (Mamane et al., 1986), as well as rounded but aspherical fly ash bonded with organic matter to form aggregates. This latter category of fly ash is thought to be associated with home heating emissions.

During the February 1993 episode, aerosol concentrations were extremely high, saturating the fine filter with layers of sulfate and carbonaceous particles (Figure 2). The coarse particle fraction was dominated by power plant fly ash and aluminum silicate minerals (resuspended soil and street dust).

Figure 2. Micrograph of a 12-h sample (fine fraction on a Teflon filter) collected February 5, 1993, at the peak of an air pollution episode in northern Czech Republic. Large terminated sulfate crystals can be observed, which may have grown in situ because of the extremely high sulfate loading of the filter. The sample composition (by mass) was approximately 53% ammonium sulfate, 32% organic carbon, 4% trace elements and elemental carbon, and 10% not accounted for.

Particles identified in the ambient samples provided evidence for additional sources: coal particles rich in pyrite, iron spheres, aluminum chips, fly ash from oil-fired boilers, incinerator emissions, and silicon-rich spheres. Biological particles such as spores and pollen were rarely observed in the ambient samples analyzed.

Receptor Modeling

Source apportionment of the major pollutants in Teplice was one of the objectives of the air monitoring program. Contributions from power plants, incinerators, automobiles, and home heating to the total fine aerosol mass measured at the monitoring site in Teplice were estimated by using chemical mass balance (CMB) receptor modeling (Watson et al., 1990).

The source signatures used in the receptor modeling calculations are shown in Table 2. The table lists data from the Ledvice power plant, home heating during the smoldering phase of combustion (Home Heating-S) and during active combustion (Home Heating-A), mobile sources, and a hospital incinerator in Teplice. As can be seen in Table 2, elemental and organic carbon dominate the fine particle emissions from

Table 2. Source Profiles Used in Receptor Modeling Calculations*

Species	Ledvice Power Plant	Home Heating-S	Home Heating-A	Mobile	Hospital Incinerator
OC	7.5%	68.0%	61.7%	53.9%	55.9%
EC	–	2.0%	6.2%	10.1%	17.6%
Al	15.6%	–	–	–	7700 ppm
Si	19.8%	823 ppm	–	2888 pm	1.7%
S	1.4%	8330 ppm	1.55%	2.72%	0.5%
Cl	–	1196 ppm	2052 ppm	3838 ppm	4.3%
K	1.8%	665 ppm	2571 ppm	1001 ppm	2.2%
Ca	4262 ppm	2048 ppm	–	1040 ppm	8660 ppm
Ti	9850 ppm	298 ppm	–	–	2060 ppm
V	337 ppm	–	–	–	–
Cr	288 ppm	–	–	–	–
Fe	4.2%	844 ppm	435 ppm	3886 ppm	5510 ppm
Zn	665 ppm	145 ppm	1936 ppm	946 ppm	9760 ppm
As	220 ppm	62 ppm	226 ppm	–	–
Se	559 ppm	25 ppm	27 ppm	–	–
Br	–	67 ppm	59 ppm	2575 ppm	282 ppm
Pb	190 ppm	60 ppm	211 ppm	9086 ppm	1051 ppm

*All signatures are based on stacked measurements of sources in the Teplice area. Home Heating-S=Smoldering phase. Home Heating-A=Active (high-temperature) phase. OC=Organic carbon, EC=Elemental carbon.

mobile sources, home heating, and the incinerator, whereas power plant emissions are dominated by crustal matter.

The CMB model formulation does not allow for the inclusion of secondary species formed by reactions occurring in the atmosphere, and most of the SO_4^{2-} found in the ambient samples is formed by the oxidation of SO_2 in the atmosphere. In order to apportion the sources of ambient secondary sulfate observed in Teplice, the contributions of power plants and home heating to ambient sulfate were scaled according to their total SO_2 emissions in the region. This was further refined by incorporating the CMB results for trace metals and carbon. The emissions scaling method (Lewis et al., 1986) uses SO_2 and particulate emissions scaled by the distance of each source from the main receptor site. Lignite-fueled power plants in the northwest CR and home heating systems were assumed to be the only sources of SO_2 in the region.

We first apportion secondary sulfate for the winter of 1993. Total annual power plant and home heating emissions of SO_2 in the Teplice region are 574,600 and 8400 tons, respectively (CHMI, 1994), as shown in Table 3. Power plant emissions during the winter months are estimated to be one-fourth of the annual emissions, or 144,000 tons. (Home heating is assumed to occur only during the winter months, so the winter emissions are the same as the annual emissions.) The impact of power plant emissions on the Teplice air shed are scaled by $1/R$, where $R = R_1/R_2$, R_1 is the distance of a power plant to Teplice, and R_2 is the distance of the Ledvice power plant to Teplice. Simple inverse scaling based on distance was chosen because channeling of emissions between the mountain ranges results in a larger contribution from upwind sources than if uniform dispersion in all directions is present ($1/R^2$ scaling). Home heating emissions from districts outside Teplice and Usti were also scaled inversely with distance according to Table 3.

Power plant and home heating contributions to ambient sulfate are calculated by scaling emissions inventory data by CMB results for trace metals and carbon according to the relation

$$(SO_4^{2-}{}_{HH}) / (SO_4^{2-}{}_{PP}) = (NSM_{HH} \times X_{HH}) / (NSM_{PP} \times X_{PP}) \qquad (1)$$

subject to the constraint

$$(SO_4^{2-}{}_{HH}) + (SO_4^{2-}{}_{PP}) = SO_4^{2-}{}_{TOTAL} \qquad (2)$$

Table 3. Emission Data for Particulate Matter and SO_2 for Major Point Sources in Northwest Czech Republic

Power Plant	SO_2, Tons/year	Particulate, Tons/year	Distance From Teplice, km	Stack Height, m
Ledvice	38200	11600	3	200
SDHZ Lovosice	3699	1054	23	80
Neratovice Spolana	7466	2151	62	200
Tusimice I	48200	3856	46	195
Tusimice II	88200	2732	46	300
Prunerov I	59700	5788	46	200
Prunerov II	163100	3984	46	300
Litvinov (Chemopetrol Z.)	55344	23055	15	100
Pocerady	110700	9448	29	200
Local Heating				
District Usti n.L.	**2378**	**1952**	**1**	
District Teplice	**2220**	**1828**	**1**	
District Most	**1494**	**1228**	**19**	
District Chomutov	**2306**	**1890**	**35**	

where SO_4^{2-} is the ambient sulfate concentration, *NSM* is the contribution to nonsulfate ambient fine particle mass (i.e., trace metals and organic carbon) derived from the CMB results, and *X* is the ratio of distance-scaled SO_2 emissions to fine particle mass emissions for home heating and power plants calculated from the emissions inventories listed in Table 3. The subscripts *HH* and *PP* in Equations (1) and (2) refer to home heating and power plants, respectively. It is assumed that the efficiency of conversion of SO_2 to SO_4^{2-} is the same for emissions from both sources.

CMB results for the Teplice fine mass data of the 1993 winter are shown in Figure 3, in which sulfate has been apportioned to power plant and home heating sources as discussed above. It can be seen that most of the fine aerosol mass during winter months is produced by home heating and power plants.

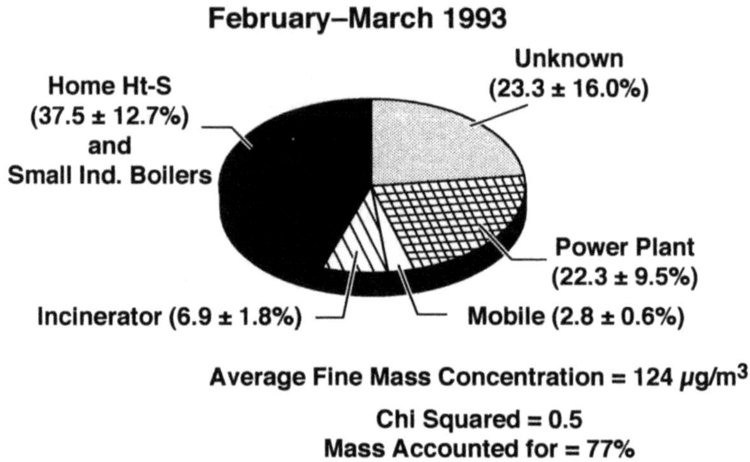

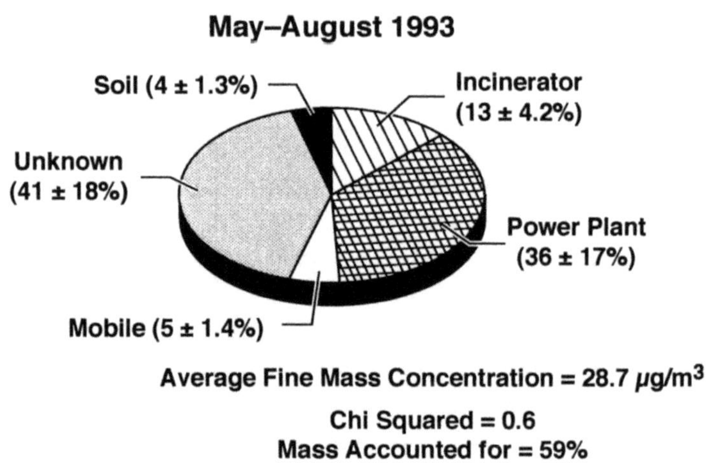

Figure 3. Chemical mass balance apportionment of Teplice fine mass. a. February-March 1993. Sulfate was apportioned between home heating (52%) and power plant (48%) sources. Water of hydration due to conditions of high sulfate and high humidity may constitute most of the unexplained mass. b. May-August 1993. All fine sulfate was assigned to power plant emissions. Water may constitute most of the unknown mass.

Home Ht-S = smoldering phase; Home Ht-A = active (high-temperature) phase.

Mobile sources account for only a few percent of the total mass. Measures of the model accuracy are shown at the bottom of the figure. Source contributions are given uncertainties equal to the standard errors calculated in the CMB model. The uncertainties assigned to home heating (active and smoldering phases of combustion) and power plant source contributions include an additional uncertainty of ±50% in the amount of secondary sulfate apportioned to these sources.

In a study by Ireson et al. (1992) the contribution of major sources (mainly power plants and glass factories) to SO_2 and particulate matter in the northwest Czech Republic during the winter was modeled with a Gaussian plume dispersion model (ISC2). Their modeled SO_2 concentrations consistently underestimated the measured SO_2 levels in the region. They also found a strong inverse correlation between wind speeds and concentrations. Thus they concluded that the contributions of small point sources to ambient concentrations were large relative to their emission rates and that the small, unmodeled sources may be the major contributors to ambient concentrations. These findings are consistent with our receptor modeling calculations.

Sulfur Dioxide-to-Sulfate Conversion

The sulfate particle size information obtained from SEM and the $SO_4^{2-}/(SO_2 + SO_4^{2-})$ ratios suggest fog-assisted particle growth and SO_2-to-SO_4^{2-} conversion by a heterogeneous mechanism, in accordance with studies by Bizjak et al. (1986, 1988) and Meng and Seinfeld (1994). Meng and Seinfeld theorized that SO_4^{2-} particles formed from heterogeneous aqueous reactions would result in particles from 0.7 to 1.5 μm in diameter, consistent with measurements in this study. In studies in Ljubljana, Yugoslavia, Bizjak et al. reported high concentrations of SO_2 and carbonaceous particles, which provided evidence for heterogeneous SO_2 oxidation. The recent work by Meng and Seinfeld suggests that Fe and Mn in fog droplets enhance the conversion of SO_2 to SO_4^{2-}. Data from the February episode in Teplice in fact indicate a relatively high efficiency for conversion of SO_2 to SO_4^{2-}, in contrast with data from a subsequent episode, which was characterized by substantially lower concentrations of fine mass, Fe, and Mn.

Conclusions

The principal objectives of this study were to assess human exposure to toxic air pollutants, to relate ambient concentrations of pollutants in the northwest CR to health risks, and to determine the major pollution sources so that air pollution control can be instituted most effectively. Results of the study are summarized below.

- The 24-h average fine particle mass loadings approached 1 mg/m^3 during periods of air stagnation in the valleys of the Ore Mountains in the northwest CR. The major aerosol constituents during these periods were sulfate and organic carbon. Concentrations of Cr, Zn, Se, As, and other trace metals were greatly elevated compared to levels measured in U.S. cities.
- Between January 29 and February 6, 1993, concentrations of SO_2 and particulate matter in Teplice approached concentrations measured in the classic London smog episode of December 5–9, 1952, which has been associated with nearly 4000 excess deaths.
- About 85% of the fine particles analyzed by SEM were sulfates or carbonaceous particles. The remaining particles were mostly fly ash associated with power plants or home heating. Coarse particle mass typically ranged from 10 to 30% of the fine mass and was dominated by power plant fly ash and soil particles.
- CMB model results indicate that residential space heating and mobile sources were the major sources of organic carbon, and power plants and space heating were the major sources of sulfate and toxic metals during the winter of 1993. Both sources burn lignite because it is the most abundant fuel in the region.
- Primary SO_4^{2-} was a much smaller fraction of the fine particle mass in the source profiles than in the ambient data, indicating that substantial oxidation of SO_2 to SO_4^{2-} occurred. Because the high solar zenith angles and high O_3 columns during winter limit the production of OH radicals, there must be efficient heterogeneous oxidation mechanisms in operation.

- The high concentrations of SO_2 and particulate matter (especially fine particles) measured in the Czech Republic may pose a health risk to people living in the region. During 1993, PM-10 mass and SO_2 concentrations exceeded the maximum 24-h values of the U.S. ambient air quality standards 20 times and 10 times, respectively. Exposure to high concentrations of SO_2 has been associated with increased respiratory problems (US EPA, 1994), and exposure to high concentrations of PM-10 particles has been associated with increased mortality.

Acknowledgement

The authors acknowledge the contributions of Radim Sram and Frantish Koetosevisch of the Czech Republic for their continued support and guidance from the beginning of this program; Bob Kellogg and David Stiles of ManTech Environmental for providing assistance in the analysis of aerosol samples; Rob Ireson, SAI, for his advice on meteorological aspects of the study; Charles Lewis and Andrew Lindstrom of EPA for their thoughtful critical review of this report; and Denise Griffin, EPA, for the preparation of the original draft and assisting with transmittal of information to our Czech scientist counterparts. Also, we acknowledge the early contributions to this project of Andrew Bond, Shaibal Mukerjee, and Teri Conner.

This paper has been reviewed in accordance with the U.S. Environmental Protection Agency's peer and administrative review policies and approved for presentation and publication. Mention of trade names or commercial products does not constitute endorsement or recommendation for use.

References

Bizjak M, Hudnik V, Hansen ADA, Novakov T (1986) Evidence for SO_2 oxidation in Ljubljana, Yugoslavia. Atmos Environ 20:2199-2204

Bizjak M, Benner WH, Hansen ADA, Hrcek D, Hudnik V, Novakov T (1988) Spatial and temporal variations of aerosol sulfate and trace elements in a source dominated urban environment. Atmos Environ 22:2851-2862

CHMI (1994) Air pollution in the Czech Republic in 1992. Czech Hydrometeorological Institute Air Quality Protection Department Prague

Davis BL, Johnson RL, Stevens RK, Courtney NG, Safriet DW (1984) The quartz content and elemental composition of aerosols from selected sites of the EPA inhalable particulate network. Atmos Environ 771-782

Ireson RG, Kellermeyer DA, Looker RE (1992) Dispersion model estimates of major source contributions to SO_2 and particulate matter concentrations in the Northern Bohemia Region of the Czech Republic. Systems Applications International San Rafael California

Lewis CW, Baumgardner RE, Stevens RK (1986) Receptor modeling study of Denver winter haze. Environ Sci Technol 20:1126-1136

Logan WPD (1953) Mortality in the London fog incident, 1952, Lancet 264:336-338

Mamane Y, Miller JL, Dzubay TD (1986) Characterization of individual fly ash particles emitted from coal and oil-fired power plants. Atmos Environ 20:2125-2135

Meng Z, Seinfeld JH (1994) On the source of the submicrometer droplet mode of urban and regional aerosols. Aerosol Sci Technol 20:253-265.

Method TO-13 (1990) Determination of benzo[*a*]pyrene (B[*a*]P) and other polynuclear aromatic hydrocarbons (PAHs) in ambient air using gas chromatographic (GC) and high performance liquid chromatographic (HPLC) analysis, Methods for determination of toxic organic compounds in air, WT Winberry, NT Murphy, RM Riggin (eds). Hayes Data Corp. New Jersey pp 370-466

Moldan B, Schnoor JL (1992) Czechoslovakia: restoring a critically ill environment, Environ Sci Technol 26:14-21

Stevens RK, Dzubay TG, Russwurm GM, Rickel D (1978) Sampling and analysis of atmospheric sulfates and related species. Atmos Environ 12:55-68

Stevens RK, Purdue LJ, Barnes HM, Ward RP, Baugh JO, Bell JP, Sauren H, Sickles JE, II, Hodson LL (1990) Annular denuders and visibility studies. Transactions - visibility and fine particles. A&WMA Publication TR-17, C Mathai (A Peer-Reviewed Publication) AWMA Transaction Series, ISSN 1040-8177 No 17 pp 120-130

Stevens RK, Pinto JP, Mamane Y, Ondov J, Abdulraheem M, Al-Majed N, Sadek M, Cofer W, Ellenson W, Kellogg R (1993) Chemical and physical properties of emissions from Kuwaiti oil fires. Water Sci Technol 27:223-233

US EPA (1994) Supplement to the second addendum (1986) to air quality criteria for particulate matter and sulfur oxides (1992). EPA-600/FP-93/002. US Environmental Protection Agency Research Triangle Park NC

Watson JG, Robinson NF, Chow JC (1990) The US EPA/DRI chemical mass balance receptor model, CMB-7.0. Environ Software 5:38

Wilkins ET (1954) Air pollution aspects of the London fog of December 1952. Q J R Meteorol Soc 80:267-271

Some results of mobile measurements of the influence of the town Sofia on air pollution in the surrounding area

Vassil Andreev
National Institute of
Meteorology and Hydrology,
1784 Sofia, Bulgaria

During the Complex Air Pollution Experiment carried out in two monthly periods (20.XI.-20.XII.1981 and 15.VI.-15.VII.1982), hereinafter conventionally named "winter" and "summer" periods, 62 and 104 air samples respectively were taken by laboratory car at different distances on the lee side of Sofia - the largest Bulgarian city and industrial center. These measurements aimed at investigating the influence that this large town has on the air pollution of the surrounding area.

Wind data are a decisive factor for this type of investigations, therefore data processing started with a test in order to ascertain that a given sample had been taken on the lee side of the town. First of all, the azimuth and the distances of the observational points have been precisely defined by maps.

Wind data analysis in observational periods has shown significant changes of wind direction with height. The most evident cases are in the layer from the ground to the height 500-700 m, reaching 45° in the cases of stronger wind, and even 180° when the speed is 2-3 m/s into the layer to 500 m. The wind direction changes were more significant in the town periphery in comparison with the central parts. These wind direction changes can lead to a wrong car laboratory route definition. Correct information on the air pollution transport in the region of Sofia can not be obtained from the surface wind data in most cases with weak wind - up to 3-4 m/s. Only in cases of stronger winds (invasion type weather) there is coincidence between the surface wind and the wind within the 300 m layer.

The wind variability can be an another cause of incorrectly determining the laboratory's route for air sampling. There are

NATO ASI Series, Partnership Sub-Series, 2. Environment – Vol. 8
Urban Air Pollution
Edited by Ivo Allegrini and Franco De Santis

days with weekly changing winds, not only with height but and in time as well. But there are also days with significant nonstationary wind velocity at different levels.

The wind observations discussed here lead to the conclusion that because of time and height variability of the wind one can select a wrong place for the air sampling.

During the Experiment the direction of route for air sampling was selected operatively with the help of the surface and upper-level wind date from the surface and pilot-balloon measurements in Sofia central meteorological station. The surface wind, however, does not always provide a correct vertical picture of the air transportation on micro - and mesometeorological scales. The upper-wind sounding does not give precise data in the lower atmospheric layers, but they are very important in such investigations. The pilot-balloon measurements are the most representative in this case, but the results have to be available at the laboratory car before starting. Because of the incorrectly selected route, some air samples proved to be defective. Special attention has been paid to the cases with air pollutant transportation from the "Kremikovtsy" Metallurgical plants towards Sofia. These plants are situated about 20km North-East of Sofia (Fig.1). The cases

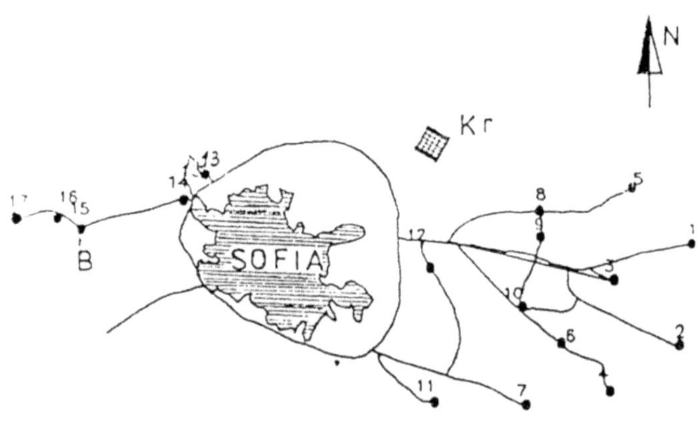

Fig.1 Scheme of the Sofia field. The black dots indicate the observation points on the lee side of Sofia, where air samples have been taken by car laboratory.

with pilot-balloon wind from "Kremikovtsy" at a level 300-600 m have been included in this group. They are shown in Table 1, where α_1 and α_2 are the surface wind directions when the air transportation is from Sofia or from Kremikovtsy respectively. **r** is the distance from Sofia centre.

Observation point		Distance r [km]	Surface wind			SO2 [µg/m3]
No	Name		Direction		Speed [m/s]	
			α_1	α_2		
1	Sarrantsy	35	W	N-WNW	4	135.27
2	Gaytanevo	33	W-WNW	NW	11	75.15
3	G.Mallina	30	W	NW	4	145.29
3	G.Mallina	30	W	NW	10	75.15
4	Bogdanlia	30	WNW	NW-NNW	9	180.36
4	Bogdanlia	30	WNW	NW-NNW	3-4	80.16
5	Tchurek	30	WSW	W	3	133.60
6	Lesnovo	25	WNW	NW-NNW	2	150.30
6	Lesnovo	25	WNW	NW-NNW	9	135.27
7	Novy chan	24	WNW-NW	NNW	10	123.58
7	Novy chan	24	WNW-NW	NNW	3	108.55
8	Iordankino	23.5	WSW-W	WNW	3	126.92
9	Stolnik	23	W	WNW-NW	7	130.26
10	Elin Pellin	22	W-WNW	NW	6	121.91
10	Elin Pellin	22	W-WNW	NW	10	145.29
10	Elin Pellin	22	W-WNW	NW-NNW	6	153.64
11	Lozen	17.5	NW	N	5	143.62
12	D.Bogrov	13.5	W	N	10	123.58

Table 1. SO_2 concentrations at single air samples taken into indicated observation points when they were on the lee side of the Sofia town (surface wind directions are α_1) or of the Metallurgical plants "Kremikovtsy" (surface wind directions are α_2)

When the wind direction is between α_1 and α_2, Sofia and Kremikovtsy jointly contribute to air pollution at a given observation point. On the other hand, the wind directions α_1 and α_2 differ a little and in the variable wind situations it is difficult to determine the main source of air pollution. The large concentrations of some gases at considerable wind speeds have led us to the conclusion that, probably, a particular

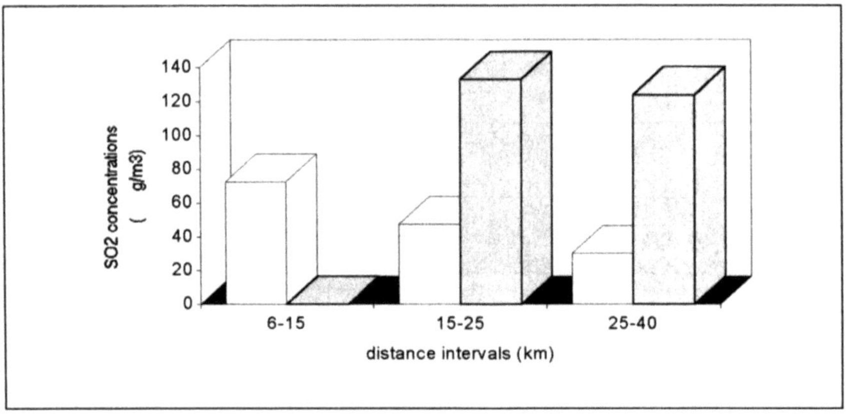

Fig. 2 . Concentration of SO2 on the lee side of Sofia (regardless of wind direction) averaged for given distance intervals from the city centre during the "summer" period. The concentrations under the influence of Sofia town are plotted white and these of Metallurgical plants "Kremikovtsy" are grey.

observation point has been placed under the direct influence of the plume of the Kremikovtsy plants. From Figures 2-4, one can see that Metallurgical plants "Kremikovtsy" can considerably pollute the Sofia field with SO2 . Its influence does not change within the boundaries of this field. The NO and NO2 pollution is more insignificant and their concentrations decrease with distance from the city centre

The city Sofia enhances to the air pollution of the its west part when there is an easterly wind. That is why, the air samples from this part have been discussed separately. The average concentrations, measured during Summer period of 1982 at different distances (r km) westerly of Sofia are shown on Table 2.

The series of observation points from N°14 to N°17 reflect the gradual penetration of the laboratory car along a given profile into the mountain situated southwest of Sofia beyond town Bankja (see Fig.1). It is interesting to see that the average SO_2 concentrations increase in the direction of the mountain, the highest one (equal to 367,4 µg/m) being measured in the village Klissura on May 7, 1982 at 11 a.m. The average concentration for this part of Sofia by the mobile sampling shown in Table 2 is equal to 129 µg/m3 i.e. about 1.5 times

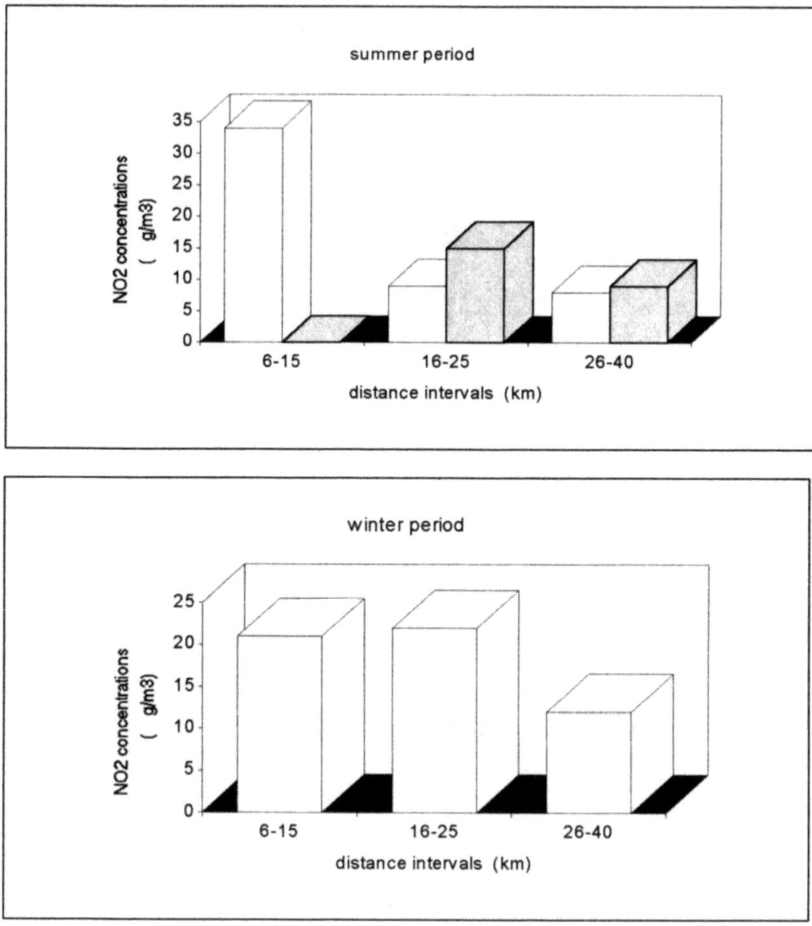

Fig 3. Same as in Fig. 2 but for NO2

higher than that in the region under the influence of the metallurgical plants. This local space maximum of SO2 concentration can be result of either local air circulations (mountain and valley winds) or from pollutant transportation from the neighbouring large towns Sofia and Pernik. The latter is situated about 20 km, south of the town Bankja. Air

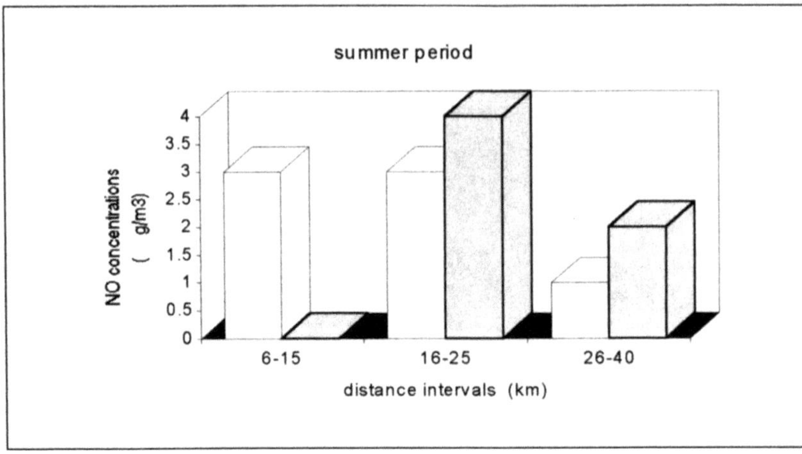

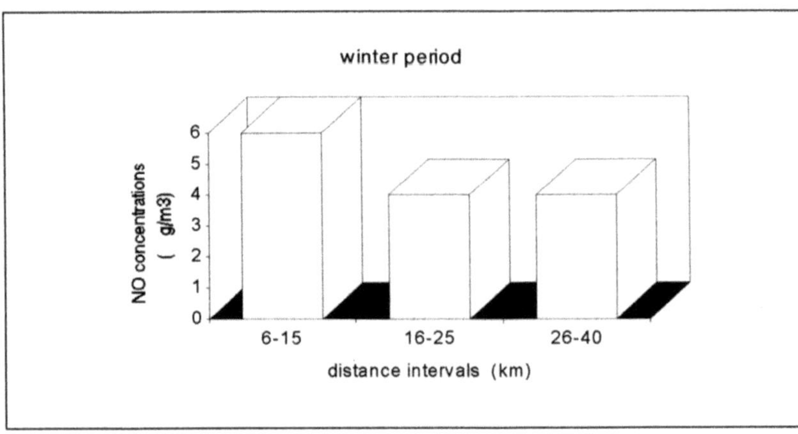

Fig 4. Same as in Fig. 2 but for NO

pollution from local sources can not be expected because there is no industry in the region and no domestic heating during the summer. It should be noted that the air samples are insufficient to draw more definite conclusions. This region requires a more detailed and separate experiment on account of its importance as a health resort.

A more general picture about the influence of the town of Sofia on air pollution in the surrounding area could be obtained not only from Figures 2 - 4 but also from Table 3.

Table 3 shows a different distribution of air pollutant

Observation point:		Distance r[km]	Elevation [m] of:		Number of air samples	Average SO_2 ($\mu g/m^3$)
No	Name		Observ point	Surroundings		
13	Obellia	6.5	540	540	2	79.33
14	Soil Institute	8.0	570	570	3	155.98
15	Bankja	15.0	700	754; 786	5	119.84
16	Verdical	15.5	650	694; 754	4	195.57
17	Klissura	19.0	750	873; 885	3	239.37

Table 2. Average SO2 concentrations in the summer period at the west part of the Sofia field measured at the different points in the cases when they were under influence of the Sofia town. The distances, **r** was measured from the city centre.

concentrations on the lee side of the town regardless of the wind direction during the summer and winter periods. In winter, the polluted gases from domestic heating, industry and transport of the towns and villages along the air transport trajectory are added on the pollution transport from the big town. In summer the domestic heating doesn't work and measurements show an evident reduction of concentration with distance on the lee side.

The investigations of some authors (e.g. see Landsberg,1981) have shown an exponential decrease of the SO_2 concentrations. We tried to verify this conclusion by our data. Acceptable results were obtained for the SO_2 and NO_2 concentrations in the summer period (Fig.5 and Fig.6). The dependence of the concentration (ln C) on the distance r, shown on Fig.5 and Fig.6, can be approximated by a linear equation in the form $\mathbf{\ln C = \ln Co - kr}$, where C and Co are the concentrations of SO_2 or NO_2 at the arbitrary distance r in the lee side of the town and at $r = 0$, respectively. It should be noted that the distances were measured from the centre of the town. On this account, the points at a distance of about 10 km reflect to the high air pollution in the town. The Figures 5 and 6 could be considered as a confirmation of the accuracy of the data selection procedure.

Interval r [km]	Average distance r [km]	Average concentrations [µg/m3] of: SO_2	NO_2	NO	H_2S
a summer period					
6-15	12	72.57	34.2	3.1	10.2
16-25	22	47.36	9.1	3.2	8.1
30-39	32	30.05	8.3	1.3	12.1
a winter period					
6-15	11	3.36	21.3	6.3	32.1
16-25	22	18.71	22.1	4.2	22.2
30-39	33	6.68	12.3	4.1	25.2

Table 3. Average concentration of some pollutants measured on the lee side of the Sofia town and grouped in given distance intervals, **r** from the city centre.

The linear approximation of the ln C dependence on the distance enables us to get an idea of the values of the quantity "radius of influence" (r_{inf}) i.e. that distance in the lee side after which no influence of the town is observed. For this purpose, the concentration - distance approximation equation has to take in the form:

$$\frac{r}{r_{inf}} + \frac{\ln C}{\ln C_0} = 1 \Leftrightarrow r_{inf} = \frac{r}{1 - \ln C_{\ln C_0}}.$$

If we use the data from Table 3 and assume that Co is equal to the average concentration at a distance 6-15 km, then by the concentrations of distances r=22 and r=32 km, respectively we can calculate from this equation that the radius of influence of Sofia is no more that 200-220 km for SO_2 and 60-80 km for NO_2. These assessments are intended for orientation only since they result from a formal extrapolation and many factors have not been taken into account such as: the local pollution sources along the air transport trajectory, the chemical

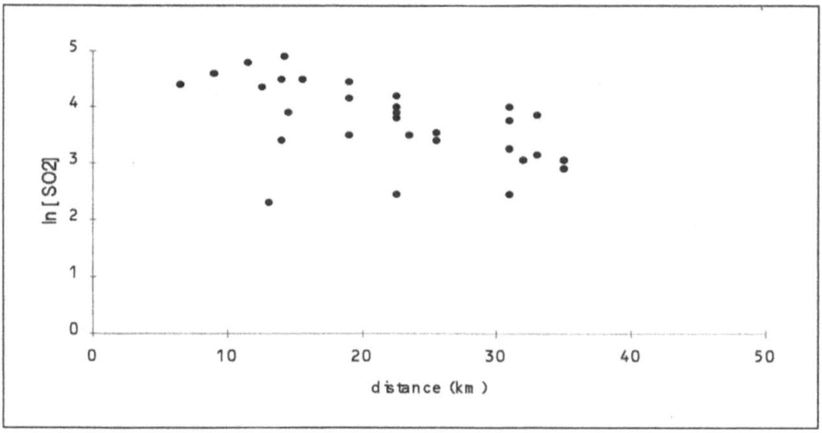

Fig.5 The logarithm of SO_2 concentrations in the air - ln[SO2], as a function of the distance from the city centre on the lee side of the town for the summer period.

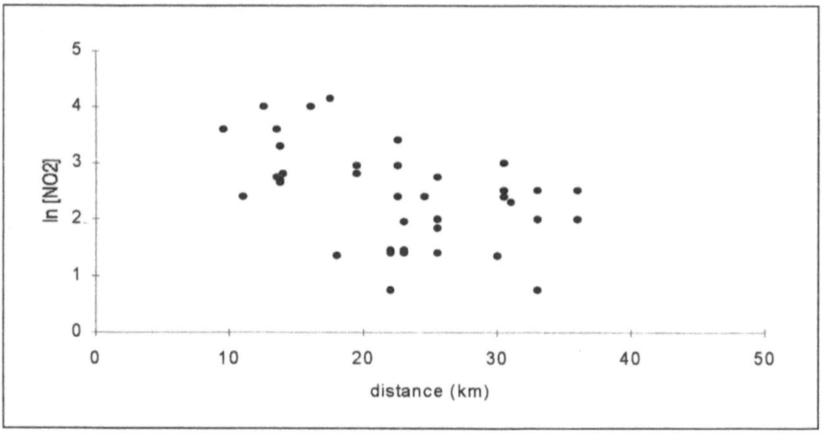

Fig. 6. Same as in Fig. 5 but for the NO_2 concentrations for the summer period.

reactions with the participation of polluted gases, the local orographic effect on air circulation keeping in mind the presence of the high mountains around Sofia. If more precise assessments are necessary it would be better to make them by a tracer experiment.

References.

Landsberg,H.E. The Urban Climate. Academic Press, New York , 1981

Non-methane hydrocarbon measurements in a road tunnel in Budapest

László Haszpra, István Szilágyi*
Institute for Atmospheric Physics
H-1675, P.O.Box 39 Budapest, Hungary
* Central Research Institute for Chemistry
H-1525, P.O.Box 17, Budapest, Hungary

Introduction

The vehicular emission of non-methane hydrocarbon (NMHC) is one of the largest anthropogenic sources of NMHC in Hungary (Molnar, 1990) and in most of the industrialized countries. Non-methane hydrocarbon plays a key role in the formation of photo-chemical air pollution, usually characterized by ozone concentration, which seriously endangers our environment and human health. The ozone forming potential of the different NMHCs differs from each other significantly (see e.g. Atkinson, 1990), while the NMHC composition of car exhaust is influenced by the use of fuel and engine type (e.g. Chang and Rudy, 1990; Gabele, 1990; Hoekman, 1992), technical condition of the vehicle, vehicle speed (see e.g. Bailey et al. 1990) and several other factors.

In Hungary the majority of cars are still of Eastern European origin. They represent the technological standard of the 70's although there have been changes recently. Due to the long-term economical decline in Hungary, the average age of the cars was about 9 years in 1990 (Balo and Lipovecz, 1992) and reached 10 years by 1993 (KSH, 1993). The condition of the majority of the cars is poor. In addition, almost one third (31.2%) of the cars are equipped with two-stroke engines (KSH, 1991) which emit less NO_x but much more hydrocarbon (Meretei and Borsi, 1990). The number of cars equipped with catalytic convertes was negligible in 1990 and only recently has it slowly increased. As a consequence of these facts the traffic emission in Hungary may differ from that measured in or estimated for the Western European countries and these differences should be taken into account in air pollution models.

NATO ASI Series, Partnership Sub-Series, 2. Environment – Vol. 8
Urban Air Pollution
Edited by Ivo Allegrini and Franco De Santis

One possible way to measure the composition of the exhaust gas is to make laboratory measurements on a few test cars. However, the cars in the actual traffic are in different technical condition and the laboratory test on a representative car fleet would be extremely expensive. In a long tunnel, the air contains a mixture of the exhaust gases and evaporated gasoline of a high number of cars. The composition of the mixture may represent the average composition of the exhaust gases and evaporation loss of the cars running under actual vehicle-usage conditions. Road-way tunnel studies have been widely used for decades to estimate vehicular emission and its composition (e.g. Lonneman et al., 1974; Chang et al., 1981; Gorse, 1984; Lonneman et al., 1986; Benner et al., 1989; Pierson et al., 1990). Several authors rely more on tunnel measurements and others on road measurement and use or suggest using, these data for the verification of emission models based on dynamometer studies (e.g. Lonneman et al., 1974; Chang et al., 1981; Gorse, 1984; Zweidinger et al., 1988; Seinfeld, 1989; Pierson et al., 1990; Doskey, 1992). For the estimation of non-methane hydrocarbon composition of the exhausts from the Hungarian car fleet and its average emission, a one-day tunnel study was performed in the downtown of Budapest in summer, 1990. To measure the consequences caused by the slow changes in the Hungarian car fleet, by the extended introduction of unleaded gasoline and the annual mandatory emission test a similar project is planned for summer, 1995, on the same site with a somewhat more extended measuring program. In this paper the methods and the results of the 1990 experiment are presented.

Experimental

The tunnel used for the present study (Fig. 1) is a 350 m long, 9.3 m wide east-west tunnel in the downtown of Budapest which can be used only by passenger cars, motorcycles and public transport buses. Here the traffic is heavy enough to produce high pollution in the tunnel and the tunnel is fairly long, so that significant mixing of the tunnel air with that outside is avoided. In the tunnel the traffic flows in one lane in each direction. The tunnel continues in a western direction in a 70 m long street canyon surrounded by buildings of 20-25 m. At the end of this street canyon the traffic is controlled by traffic lights.

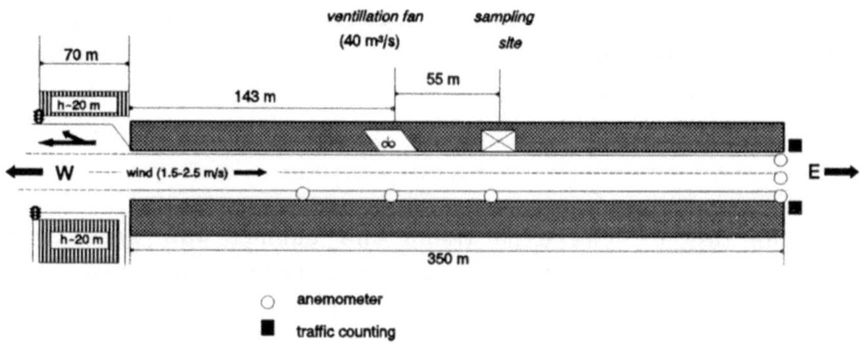

Figure 1: The sketch of the tunnel used in the study

For the measurement of the air exchange three anemometers were located at the eastern entrance which was the outflow side at the time of the experiment, to measure the air flow through the tunnel. Two anemometers, one on each side, were located one above the sidewalk at about 5 m height, and one was hung down the ceiling of the tunnel. The height of this latter anemometer was about 9 m above the street level.

The air samples for hydrocarbon analyses were collected near the middle of the tunnel at a height of 1.5 m. One spot sample was taken every hour in evacuated Pyrex glass bulbs of 1 litre fitted with teflon valves. A total of 16 valid samples were collected during the experiment between 5 a.m. and 11 p.m. For the analysis a HP5840A gas chromatograph equipped with a 30 m long Pyrex glass capillary column of 0.25 mm internal diameter and a FID detector were used. The total non-methane hydrocarbon concentration was determined from the sum of peak areas using a calibration factor determined experimentally. For further details of the analytical method see Haszpra et al. (1991).

Monitor Labs carbon monoxide (Model ZRC3) and nitrogen dioxide (Model 8840) monitors were also located at the sampling site. By means of them, in addition to the non-methane hydrocarbon species, nitrogen oxide and carbon monoxide emissions could also be determined.

For the determination of the emission per car the number and speed of the cars crossing the tunnel were necessary. The traffic was counted manually, recording the cars equipped with two-stroke engines separately. The speed of the traffic was followed by a pace car.

Results

On the day of the experiment the tunnel was ventilated by a westerly wind (1.5-2.5 m/s) and a longitudinal fan pressing 40 M^3/s air into the tunnel supported the natural ventilation. During the experiment of 18 hours 30340 vehicles crossed the tunnel. 96.9% of them were passenger cars, 2.5% of them were public transport buses, while the contribution of motorcycles was only 0.5%. The ratio of the cars equipped with two-stroke engines, was only 13.6%, which was significantly lower than it was expected from the statistical data (31.2%). It indicates that the population statistics alone are not suitable for the determination of the total emission. One possible explanation for this observation is that these cars are less used than those equipped with four-stroke engines. The number of cars equipped with Diesel-engines, was not counted but in the Hungarian car fleet it was only 0.75% at the time of the measurements.

The tunnel worked at its full capacity (approx. 1000 vehicle/hour/direction) almost all day, however the speed of the traffic changed with the actual driving conditions outside the tunnel. As a consequence the number of cars in the tunnel varied in a wide range causing significant variation in the air pollution.

For the determination of emission factors from tunnel measurements there are several mathematical models (e.g. Chang et al., 1981; Chang and Rudy, 1990; Pierson et al., 1990). In our model we assumed that there is an instantaneous mixing of emitted pollutants within the tunnel cross-section and the concentration at a given point does not depend on time (steady-state approximation). The deposition velocity of the pollutants measured are low, therefore, we neglected this process in the model. Taking into account the short transport time (0-2 minutes) from the sources to the measuring site, we did not deal with the chemical transformations either. As the $NO \rightarrow NO_2$ conversion is rather fast the emission factor is determined only for their sum.

In our case, where the natural ventilation through the western entrance was also supported by the ventilation fan, the concentration at the measuring site can be given in the following way:

$$c = c_0 + \frac{en}{Av + W}$$

where c [$\mu g\ m^{-3}$] is the concentration at the measuring site, c_0, [$\mu g\ m^{-3}$] is the concentration in the outside air pressed into the tunnel by natural and forced ventilation, e [$\mu g\ s^{-1}$] is the emission of a single car, n is the number of cars in the tunnel upwind of the measuring site, A [m^2] is the cross section of the tunnel, v [$m\ s^{-1}$] is the speed of the natural airflow at the tunnel entrance, while W [$m^3\ s^{-1}$] is the air volume pressed by the ventilation fan into the tunnel. From this equation e can be expressed. As the average speed of the traffic was also measured the emission can be expressed in the more conventional g/km unit. The number of cars upwind of the measuring site cannot be counted easily. For its determination a mathematical model was compiled describing the locations of the cars in time. The results of the traffic counting and speed measurements were used as input data. Assumptions were made on the average acceleration and deceleration of the cars and on the average reaction time of the drivers. The model describes the real situation well as long as the average speed is above 30 km/h, but at lower speeds it increasingly deviates from the reality. The reason for this is that in a traffic jam, the actual traffic conditions produced, strongly depend on the individual behaviour of the drivers which cannot be easily modelled. Therefore, the emission factors are calculated only for the 30-60 km/h speed range.

The NMHC composition depends on the average speed of the cars. With increasing speed the relative contribution of aromatics (other than benzene) increases, while this is balanced by the decreasing shares of C_5-C_6 hydrocarbons (for further details see Haszpra and Szilagyi (1994)). However, no change can be found in the relative contributions of alkenes (ethene, propene, butene) which are quite sensitive to the changing combustion conditions (see e.g. Bailey et al., 1990). A possible explanation for this phenomenon is that the drivers use different gears in the speed range observed,depending on the construction of their cars. The effect of a fast revolving engine in a low gear may be compensated by another car of different construction (or different driver's behaviour) which reaches the same speed in higher gear with less engine revolution per minute.

In Table I it can be seen that in Budapest the relative contribution of light molecular weight hydrocarbons is lower, while that of pentane and hexane species is higher than in the other cities. This means that the tunnel air in Budapest is relatively rich in C_5-C_6 species which are the main alkane components of gasoline and gasoline vapour. In the case of a working engine the evaporative loss is only a few percentage of the total emission, so the NMHC composition of the tunnel air is primarily determined by car exhaust.

Based on the concentration measurements in the tunnel and on the traffic model described above the TNMHC, CO and NO_x, emission of a hypothetic average Hungarian car in the 30-60 km/h speed range are 4.1 g/km, 28.5 g/km and 1.1 g(NO_2)/km, respectively.

For the calculations the approximate downtown concentrations were assumed as background concentrations.

Actually, the accuracy of the background values is not crucial because the concentrations in the tunnel are roughly a magnitude higher than outside.

Table I: Emission profiles expressed in wt% of TNMHC

	Chicago[1]	Lincoln Tunnel[2]	Raleigh[3]	Atlanta[4]	Budapest[5]
ethene	6.98	9.51	7.15	4.51	7.06
ethane	1.81	--	1.82	1.79	1.34
propene	2.60	2.85	2.51	2.00	1.45
propane	1.00	--	0.88	1.17	0.21
i-butane	1.93	1.78	--	1.26	0.32
n-butane	10.79	4.61	1.75	4.57	2.05
i-pentane	7.31	7.11	5.89	9.35	7.30
n-pentane	3.71	3.40	2.82	2.86	4.57
2-Me-C5	3.21	--	--	2.62	4.13
3-Me-C5	1.69	1.72	--	1.56	2.80
n-hexane	2.08	1.74	--	1.18	3.26
benzene	3.35	4.70	3.23	2.63	2.56
toluene	5.61	7.06	6.59	6.50	6.18
ethylbenzene	1.02	--	5.32	1.28	1.34
m/p-xylene	3.52	3.79	4.43	4.33	3.87
o-xylene	1.27	1.73	1.41	1.67	1.53

[1] Doskey et al. (1992); [2] Lonneman et al. (1986); [3] Zweidinger et al. (1988); [4] Lonneman et al. (1991); [5] this study

Using an early report of Veldt (1986) summarizing emission measurements in West Germany, the United Kingdom, the Netherlands and Switzerland the emission of the Hungarian car fleet can be compared with that of Western European (Table II).

In the speed range mentioned above the TNMHC emission factor for an average Hungarian car is 2-4 times higher than that suggested for engine size below 1400 CM^3 (typical in Hungary) for 1985 in Western Europe (1.1-1.9 g/km including approx. 10% methane). This difference is real, taking into account the emission of the cars equipped with two-stroke engines and their proportion in the Hungarian car fleet. The carbon monoxide emission of the cars is also significantly higher than in Western Europe (4.4-8.8 g/km for 1985, 6.5-12.3 g/km for 1980), which reflects the outdated technical standard and poor maintenance of the aged engines in Hungary.

Table II: Emission factors for non-methane hydrocarbons (mg/km)

	Budapest[1]	Stevenage[2]	Western Europe[3]
methane	--	80	114
ethene	278	123	91
ethane	67	15	18
propene	49	61	46
propane	3	4	3
1-isobutene	22	11	18
2-butene	28	30	15
i-butane	6	30	
n-butane	80	67	
i-pentane	297	92	
n-pentane	189	32	
2-methyl-pentane	165		
3-methyl-pentane	113		
n-hexane	127		
benzene	99	101	66
toluene	260	237	162
ethyl-benzene	55	68	27
m/p-xylene	155	212	129
o-xylene	62	84	
THC	40844	1941	1500

[1] this study; [2] Bailey et al. (1990) [3] calculated from the data of Veldt (1986); [4] total non-methane hydrocarbon

The average age of the car fleets in most Western European countries is almost half of that of the Hungarian one. The Hungarian CO emission factor may be close to that measured in Mexico city (Beaton et al., 1992) having also an environmentally uncontrolled, aged car fleet, as far as it can be estimated from the data given in g/gal unit.

The high number of cars with two-stroke engines and the less effective engines in general cause the NOX emission of an average Hungarian car to be somewhat lower than its Western European counterpart (1.5-1.7 g/km in NO_2).

Since the measurements presented in this paper have been carried out significant changes have happened which may influence traffic emission: (a) the average age of the Hungarian car fleet has increased further; (b) the number of cars equipped with two-stroke engine has decreased due to the cessation of their production; (c) the use of unleaded gasoline has become more extended and the lead content in leaded gasoline has decreased which may influence the traffic emission through the modified average composition of the gasoline; (d) the number of cars equipped with catalyzers and Diesel-engines, is slowly but steadily increasing; (e) mandatory yearly emission tests have come into force. All these changes indicate valid reasons for repeating this relatively cheap study in the near future.

Acknowledgements: The authors express their appreciation to L. Bozo, L. Horvath, A. Molnar, K. Molnar, G. Szilagyi, L. Takacs (Inst. for Atmospheric Physics) and T. Turanyi (Central Research Institute for Chemistry) for the preparation and in-site completion of the experiment, to T. Weldinger and his students (Dep. of Meteorology, Eotvos Lorand University, Budapest) for the meteorological measurements and traffic counting, as well as to J. Fazekas and L. Rabai (municipal Public Services Enterprise, Budapest) for providing the sampling site and technical conditions for the experiment. The study was partially sponsored by the National Committee for Technological Development, Hungary, as part of an EUREKA/EUROTRAC/TOR contribution.

References

Atkinson R. (1990) Gas-phase tropospheric chemistry of organic compounds: a review. Atmospheric Environment 24A, 1-43.

Bailey J. C., Schmidl B. and Williams M. L. (1990) Speciated hydrocarbon emissions from vehicles operated over the normal speed range on the road. Atmospheric Environment 24A, 43-52.

Balo Gy. and Lipovecz I. (editors) (1992) Tények könyve 192. Racio Kiadó kft., Budapest.

Beaton S. P., Bishop G. A. and Stedman D. H. (1992) Emission characteristics of Mexico City vehicles. J. Air & Waste Management Association 42, 1424-1429.

Benner B. A. Jr. , Gordon G. E. and Wise S. A. (1989) Mobile sources of atmospheric aromatic hydrocarbons: A roadway tunnel study. Environ. Sci. & Technol. 23, 1269-1278.

Chang T. Y., Modzelewski S. W., Norbeck J. M. and Pierson W. R. (1981) Tunnel air quality and vehicle emissions. Atmospheric Environment 15, 1011-1016.

Chang T. Y. and Rudy S.J. (1990) Ozone-forming potential of organic emissions from alternative-fuelled vehicles. Atmospheric Environment 24A, 2421-2430.

Doskey P. V., Porter J. A. and Scheff P. A. (1992) Source fingerprints for volatile non-methane hydrocarbons. J. Air Waste Manage. Assoc. 42, 1437-1445.

Gabele P.A. (1990) Characterization of emissions from a

variable gasoline/reethanol fuelled car. J. Air Waste Manage. Assoc. 40, 296-304.
Gorse R. A. Jr. (1984) On-road emission rates of carbon monoxide, nitrogen oxides, and gaseous hydrocarbons. Environ. Sci. & Technol. 18, 500-507.
Haszpra L. and Sziladgyi I. (1994) Non-methane hydrocarbon composition of car exhaust in Hungary. Atmospheric Environment 28, 2609-2614.
Haszpra L., Sziladgyi I., Demeter A., Turdnyi T. and Berces T. (1991) Non-methane hydrocarbon and aldehyde measurements in Budapest, Hungary. Atmospheric Environment 25A, 2103 -2110.
Hoekman S. K. (1992) Speciated measurements and calculated reactivities of vehicle exhaust emissions from conventional and reformulated gasoline. Environ. Sci. & Technol. 26, 1206-1216.
KSH (1991) Magyar Statisztikai Evkonyv. Kozponti Statisztikai Hivatal, Budapest.
KSH (1993) Statisztikai Havi Kozlemenyek/Monthly Bulletin of Statistics 1993/4. Kozponti Statisztikai Hivatal, Budapest.
Lonneman W. A., Kopczynski S. L., Darley P. E. and Sutterfield F. D. (1974) Hydrocarbon composition of urban air pollution. Environ. Sci. & Technol. 8, 229-236.
Lonneman W. A., Seila R. L. and Meeks S. A. (1986) Non-methane organic composition in the Lincoln Tunnel. Environ. Sci. & Technol. 20, 790-796.
Lonneman W. A., Seila R. L., Daughtridge J. V. and Richter H. G.(1991) Results from the canister sampling program conducted during the 1990 Atlanta precursor study. Paper 91-68P.2, presented at the 84th Annual Meeting and Exhibition of the Air & Waste Management Association, Vancouver, June 1991.
Meretei T. and Borsi Z. (1990) Verminderung der Schadstoff Emission von PKW-zweitaktmotoren durch Katalizatoren Erfahrung in Ungarn. Vortrag am Kolloquium "Die Zweitaktmotor im Kraftfahrzeug". TU-Berlin, 28 February 1990.
Molnar A. (1990) Estimation of volatile organic compounds (VOC) emissions for Hungary. Atmospheric Environment 24A, 2855 -2860.
Pierson W.R., Gertler A.W. and Bradow R.L. (1990) Comparison of the SCAQS tunnel study with other on-road vehicle emission data. J. Air Waste Manage. Assoc. 40, 1495-1504.
Seinfeld J. H. (1989) Urban air pollution: State of the science. Science 243, 745-752.
Veldt C. (1986) Emission from road transport. Discussion paper for the OECD Workshop on Comparison of Emission Inventory Data, Apeldoorn, the Netherlands, October 22-24, 1986.
Zweidinger R. B., Sigsby J. E. Jr., Tejada S. B., Stump F. D., Dropkin D. L. and Ray W. D. (1988) Detailed hydrocarbon and aldehyde mobile source emission from roadway studies. Environ. Sci. & Technol. 22, 956-962.

Applications of Highly-Resolved Size-Spectra for Source Attribution, Growth, and Deposition of Urban Aerosol Particles Bearing Various Elements

J. M. Ondov, F. Divita, Jr., T. L. Quinn, M. Han
Department of Chemistry & Biochemistry, University of Maryland, College Park, MD 20742

INTRODUCTION.

Knowledge of composition and size of atmospheric aerosol particles is needed to elucidate their sources, atmospheric transformation processes, contributions to visibility reduction, and respiratory and environmental deposition (Heintzenberg and Covert, 1987; Wu et al., 1994). In a study conducted at Deep Creek Lake (Dodd et al., 1991), a rural recreational area in the Appalachian mountains of western Maryland, we used instrumental neutron activation analysis (INAA) and X-ray fluorescence (XRF) to determine the concentrations of >26 elements in size-fractionated aerosol samples collected with a micro-orifice impactor (MOI). The results showed that elemental composition vs. particle size spectra contain a wealth of fine structure from which information on origin and changes associated with atmospheric processing might be inferred. For example, we observed as many as four discrete aerosol peaks in the elemental size spectra of a single MOI sample. These included narrow peaks with modal diameters of 0,1 to 0,2 μm indicative of "fresh" accumulation aerosol from HTCSs and sulfur-containing peaks at larger diameters, representing "regional sulfate" aerosol, i.e., a more highly aged aerosol that accounts for much of the line-particle mass in ambient aerosol of the Northeastern U.S.

At DCL, we studied a rural site influenced by emissions from utility coal combustion and motor vehicles. Complex mountain meteorology and the multiplicity of similar sources precluded definitive assessment of source receptor relationships with various peaks in the spectra. However, it

NATO ASI Series, Partnership Sub-Series, 2. Environment – Vol. 8
Urban Air Pollution
Edited by Ivo Allegrini and Franco De Santis

was clear that modal diameter and particle composition might indicate age or degree of atmospheric processing, and might be used to identify HTCSs in *urban and industrial* areas. Furthermore, both of these parameters are important in estimating deposition fluxes of toxic substances which are substantial for the Great Lakes and mid Atlantic coastal waters of the United States. Although both dry and wet deposition processes appear to be important, dry deposition fluxes of various trace elements and heavy metals remain highly uncertain, in part, because of their strong dependency on largely unavailable composition vs particle data (Ondov et al., 1994; Wu et al., 1994) and knowledge of the changes in particle size associated with hygroscopic growth.

More recently, we have collected and analyzed.size-segregated aerosol particles in two urban areas, i.e., Washington, DC, a relatively simple airshed, influenced by emissions from motor vehicles, utility coal combustion, municipal incineration, and regionally transported sulfate aerosol; and Camden, NJ, an area influenced by heavy industry and municipal incineration. One goal of these studies was to confirm the observations made in the more limited DCL study and to characterize changes in elemental spectra associated with temporal fluctuations in RH. An additional study was undertaken at two rural, overwater sites along the middle Chesapeake Bay to evaluate the importance of temporal changes in RH on dry deposition of contaminants to Bay surface waters. Herein, we present a summary of pertinent results of these studies, focussing on data developed for three primary sites, i.e., Camden, N.J., College Park, MD, and the Chesapeake Bay Laboratory. The urban and Bay aerosol studies are described in detail by Divita (1993) and Quinn (1994), respectively.

METHODS

Sampling. Size-segregated samples in these studies were collected with 8- or 9-stage microorifice impactors (MOI) for from 8 to 22 h at a nominal flow rate of 30 L m^{-1}. Along with an integral backup filter, each impactor sample provided up to 10 discrete particle fractions with midpoint diameters estimated as follows: 21.2, 6.9, 2.4, 1.3, 0.74, 0.40, 0.22,

0.12, 0.070, and 0.033 µm. Gelman "Teflo" 37-mm dia., 2-µm-pore Teflon filters were used as impaction substrates on all stages and as the backup filters. To reduce particle bounce, the first three stages were coated with approximately 2 mg of dimethylpolysiloxane (Dow Corning 200 fluid). Laboratory and field blanks were collected during each study by exposing clean, unused impaction substrates in the laboratory during sample preparation and at the field sites during sample changes, respectively.

In the Washington, DC, study samples were collected daily, from 7 a.m. to 6 p.m. (EST) between 18 August and 26 September using 8-stage MOIs. College Park samplings were accompanied by simultaneous collections with a second MOI at an air-monitoring trailer maintained by the Fairfax County Department of Air Pollution Control at the Marshall Road Elementary school in Vienna, VA. At Camden, NJ, samples were collected on the roof of an air quality trailer operated by the New Jersey Department of Environmental Protection, from 16 October to 11 November, also using an 8-stage MOI. Additional MOI samples were collected simultaneously at the Belmont water treatment plant, a suburban, northwest Philadelphia site. Wind speed and direction, and RH were recorded on site with our own meteorological stations. Relative humidity typically followed a strong diurnal cycle, ranging from a high of 80 to 97% at the onset of sampling to a low of from 40 to 65% in the late afternoon.

Overwater sampling was conducted, between December, 1992, and July, 1993, on a pier 60 m from the shore and 6 m from the water surface at the University's Chesapeake Biological Laboratory (CBL), located on the Patuxent River, about 80 km south of Washington, DC. A series of 13 samples were collected at average RH ranging from 41 to 95%. Additional samples (Quinn, 1994) were collected at the Natural Resources Police Facility in Matapeake State Park and at the Elms Environmental Education Center (ELMS) located about 20 km south of CBL.

Analyses. Micro-orifice impactor samples from the various sites were selected for elemental analysis on the basis of meteorological conditions and analyzed for up to 44 elements by INAA using techniques described by Ondov et al. (1995). Sample-to-blank (S/B) ratios, defined as the ratio of the elemental mass (determined before blank correction) to that of the blank, were typically ≥ 10 for at least 70% of all of the impaction substrates analyzed in these studies. Uncertainties in the analyses vary, depending on the element determined and sample mass, but are typically <10% for Al, As, Br, Cr, Mn, Sb, Se, V, and Zn; and <15% for Fe and S on peak stages (Ondov et al., 1990).

Distribution parameters were determined from the elemental-size spectra using the computer program of Dzubay and Hasan (1990) which uses impactor efficiency curves and observed stage loadings to fit the sum of up to three lognormal distribution functions, each with its own amplitude, mass median aerodynamic diameter (mmad), and geometric standard deviations (σ_g). To obtain high accuracy in the fitted parameters, we fit elemental spectra only when the elemental S/B exceeded 5 and the spectrum contained at least three consecutive stages for which the relative uncertainty in the mass concentration was < 33%. Data for As, Br, Fe, I, Mn, S, Se, Sb, V, and Zn usually met these conditions. Nonparametric parameters, i.e., geometric mean aerodynamic diameter (gmmad) and σ_gs were also determined (Ondov et al., 1995) as were modal diameters (i.e., the midpoint diameter for the stage associated with the maximum in the frequency distributions. The former usually agreed with parametric determinations to within 10%.

RESULTS AND DISCUSSION

General Features of Size Spectra. The general character of the size spectra are illustrated in plots of spectral data shown for CBL in Figure 1, wherein the average elemental concentrations of three representative elements determined for each stage of 5 MOI samples collected at RH<60% were normalized to the logarithmic bin width and plotted vs midpoint diameter. As indicated in the figure, the spectra clearly show both super- and submicrometer aerosol modes. The former, represented by Al, consistently occurred in the 3.2 - to 15-μm interval at a modal diameter of 6.9 μm and contain large fractions of the aerosol loadings of elements with strong crustal, e.g., Al, Ca, Sm, Ce, La, Sc, Ti, Nd, Fe, and Mn, or marine association, e.g., Na and K.

We note that the crustal elements also have substantial mass fractions in submicrometer particle due to contributions from HTCSs.

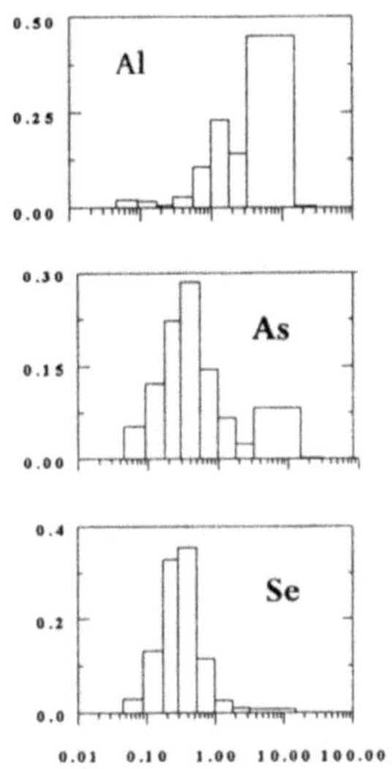

Figure 1. Averaged spectra for CBL aerosol collected at <60% RH.

In contrast to elements associated with soil, lime, and marine aerosol, As, Br, I, S, Sb, Se, V, Zn, i.e., noncrustal elements, were predominately associated with submicrometer particle populations with modal diameters of 0.2 or 0.4 µm in the averaged spectra. Note that individual spectra and mmads for these elements often differed significantly in the same sample. This suggests that aerosol particles advected to the Bay remain suspended as separate particle populations over a scale of 50 km. Despite their strong association with submicrometer aerosol modes, significant fractions of their masses were associated with the supermicrometer aerosol mode, e.g., mass fractions associated with the 6.9-µm interval were: As, Br, Zn, and Sb, 8 to 11%; 13±10%, and V, 16±11. As shown in Figure 1, very little Se was associated with the 6.9-µm mode.

Detection of an Incinerator Plume at Camden, NJ. Greenberg et al. (1978) have shown that particles emitted from municipal incinerators are highly enriched in Zn, Cd, Pb, Sb, Ni, Cr, and

Na, K, and Cl, the major anion, and are readily identified on this basis. According to Olmez et al. (1988), fine (<2.5 μm) particles emitted from the Philadelphia incinerator have the following composition: Cl, 14%; Na, K, and Zn, 4 to 6%; Pb and Fe, 1.7±0.5%; Cr, 0.1±0.05%, and Cd, 0.08±0.03%. As shown in Figure 2, wherein we plot spectra for Cd, Zn, Cr, and Na, fresh accumulation aerosol from the municipal incinerator located 5 km northwest of the Camden site is readily identified in size spectra developed for the site on the 27th of October, 1990, when the prevailing wind direction was predominately from the northwest. In each of the spectra, a distinct accumulation aerosol mode is evident at a modal diameter of 0.22 μm, however, Na, Cl, and Fe (not shown) also have substantial amounts of mass in 0.8 μm particles. Observation of the latter, coarse- particle mode, is consistent with in-stack measurements by Kauppinen et al. (1986). Fine-particle mass ratios for these elements relative to Zn, i.e., ~1 for Na and K, and ~0.1 for Cr and Cd; and ~0.5 for Fe, are in good agreement with those reported by Olmez.

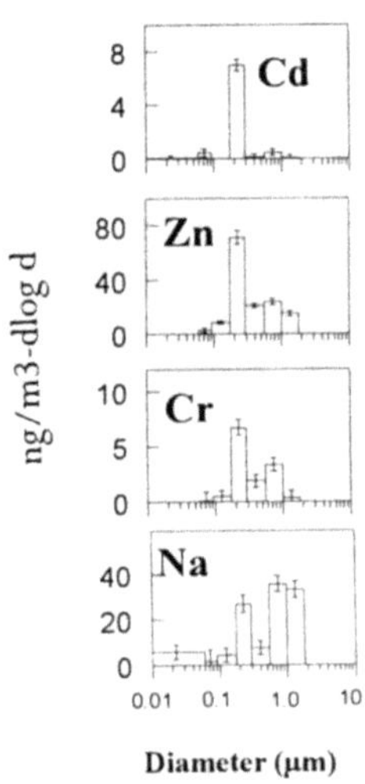

Figure 2. Size spectra for marker elements in aerosol collected at Camden, NJ, 5 km downwind of a municipal incinerator

As described by Divita, Jr. (1993), CMB calculations indicate that fine aerosol mass attributed to incinerators at Camden on this date, i.e., 300 ng m^{-3}, was nearly 3-fold greater than the amount attributed to incineration in the MOI sample simultaneously collected at the upwind site in Northwestem Philadelphia. This clear observation of the plume of a major HTCS is important in that such measurements could obviate the need, for expensive, cumbersome, and often impractical in-stack measurements.

Particle Size vs. RH at College Park, MD. The influence of RH on the mean size of particles bearing Sb, an element whose submicrometer spectra are typically unimodal, is shown in Figure 3, wherein by convention, RH is plotted against mean aerodynamic diameter. The latter is the average of the mmad and gmmad, or the gmmad when a fit could not be obtained. Most of the data were collected at College Park, MD. Data for samples collected when the average RH exceeded 84%, e.g., during rain, were excluded from the plot because at such high RH much of the mass is likely to have grown outside the measurement range (i.e., >1.3 μm). Plots for As, Se, and Zn, i.e., other elements with unimodal submicrometer size spectra were similar.

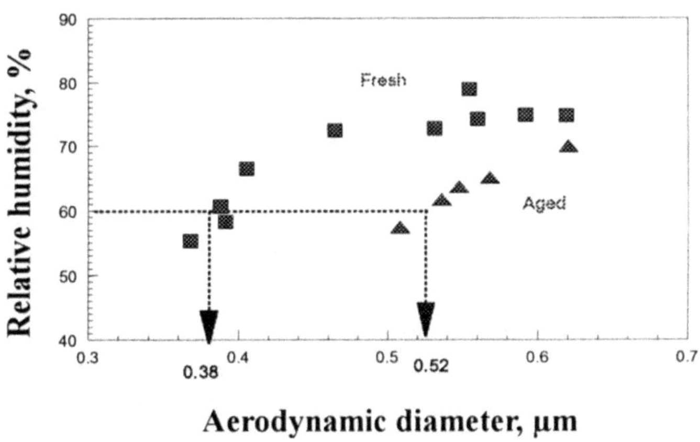

Figure 3. Relative humidity vs mmad for aerosol particles bearing Sb at College Park, MD.

An outstanding feature of Figure 3 is that data clearly fall on two separate curves. As discussed by Ondov et al. (1995), points on the lower curve represent samples influenced by westerly winds, including one sample which was collected in Vienna, about 30 km westsouthwest of College Park. In contrast, nearly all of the data for samples influenced by winds from other directions, (mainly SW, NW and NE) fall on the upper curve. Thus, at the same RH, aerodynamic mass mean diameters for samples influenced by northeasterly, northwesterly, or southwesterly winds were uniformly smaller than those influenced by more westerly winds. For example, at 60% RH gmmads determined from the two curves were Sb, 0.39±0.03 and 0.53±0.04; As, 0.30±0.03 and 0.46±0.04; Se, 0.33±0.06 and 0.54± 0.04; and Zn, 0.39±0.06 and 0.53±0.08; respectively. Major sources of these elements lie within 50 km of College Parkin SW, NW and NE directions, including CFPPs located 20 and 45 km to the southwest and northwest, respectively; several sources, including two large municipal incinerators, a blast furnace, and two CFPPS, in the heavily-industrialized southeastern quadrant of Baltimore, about 50 km to the northeast; and two incinerators lying 20 km to the southwest of College Park. In contrast, the closest major source lying to the west of the Washington, DC, area is the CFPP at Mount Storm 180 km away across mountainous terrain. Furthermore, emissions from CFPPs in the Ohio River Valley >300 km to the west account for major fractions of the regional sulfate aerosol (and associated elemental components) observed at College Park. We hypothesize, therefore, that the lower growth curve represents older, more highly-processed, regional sulfate aerosol transported from the Ohio River Valley, whereas the upper curve represents more local sources.

This hypothesis is supported by the sulfur data. Excluding data for mixed wind directions, sulfur concentrations in the westerly samples were, on average, 4-fold greater than those for the other directions, and increased, more-or-less, monotonically with increasing size and RH, i.e., from 1200 to 8700 ng/m^3. This trend was also evident in the upper curve, at RH>70% (i.e., where S concentrations ranged from 1700 to 9300 ng/m^3), but not at lower RH (where S concentrations were only ≈ 900 ± 100 ng/m^3). It appears, therefore, that hygroscopic growth associated with temporal RH fluctuations is accompanied by the deposition/conversion of atmospheric sulfur and that sulfur deposition is associated with RH>60 or 70%, depending on the origin of the aerosol. It may be significant that the onset of S deposition occurs in the region of deliquescence of atmospheric aerosol particles. In any event, all of this suggests that particle growth is neither linear nor reversible.

Effect of High RH Episodes on Particle Deposition. As discussed above, average RH during collection of CBL samples ranged from 42 to 95% and, although RH was often quite variable during their collection periods, hygroscopic growth of elements bearing the various elements is evident in their corresponding size spectra and distribution parameters. As illustrated in Figure 4, the size of submicrometer particles bearing As, S, Sb, Se, V and Zn, increased continuously with increasing RH by from 40 to 400%.

To estimate the significance of temporal changes in RH on dry deposition, elemental size spectra derived from CBL samples were used in an adaptation of the Williams model (Quinn, 1994) along with half-hourly meteorological observations made over a 9-month period at northern Bay buoy station (Howell Point). As discussed by Quinn et al., (1992) theory and experiments suggest that dry particle deposition is a strong function of particle size and wind speed.

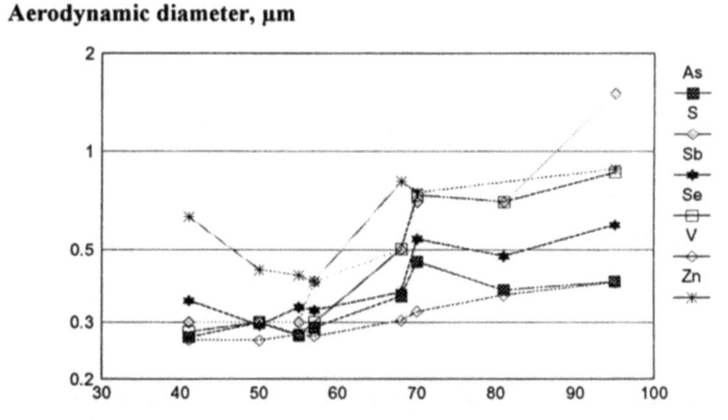

Figure 4. Aerodynamic size of aerosol bearing various elements versus mean RH during sampling overwater at the University of Maryland's Chesapeake Bay Laboratory.

To incorporate these effects into deposition velocity calculations, the Howell Point data were parsed into a 104-bin, combined, RH/U frequency distribution matrix containing 13 U and 8 RH bins. The Williams model was run for each of the bins and the resulting set of 104 deposition velocity-(V_d)-vs-size curves were weighted by the observational frequencies and combined to produce a single, weighted V_d curve. Note that to account for particle growth, the curves were calculated on the basis of the "wet" particle diameter inferred nonparametrically from the MOI spectra obtained at various RH and, parametrically, using a growth function for urban sulfate particles developed by Koutrakis et al. (1989). Separate, weighted, V_d curves were developed i) the full RH/U matrix, ii) the RH frequency distribution and average U, iii) the U frequency distribution and average RH and iv) average values of U and RH for the 9-month period. As reported by Quinn (1994), the minimum in the deposition velocity curve for average conditions (0.005 cm s^{-1}) occurs at 0.4 μm and lies 2.5-fold lower than the curve derived from the full RH/U matrix. The curves differ 3.7-fold at for 1-μm particles. Thus, while others have speculated about the importance of particle growth in RH gradients over water, results reported herein clearly show that temporal changes in RH should not be neglected in studies of fine particle dry deposition.

ACKNOWLEDGEMENTS

This work was funded by the US Department of Energy, US Environmental Protection Agency, and Maryland Department of natural Resources under contract numbers, DE-AC22-83PC61253, R-815768-01-0, and CB-92-006-002/CB93-002-004, respectively.

REFERENCES

Divita F, Jr. (1993) Size distributions and sources of submicrometer atmospheric particles in Washington, DC, and Philadelphia, PA. Ph. D. Thesis, University of Maryland, College Park

Dodd JA, Ondov JK Tuncel G (1991) Multimodal size spectra of submicrometer particles bearing various elements in rural air. *Environ. Sci. TechnoL, 25:890-903*

Dzubay TG, Hasan H (1990) Fitting Multimodal Lognormal Size Distributions to Cascade Impactor Data. *Aerosol Sci. Technol, 13,144-150.*

Greenberg RR, Zoller WH, Gordon GE (1978) Composition and size distributions of particles released in refuse incineration. *Environ. Sci. Technol 12:566-573*

Heintzenberg J., Covert, DS (1987) Chemically resolved submicrometric size distribution and internal mixing of the Arctic haze aerosols. *Tellus, B39:374-382*

Koutrakis, P, Wolfson JM, Spangler JD, Stem, B, Franklin CA (1989) Equilibrium size of atmospheric sulfates as a function of the relative humidity. J *Geophys. Res.,* 94:6442.-6448

Olmez I, Sheffield AE, Gordon GE, Jouck JE, Pritchett LC, Cooper JA, Dzubay TG, Bennett RL (1988) Compositions *JAPCA 38:1392-1402*

Ondov, JM, Dodd, JA, Tuncel, G (1990) Nuclear Analysis of Trace Elements in Size-Classified Submicrometer Aerosol Particles from a Rural Airshed. *Aerosol Sci. Technol. 13:249-263.*

Ondov JM, Divita J Jr., Suarez A. (1995) Size-Spectra and Growth of Particles Bearing as, Se, Sb, and Zn in Washington, DC, Area Aerosol by Instrumental Neutron Activation Analysis. *J Radioanalyt. Nuc. Chem.,* in press.

Ondov, JM, Quinn TL (1995) Influence of Temporal Changes in Relative Humidity of Size and Dry Depositional Fluxes of Aerosol Particles Bearing Trace Elements. Manuscript submitted for publication in a special edition of the Journal of the Society for Environmental Toxicology and Chemistry, in press

Quinn TL, Ondov JM, Holland, JZ (1992) Dependence of deposition velocity on the frequency of meteorological observations for the Chesapeake Bay. J *Aerosol Sci.* 23:S973-S976.

Quinn, TL (1994). Ph.D. Thesis, University of Maryland, College Park

Wu, Z.Y., Han, M., Lin, ZC, Ondov, JM (1994) Chesapeake Bay Atmospheric Deposition Study: Year 1: sources and dry deposition of selected elements in aerosol particles. *Atmos. Environ., 28:1471-1486*

Use of Scanning Electron Microscopy for Particle Characterization and in Support of Receptor Modelling

Yaacov Mamane
Environmental and Water Resources Engineering
Technion,
Haifa 32000,
ISRAEL

1. *Introduction*

The basis for an effective air pollution control lies in the knowledge of source contribution to air quality as measured at the receptor site. In the past, dispersion models and emission inventory estimates were used almost solely by regulatory agencies to evaluate source impacts. Unlike dispersion models that are based on pollutant emission rates and plume dispersion parameters to estimate source contribution, receptor models decode the chemistry, variability, or the morphology of individual particles to provide source apportionment. Receptor models can be grouped into three major classes:

(a) Models that are based on regression analysis of the particulate mass or component features (the chemical mass balance models, Watson et al., 1990).

(b) Models that use statistical analysis of aerosol composition variability (factor analysis or multiple linear regression);

(c) Models that use individual particles properties gathered by optical and electron microscopy techniques.

One of the most widely used receptor model is the chemical mass balance (CMB) model. CMB models, however, could not be applied to determine the impact of sources of similar composition. For example, fly ash, soil particles or road dust may have similar mineral composition (mostly silicates of aluminum and varying concentrations of iron and calcium). Other

NATO ASI Series, Partnership Sub-Series, 2. Environment – Vol. 8
Urban Air Pollution
Edited by Ivo Allegrini and Franco De Santis

collinear sources include particles of carbon sources such as carbonaceous, diesel and biological particles. Scanning electron microscopy (SEM) coupled with energy dispersive X-ray analysis (EDX) can distinguish individual coal fly ash particles from minerals, and carbonaceous from biologicals on the basis of their different shape and surface structure.

Individual particle analyses by microscopy provide information, not only on particle composition and morphology, but also on particle size distribution. These date provide a sound estimate of particle volume and density to be used for calculations of mass concentrations. Microscopy also provides direct information on aerosol sources that impact a given airshed, and any transformations that may occur during transport. For example it provides information on secondary sulphate that were formed in the atmosphere. CMB models and most of the statistical receptor models, cannot apportion sources of secondary aerosols.

Microscopy may be used as one of the receptor models available, or preferably may be used to enhance CMB source resolution. It is the purpose of this review to present the use of microscopy in receptor model studies, and in particle characterization.

2. *Experimental Methods in Microscopy*

2.1 *Particle Sampling*

Dichotomous samplers have been used in receptor modelling studies to collect size-fractionated atmospheric particles on a pair of filters. Samples collected in such devices are well suited for analysis of total mass and bulk composition but are poorly suited for analysis of individual particles by SEM, a technique that requires particles to be spaced a few diameters apart on the filter surface. This is a consequence of the typical size distribution of atmospheric aerosol that causes fine particles to be collected too densely while coarse particles are collected too sparsely on filters with the same area. One way to collect adequately spaced fine and coarse particles would be to use a larger deposit area for the fine fraction. An easier way to avoid pileup is to divert a portion of the airflow from the fine filter. Figure 1 shows a virtual impactor modified for use with electron microscopy. The inlet flow rate F is

$$F = F_c + F_{f1} + F_{f2}$$

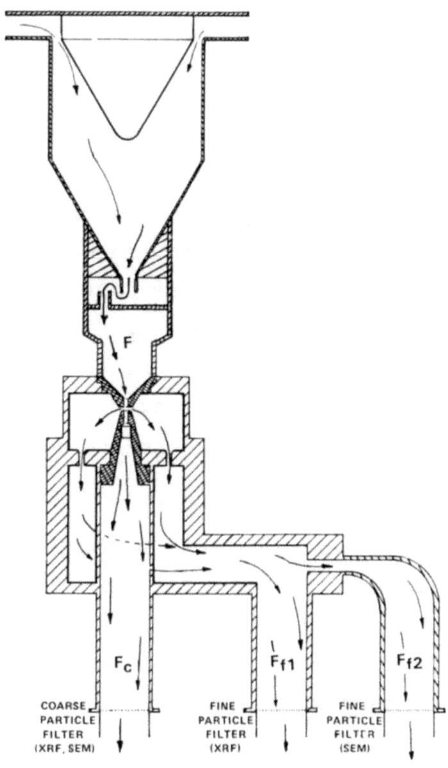

Figure 1: Modified dichotomous sampler to be used with electron microscopy. The fine fraction portion was modified by diverting ~ 10% of the flow to a second fine filter. A similar configuration is being used by the VAPS.

Where F_c and F_{f1} are the flow rates through the coarse and fine filters, respectively, and F_{f2} is the diverted flow rate (about 10% of F) of the fine fraction to be used for microscopy.

Similar considerations were used by Stevens et al. (1993) to develop the Versatile Air Pollution Samplers (VAPS). In receptor modelling studies it is recommended to collect all aerosol sources, resuspended soil, and ambient samples with dichotomous samplers and analyze the samples for mass and elemental composition by X-ray fluorescence or instrumental neutron activation techniques (Dzubay et al., 1988).

2.2 *Particle Analysis*

A subset of the filters collected for a receptor modelling study may be analyzed by scanning electron microscopy. For SEM analyses a 4 x 4 mm segment of each filter is cut and coated with carbon, and then viewed in the microscope. Several hundred particles in random fields of view are sized, and analyzed for elemental composition. Particles are identified by shape, surface structure, inhomogeneity in appearance and elemental content. During the SEM observation, each particle is assigned to one of the following categories: (a) Minerals - nonspherical particles of soil origin, such as calcite ($CaCO_3$), dolomite ($CaMg(CO_3)_2$), quartz (SiO_2), gypsum ($CaSO_4$), and more complex mixtures of alkaline earth silicates containing elements such as Na, Mg, Al, Si, K, Ca, Ti, and Fe; (b) Biologicals - particles defined by characteristic morphology and low X-ray count rate and low peak to background ratio. They include pollen, spores, plant and animal fragments; (c) Oil fly ash - includes lacy, S-rich spheres, smooth alumino-silicate spheres which contain either V or Ni and sulphates or carbonaceous particles rich in V or Ni; (d) Coal fly ash - includes mostly smooth spheres of alumino-silicate matrix; (e) Particles of industrial origin; (f) Sulphates; and (g) Elemental C and others.

3. *Application of Microscopy to Particle Characterization*

Observation of individual particles by scanning electron microscopy provides unique information on the particle make-up, and often allows to infer its origin or mechanism of formation, as well as atmospheric reactions that took place on the particle surfaces.

3.1 *Formation of nitrates on sea-salt particles*

In contrast to sulphate particles which are confined mostly to the submicrometer size range, nitrates have been found to be spread over a larger range of sizes. Ammonium nitrates were identified in the fine particle mode and nitrates associated with sodium were found in the coarse mode. Scanning electron microscopy and a unique microspot technique were used to study the presence of individual nitrate particles in the coastal city of Haifa (Mamane and Mehler, 1987, and references therein).

Figure 2 is a typical photomicrograph of particles collected in Haifa on Stage 3 of a Casella Impactor (0.7 to 3 μm) on a summer day. The sample was treated with nitron to allow the identification of nitrate particles. Almost all the particles in this size range reacted with nitron, forming characteristic reaction spots of $NaNO_3$ particles. Sulphates dominated the smaller size range (Casella Impactor Stage 4), 0.3 - 0.8 μm. This study followed by laboratory chamber experiments (Mamane and Gottlieb, 1992) pointed at the main pathway of nitrate formation in Haifa:

$$HNO_3(g) + NaCl(s) \rightarrow NaNO_3\ (s) + HCl(g).$$

During the summer months in Haifa, photochemical production of HNO_3 and sea breeze transport of NaCl particles (in the micrometer range), are responsible for $NaNO_3$ formation in the micrometer size range.

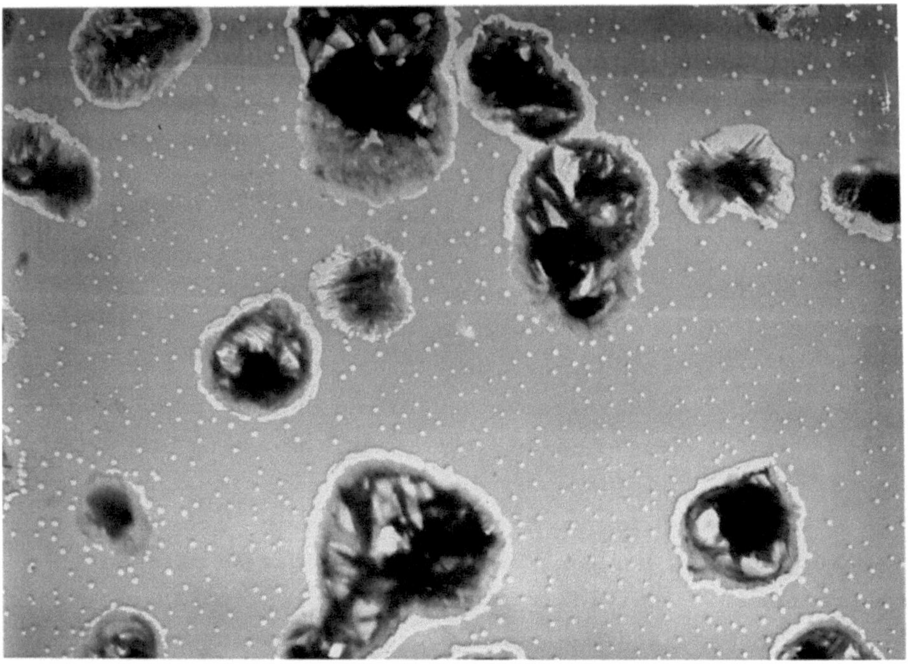

Figure 2. Typical photomicrograph of micrometer particles collected in the coastal city of Haifa on a summer day. The sample was treated with nitron. Almost all particles in this stage have the typical "fiber" appearance, indicative of sodium nitrate particles.

3.2 *Characterization of sulfate particles*

Analysis of individual submicrometer particles by the author, either in the plume of coal fired power plants or in rural and urban sites showed definitely that sulfate particles dominate the fraction below 0.5 or 1.0 μm. In the micrometer size range (> 2.5 μm) sulfate presence is limited. When particles viewed with microscopy, it was often found that some of the sulfates were formed through heterogeneous reaction on the surfaces of fly ash, minerals or pollen and spores (Mamane et al., 1992). Figure 3 shows an example of a spherical particle surrounded by a thin layer of sulfate. Elemental analysis by EDX showed that the spherical particle was a coal fired fly ash particle, while the surrounding layer, analyzed by the $BaCl_2$ microspot method, was found to contain sulfate. Thus microscopy techniques provided direct evidence on the heterogeneous reaction of sulfur oxides with fly ash and other mineral particles.

Figure 3. A photomicrograph of fly ash particle that had reacted with sulfur oxides to form a layer of sulfate on its surface. The dark spherical particle is the fly ash and the ring surrounding it is the reaction spot formed by the sulfate layer on the fly ash.

4. *Application of Microscopy to Source Apportionment*

4.1 *The Phildelphia Aerosol Airshed Study*

During the summer of 1982, the U.S. Environmental Protection Agency conducted a major field study to evaluate the ability of receptor models to determine the impacts of emission sources on air quality. Bulk (Dzubay et al., 1988) and individual particle analyses of the same samples were used to apportion the particulate mass into source categories.

Details on the sampling and analytical procedures are given elsewhere (Dzubay et al., 1988; Dzubay and Mamane, 1989; Mamane and Dzubay, 1988). For electron microscopy analysis a few samples from the Camden (industrial) and North Airport (rural) sites were studied. Particles were observed in the SEM-EDX system for size, appearance and elemental content. They were assigned tentatively to a defined category such as biologicals, minerals, industrial sources and others.

The SEM-EDX data were used to calculate the mass concentration C based on the particle counts, an assumed density, the filter and sampling parameters, using the following equation:

$$C_j = 4\pi/3\left(\sum_i N_{ji}\,(d_i/2)^3\right)\rho_j \cdot A_f/A_{SEM} \cdot 1/ft$$

where C_j is the mass concentration of particles from category j that deposited on the filter, f is the flow rate of the dichotomous sampler ($16\ell\ \mathrm{min}^{-1}$), t is the sampling time (12h), N_{ji} is the number of particles of the j group and size range i, d_i is the average diameter for size range i, ρ_j is the assumed group density, A_{SEM} is the filter area scanned in the microscope and A_F is the filter deposit area. Two assumptions were made: the particles were approximated by a volume of an equivalent sphere, and based on the particles elemental composition, a density was assumed. A density of 2.6 g cm^{-3} was assigned for alumino-silicates, fly-ash, Ca-rich particles and chlorides. Organic particles and biologicals were assumed to have a density of 1.0 and 1.3 g cm^{-3}, respectively. The overall error was estimated not to exceed 30-40% including counting statistics.

TABLE 1. Particle count by category and size in Camden (Philadelphia), and SEM estimated mass concentration for the 12-h sample that began on 21 July, 1800, EDT

Catagories*	Particle diameter (μm)						Mass Conc.
	<1.5	<2.1	<3.1	<5.1	7.1	<10	(μg m^{-3})
Minerals:							
Alumino-silicates	1	20	23	18	4	1	
Quartz		5	9	9	4		
Calcite/dolomite and other Ca		8	5	3			8.5 (Alumino-silicates to S-rich)
S-rich		3	1				
Cl-rich		2	2	6	1	1	2.1
Biological (spores and pollen)		4	11	9	4	1	2.0
Organics		2	3	7			0.3
Coal fly-ash	6	17	7	4			0.7
Fe-rich**	2	2	3	1			0.3
Al-rich							
Pb-rich: cars**	3	1	3				0.2
others	1	1	1				0.1
Incinerators	3	13	10	10	1		1.8
Other Zn particles			1		1		0.4

* The area scanned on 21 July, 1800 filter was 47,300 μm^2 and total of 258 particles were analyzed.

**Density for iron and lead particles was assumed 4 and 6 g cm^{-3}, respectively.

Table 1 is an example of the electron microscopy analysis for the 21 July, 1800 (night) coarse particle sample. Six fields of view were scanned and 258 particles were individually analyzed. Based on particle counts, minerals (some of them rich in chlorine), biologicals, incinerator particles and coal fly-ash are the most abundant particles. Table 1 also lists the SEM estimated mass concentration: 16.4 (±30%) μgm^{-3} as compared with 14.7 measured by β-gauge.

The fine (<2.5 μm) particles collected in Camden on 21 July, 1800 (night sample) are made of sub-μm aggregates. Individual analysis of 299 fine particles revealed that 30% of the particles contained S, K and Zn; 10% were minerals rich in S, K, Zn; 18% were organic; 16% were minerals rich in S, K, Zn; 18% were organic; 16% contained S, K, Zn and Cl; 8% contained S and K; 7% 'pure' sulfur particles, 7% coal fly ash; and 4% were Pb particles. These particles are the result of fine incinerator particles, soluble chlorides of Zn and K, that reacted with submicrometer fine sulfates (Mamane, 1988).

The estimated SEM mass concentrations were compared with the beta gauge measured mass on the parallel Teflon samples. Agreement is very good; the slope on the regression line is almost 1.0 and the linear correlation coefficient is 0.9. Figure 4 compares the mineral sources as obtained by the SEM analysis and by the CMB receptor model. Agreement for the apportionment of this source is very good with a slope of 1.1 and correlation coefficient of 0.9. Thus the assumption made in the SEM-EDX estimates, and the procedures chosen for classfication by size and category are satisfactory (Mamane, 1990).

4.2 *Czech Air Quality Monitoring and Receptor Modelling Study*

Air pollution measurements in the Czech Republic from 1992 provided data on ambient concentrations of sulfur dioxide, polynuclear aromatic hydrocarbons, fine and coarse particle mass, elemental compposition, and aerosol acidity. The primary

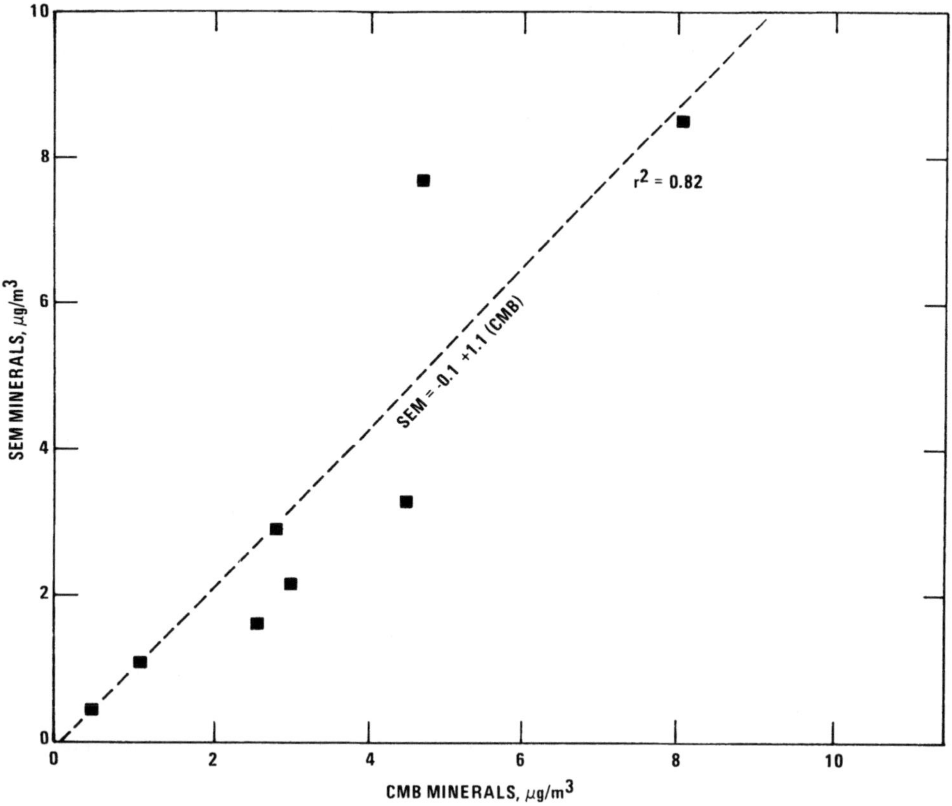

Figure 4. SEM estimated mineral mass concentration in the coarse fraction versus CMB estimates of the soil component, based on the Phildelphia aerosol data.

monitoring sites were located in Teplice in northwestern Bohemia and Prachatice in southern Bohemia. Emissions from power plants, residential space heating, local factories, and motor vehicles were characterized. The ambient monitoring and source characterization data were then used in receptor modelling calculations, which indicated that home heating and power plant emissions were responsible for up to 80% of fine particle mass concentrations observed during winter inversion episodes. Sample data obtained from mobile source and incinerator emissions and from windblown dust indicated that these sources contributed to the balance of the inhalable paricles (Stevens et al., 1993b).

SEM analysis of the coarse fraction of several filters supported the CMB results. A large fraction of the mineral elements were associated with fly ash particles. Table 2 summarizes the SEM analysis of the coarse fraction particles collected on March 2, 1992, in Teplice (sample N0015). The data in Table 2 clearly show the extremely high concentration of fly ash particles in the sample (46%). The rest of the particles were made of alumino-silicate minerals and gypsum (some of which may be formed through reaction of calcites with anthropogenic sources). Not all the fly ash particles were spheres or cenospheres, as is found in coal power plants emissions (Mamane et al., 1986). Some Teplice fly ash particles were round, not strictly spherical particles, and bonded with organic matter to form aggregates. The latter may be the results of combustion at lower temperatures, which would produce more condensable organic particles and nonspherical fly ash. They may be associated with home heating emissions.

The fine fraction as observed by SEM was dominated by sulfates and carbonaceous particles. fly ash particles were also found in large numbers. The majority, 85%, of the particles analyzed were submicrometer sulfates or carbonaceous particles. The rest were Si-rich round or spherical particles, apparently fly ash. Other samples collected during the winter of 1992 produced similar results.

SEM analysis provided information on specific sources (in addition to soil minerals, power plants, fine sulfates, and carbonaceous particles). Components found in the ambient samples included coal particles (rich in pyrites - Fe or S), iron spheres, Al chips, oil fly ash, and silicon-rich spheres. Biological particles such as spores and pollen were almost absent in the coarse fraction.

Table 2. Particle count by size and category obtained from SEM analysis of ambient sample in Teplice (Czech Republic) 2 March 1992

Catagories	Particle size (μm)						Total*
	1-5-2.1	<3.1	<5.1	<7.1	<10	<15	
Minerals:							
Alumino-Silicate	28	22	15	1	1		67
Quartz	5	3	2				10
Ca + S	12	11	2				25
Other	1	1					2
Organics:							
Soot	7	2					9
Other	11	13	6	1	1		32
Fly ash	82	37	14	1			134
Fe-rich	4	2					6
S-rich	3	1					4
Other	1						1

* 290 particles were analyzed manually by SEM.

5. *References*

Dzubay T.G., Stevens R.K., Gordon G.E., Olmez I., Sheffield A.E. and Courtney W.J. (1988) A composite receptor method applied to Philadelphia aerosol. *Envir. Sci. Technol.* 22, 46-52.

Dzubay T.G. and Mamane Y. (1989) Use of electron microscopy data in receptor models for PM-10. *Atmospheric Environment*, 23, 467-476.

Mamane Y., Miller J.L. and Dzubay, T.G.(1986) Characterization of individual fly ash particles emitted from coal-and oil-fired power plants. *Atmospheric Environmement* 20, 2125-2135.

Mamane, Y. and M. Mehler (1987) On the nature of nitrate particles in a coastal uban area, *Atmospheric Environment*, 21, 1989-1994.

Mamane Y. and Dzubay, T.G. (1988) fly ash concentrations in Philadelphia aerosol determined by electron microscopy. *J. Wat.Air Soil Pollut.* 37, 389-405.

Mamane Y. (1988) Estimate of incinerator conribution to Philadelphia aerosol – I.Source analysis. *Atmospheric Environment* 22, 2411-2420.

Mamane, Y. (1990) Estimate of municipal refuse incincerator contribution to Phildelphia aerosol using single particle analysis - II. Ambient measurements, *Atmospheric Environment*, 24B, 127-135.

Mamane, Y. and J. Gottlieb (1992) Nitrate formation on sea-salt and mineral particles - A single particle approach, *Atmospheric Environment*, 26A, 1763-1769.

Mamane, Y., T.G. Dzubay and R. Ward (1992) Sulfur enrichment of atmospheric minerals and spores, *Atmospheric Environment*, 26A, 1113-1120.

Stevens, R.K., J.P. Pinto, Y. Mamane, J. Ondov, M. Abdulraheem, N. Al-Majed, M. Sadek, W. Cofer, w. Ellenson. and R. Kellogg (1993) Chemical and physical properties of emissions from Kuwaiti oil fires, *Water Sci. Technol.*, 27, 223-233.

Stevens, R.K., J.P. Pinto, T.L. Conner, R. Willis, R.A. Rasmussen, Y. Mamane, G. Casuccio, I. Benes, J. Lenicek, P. Subrt, J. Novak, and J. Santroch (1993) Czech air toxic study (CATS) - Project summary, paper 93-RA-128A.03, 86th Annual Meeting, Air Waste Management Assoc., June 13-18, 1993, Denver, Colorado.

Watson, J.G., N.F. Robinson, and J.C. Chow (1990) The US EPA/DRI chemical mass balance receptor model, CMB-7.0, *Environ. software*, 5, 38.

Vehicular Contributions to Primary Airborne Particulate Matter and Urban Air Quality

Andrew Deacon and Roy M Harrison
The University of Birmingham
Institute of Public and Environmental Health
School of Chemistry
Edgbaston
Birmingham B15 2TT
United Kingdom

Introduction

Previous studies have highlighted the importance of vehicles as a contributor to airborne particulate matter. For example, the Quality of Urban Air Review Group described results from a study of urban particulate matter carried out by London Scientific Services (LSS, 1990). This study showed that in central London, roadside levels of particulate matter were 2-3 times higher than at a nearby "urban background" monitoring site, indicating road vehicles as a major contributor to particulate matter. QUARG (1993a) also analysed carbon monoxide and PM_{10} data for the central London Enhanced Urban Network site for a part of 1992. As a correlation coefficient of only 0.3 was found it was concluded that traffic might be only a minor influence on particulate matter concentrations. It was noted that although there were large day to day variations in PM_{10} concentration, the variations in traffic activity were relatively small. QUARG concluded that episodes of low rainfall and high windspeed would exert a strong influence on the concentration of particulate matter, as these conditions would lead to much resuspension of deposited dusts. Traffic has long been recognised as a major source of carbon monoxide, especially in urban areas. Emission inventories for London for 1991 (QUARG, 1993b) show that vehicles are responsible for 98.4% of CO emissions. CO therefore may be use as a good indicator of traffic activity. Through the use

NATO ASI Series, Partnership Sub-Series, 2. Environment – Vol. 8
Urban Air Pollution
Edited by Ivo Allegrini and Franco De Santis

of regression analysis, changes in PM_{10} from background to mean conditions have been calculated and attributed to traffic. These results are compared with previous work on urban particulate matter, and with other studies currently being undertaken at the University of Birmingham.

Method

Data was analysed for two periods; Summer (June, July and August 1993) and Winter (December, January and February 1993-94). It was felt that a seasonal split in the data might help to isolate any meteorological influence on the relationship between the two pollutants. Hourly data were taken from UK Enhanced Urban Network sites, all of which are situated in central urban background locations. Seven sites were used in all : London, Birmingham, Edinburgh, Cardiff, Newcastle, Leeds and Bristol. Initially, average diurnal plots were produced to assess the relationship between carbon monoxide and PM_{10}. These were then converted to regression analyses, and the numerical relationship and correlation coefficient between the two pollutants was obtained.

It was then necessary to estimate the UK background carbon monoxide level. Novelli et al (1994) give a northern hemisphere background CO figure of 0.125 ppm based on measurements at 27 locations in northern and southern hemispheres with records of 2 to 5 years. However, Derwent et al (1994) give a background CO figure of 0.128 ± 0.004 ppm for Macehead on the Atlantic coast of Ireland, based on measurements made over three years (1990-1992). Derwent et al also identified a distinct seasonal cycle in the background CO concentration, with highest levels in months January to March, and lowest levels in months July to September. It was therefore decided to seasonally adjust the CO baseline used in this study. A figure of 0.10 ppm was used during the Summer and 0.14 ppm during the Winter, based upon the data in the Derwent et al (1994) paper. This level was set as that associated with zero traffic. Through correlation with CO, the change in PM_{10} from that occurring at background CO to mean conditions could be calculated, and then attributed to traffic.

The results obtained were then compared with data collected in Birmingham as part of another project on source apportionment of atmospheric particles. Again a seasonal split in the data was observed with sampling campaigns being undertaken in May 1993 and January 1994. Air samples were collected from a road tunnel in the city centre and from the University campus, located about 3 miles south of the city centre. Pollutants measured include Total Suspended Particulate (TSP), PM_{10}, Black Carbon (BC) and Organic Carbon (OC). Using this data the relationship between black carbon and TSP was analysed both in the ambient samples and in those from the road tunnel. By using the ratio of BC to TSP in the road tunnel as an indicator of the vehicular source composition, an assessment of the vehicular contribution to TSP at the University air sampling site was made.

Results

Table 1 shows the correlation coefficients between CO and PM_{10} (R^2) for all sites used during the study. The correlation coefficients between the two pollutants are higher in Winter than in Summer at all sites. All correlations are significant at the 1% level, indicating a strong relationship between CO and PM_{10}. The highest correlation coefficient is found in Birmingham during the Winter, whilst the lowest is found in Cardiff during the Summer.

Table 1 - Correlation between CO and PM_{10} (R^2 value) for all sites

Site	Summer	Winter
Birmingham	0.69	0.90
London	0.83	0.88
Cardiff	0.35	0.49
Edinburgh	0.58	0.86
Newcastle	0.41	0.74
Leeds	0.58	0.65
Bristol	0.61	0.86

Some examples of the diurnal variation in CO and PM_{10} are shown in Figure 1. This shows a typical urban pattern of peaks in CO concentration at around 07.00 - 09.00 and again at 17.00 - 19.00 associated with rush hour traffic. However, for both cases shown (Birmingham and London) the diurnal variation in PM_{10} appears to vary between seasons, exhibiting only one clear, morning rush hour associated peak in concentration during Summer, with an extended evening peak in concentration, whilst in Winter, two clear rush hour associated peaks in PM_{10} concentration can be seen.

Having carried out regression analysis the traffic contribution to PM_{10} could be assessed, and is shown in Figure 2. This graph shows that in Birmingham, London, Edinburgh and Bristol, the traffic contribution to particulate matter is higher in Winter than Summer. The highest traffic contribution can be found in London and the lowest in Edinburgh. When these results are averaged by season it is found that the average contribution of traffic to PM_{10} in Summer is 39.4% and in Winter 40.4%.

A plot was also produced relating the seasonal average traffic contribution to CO and PM_{10} (Figure 3). This shows a reasonable correlation between the two series (r=0.8). The traffic contribution to CO was calculated by subtracting the seasonal background value from the seasonal average CO level, whilst the seasonal average traffic contribution to PM_{10} was obtained from the regression analysis. This graph shows London to be well above the best fit line, and this may be due to the high proportion of diesel taxis and buses in use there.

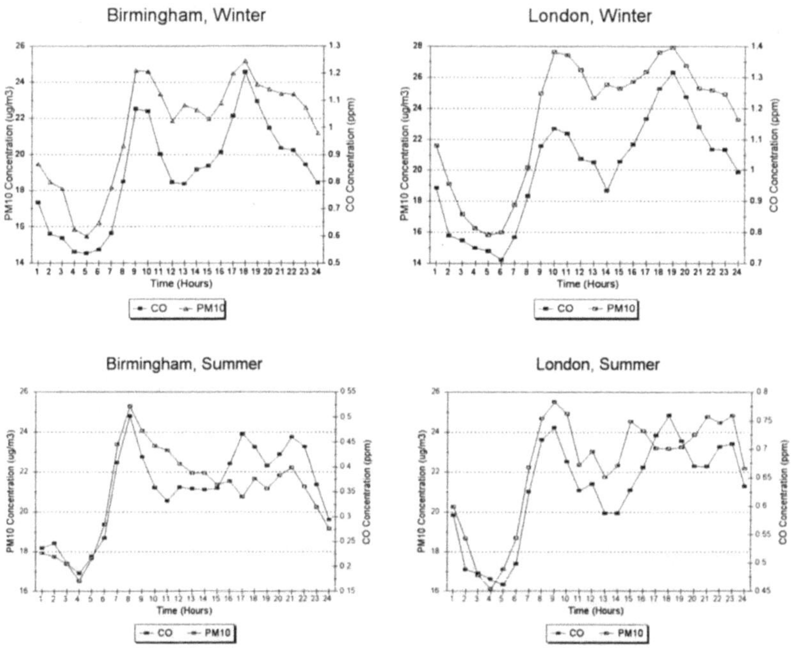

Figure 1 - Diurnal Variation in CO and PM_{10} for London and Birmingham for Summer and Winter Sampling Periods

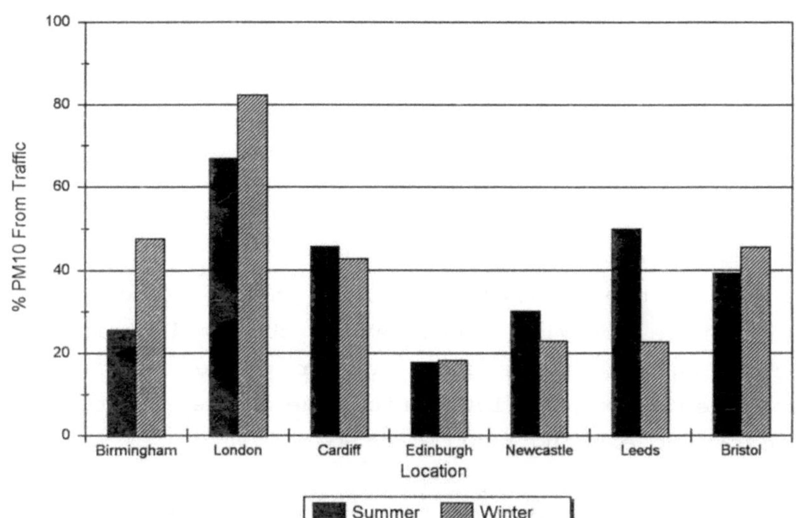

Figure 2 - Seasonal Average Traffic Contribution to PM_{10}

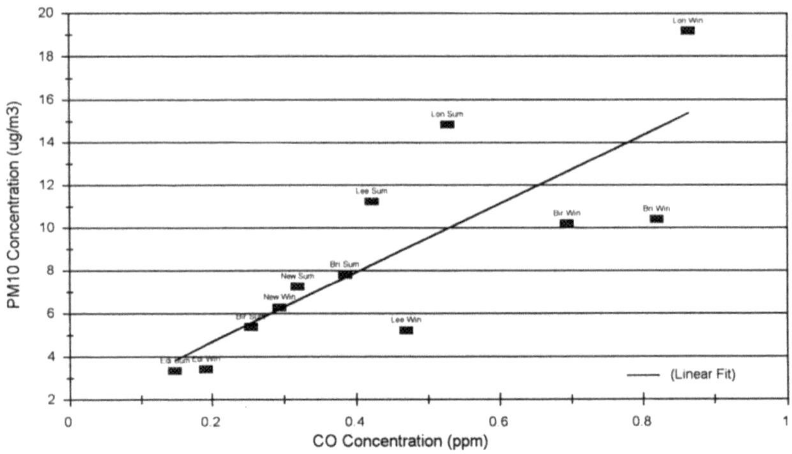

Figure 3 - Seasonal Average Traffic Contribution to CO and PM10 Concentrations

These results may then be compared with those obtained from analysis of TSP data collected in Birmingham. Figure 4 shows the relationship between BC and TSP at the University site for the Winter monitoring, Figure 5 shows the corresponding pattern for the Summer sampling period. A much larger variation around the best fit is observed during the Summer (R^2 = 0.023), whilst for the Winter the correlation between these two pollutants is improved (R^2 = 0.86). Figure 6 shows the calculated traffic contribution to TSP during both Summer and Winter monitoring campaigns. The graph shows the day to day variation in the contribution of traffic to TSP.

Discussion

It appears from the diurnal variation in CO (Figure 1) that different traffic activity patterns are observed during Winter and Summer. During Winter there appear to be two clear rush hour associated peaks, whilst during the Summer there appears to be an extended evening traffic peak.

The results highlight some of the siting problems which were encountered. The monitoring site at Newcastle has recently

Figure 4 Relationship between between Black Carbon and TSP University Site, Winter

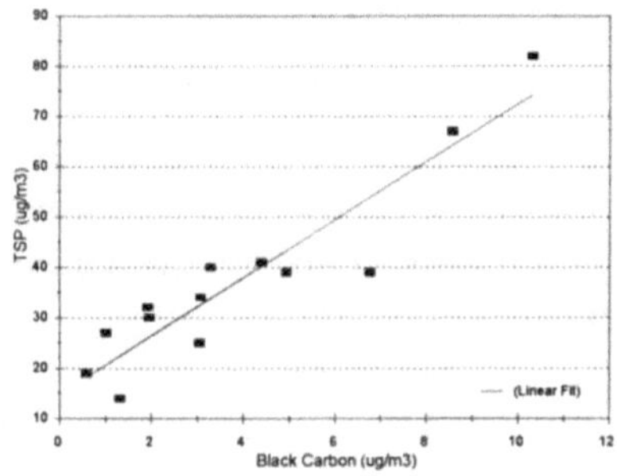

Figure 5 Relationship between Black Carbon and TSP University Site, Summer

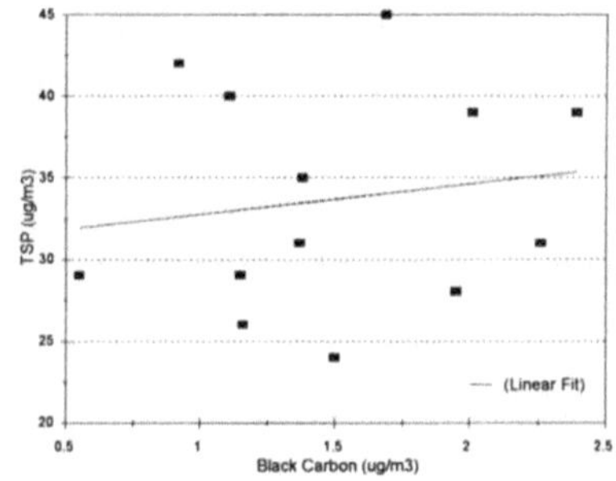

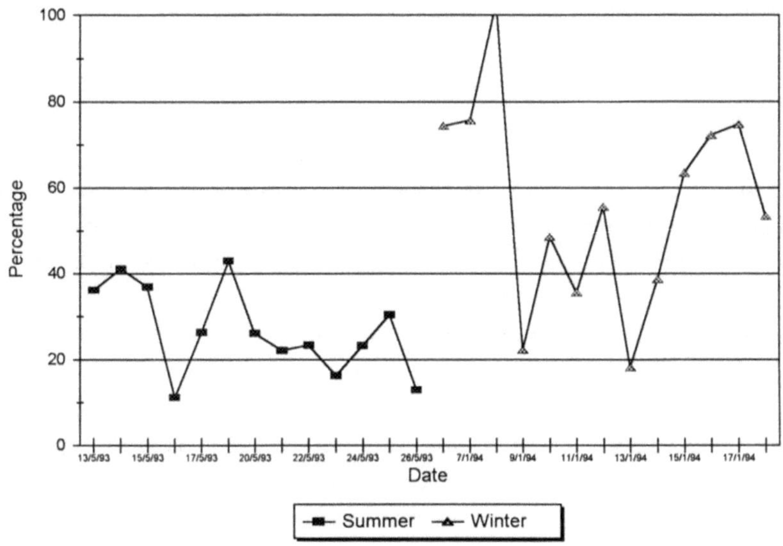

Figure 6 - Vehicular Contribution to TSP at the University Site

been found to have been located too near an Heavy Goods Vehicle depot which would undoubtedly have an influence on the PM_{10} concentration. The site has recently been moved, but it is thought that data may have been affected for the whole of the sampling period used during this study. The site at Cardiff has been found to be under the influence of construction and demolition activity, again a factor which may

influence the PM_{10} concentration through the increased concentrations of building dust present in the atmosphere.
The pattern observed in the plot of traffic contribution to PM_{10} with most sites showing higher contributions from traffic in Winter than Summer is thought to be due to the formation of photochemical PM_{10} in the Summer, through the action of sunlight on SO_x and NO_x to form sulphate and nitrate aerosols. This would have the effect of reducing the proportionate primary vehicular contribution to PM_{10}.

The national inventory for PM_{10} is shown in Table 2. This shows road transport to be responsible for 27% of PM_{10} emissions. However this figure is likely to represent an underestimation of the situation in urban areas where transport emissions are of increased importance relative to the national average.

The figure of a 40% average traffic contribution to PM_{10} can be compared with results from previous studies. Horvath et al (1988) added tracer to diesel sold in Vienna, and found that the mass concentration of diesel particles in the atmosphere varied between 5 and 23$\mu g/m^3$. A background diesel particle concentration of 11$\mu g/m^3$ was found, and this value increased by 5.5$\mu g/m^3$ per 500 diesel vehicles per hour passing near the sampling location. It was found that diesel particles constituted 12.2 - 33% of the total aerosol mass. Ball (1984) reported the results of a study carried out by the Greater London Council correlating lead and particulate matter at a rooftop site in central London, which pointed to motor vehicles as a major contributor to particulate matter.

The analysis of the data collected in Birmingham also serves to reinforce these findings. The relationship between black carbon and TSP appears to be very seasonally dependent, but this may be due to the low carbon loadings found on the filters during the Summer sampling campaign and hence the low spread of concentrations. The vehicular contribution to TSP is found to be an average of 29.4% during Summer and 60.9% during Winter, giving an overall average of 45.15%. The very high values observed at the start of the Winter sampling campaign were found to be due to adverse meteorological conditions prevailing at the time (see Figure 7). The sampling site was approximately 50 metres from a busy road, and the graph demonstrates the effects of very low windspeed on the dispersal of pollutants from this source.

Having calculated the vehicular contribution to PM_{10} it is then possible to further divide this figure into petrol and diesel contributions using emission inventory data. Table 2 shows the UK national emission inventory for PM_{10}. Ignoring the motorcycle contribution (<1%) then diesels are responsible for 16/26 = 62% of vehicular PM_{10}. Applying this to the figures obtained earlier it is possible to estimate that diesel vehicles alone are responsible for 26% of primary PM_{10} emissions in UK urban areas.

Table 2 - Source contributions to total UK emissions, 1991 (%)

Road Vehicles	**Type**	**PM_{10}**
Cars	Petrol	9
	Diesel	1
	Total	10
Light Goods Vehicles	Petrol	1
	Diesel	1
	Total	2
Heavy Goods Vehicles	Large	5
	Small	7
	Total	12
Buses		2
Motorcycles		<1
Exhaust Emissions		27
Total Road Vehicles		27

Source : QUARG (1993b)

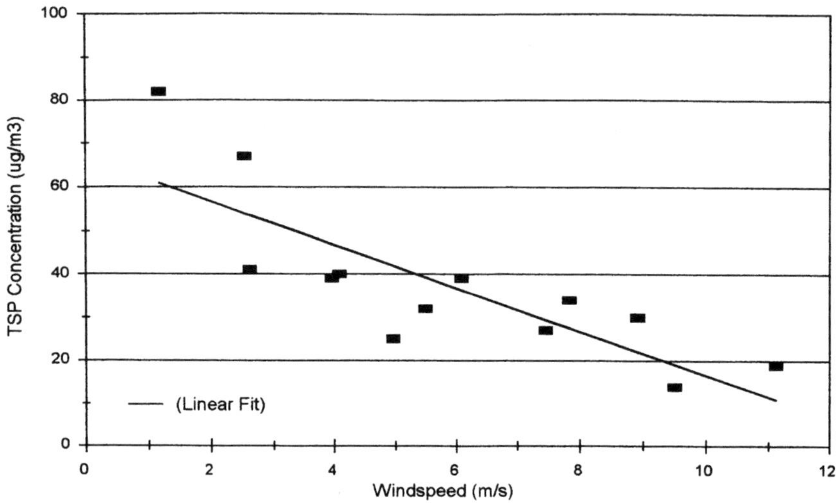

Figure 7 - Relationship between windspeed and TSP concentration for the Winter sampling period at the University site.

Conclusion

Our data analysis using CO as a tracer of road traffic indicates that on average traffic generates 40% of PM_{10} in urban areas in the UK, but there are distinct seasonal and locational differences in this contribution. Work carried out on Total Suspended Particulate and Black Carbon data collected in Birmingham supports these results, showing that on average traffic contributes 45% of TSP in a smaller data set. These results concur with those from other studies, pointing to vehicles as major contributors to urban particulate matter. At present however this work only assesses the primary contribution of vehicles to PM_{10}, and work is currently underway to make an additional assessment of the secondary contribution of vehicles to particulate matter.

Acknowledgements

The authors are grateful to the UK Department of the Environment for financial support of this work, and to Professor C.A. Pio and Mr M. Castro (University of Aveiro, Portugal) for the carbon analyses.

References

Ball D. J. (1984) Environmental implications of increasing particulate emissions resulting from diesel engine penetration of the European Automotive Market. Sci. Tot. Environ. 33: 15-30.

Derwent R.G., Simmonds P.G., Collins W.J. (1994) Ozone and carbon monoxide measurements at a remote maritime location, Mace Head, Ireland, from 19990 to 1992. Atmos. Environ. 28: 2623-2637.

Horvath I., Kreiner I., Novek C. Preining O. and Georgi B. (1988) Diesel emissions in Vienna. Atmos. Environ. 22: 1255-1269.

LSS (1989) London Air Pollution Monitoring Network Fourth Report LSS, London

Novelli P.C., Masarie K.A. , Tans P.P., Lang P.M. (1994) Recent changes in atmospheric carbon monoxide. Science: 263: 1587-1590.

Quality of Urban Air Review Group (1993a) Urban Air Quality in the United Kingdom. QUARG, London

Quality of Urban Air Review Group (1993b) Diesel vehicle emissions and Urban Air Quality. QUARG, London.

Receptor Methods for VOC Source Apportionment in Urban Environments

Charles W. Lewis
Atmospheric Research and Exposure Assessment Laboratory
U.S. Environmental Protection Agency
Research Triangle Park, North Carolina
U.S.A.

Introduction

Receptor modeling provides quantitative estimates of the impacts of sources on ambient air, primarily from inter-relationships of the ambient concentrations of individual air pollutant species measured at a receptor site. In contrast to dispersion modeling, receptor modeling employs only minimal meteorological and emissions inventory information. Its ultimate use is to arrive at minimum cost emissions control strategies that will be effective in reducing the air pollutants of interest. The development of receptor modeling methods began about 20 years ago, to provide answers to questions about the quantitative contributions made by potential sources to particulate air pollution. Since that time particulate receptor modeling has expanded from its initial focus on determining the sources of measured ambient particulate mass concentration, to the sources of particulate-related parameters such as visibility degradation (Lewis et al., 1986), mutagenicity (Lewis et al., 1988) and tumorigenicity (Lewtas et al., 1992). The success of these methods for particulate problems has naturally led to their consideration in the volatile organic compound (VOC) area. By definition VOC excludes CH_4 and the "inorganic" compounds CO and CO_2.

There are two general reasons why determining the source contributions of VOC is of interest in urban air pollution: (1)

NATO ASI Series, Partnership Sub-Series, 2. Environment – Vol. 8
Urban Air Pollution
Edited by Ivo Allegrini and Franco De Santis

the toxicity of some VOC species, and (2) their contribution to tropospheric ozone formation through photochemical reaction with oxides of nitrogen (NO_x). Although both are of importance, the work discussed in this article is motivated more by the second reason than the first. The article will summarize current work on the source apportionment of VOCs which is being sponsored by the United States Environmental Protection Agency's Atmospheric Research and Exposure Assessment Laboratory. The work has three components: (1) chemical mass balance, (2) extracting source profiles from ambient data, and (3) ^{14}C determination of biogenic VOC emissions. Each is illustrated by recent results derived from ambient VOC data collected in Atlanta, Georgia, USA, an ozone non-compliance area.

Chemical Mass Balance (CMB)

The CMB method has become the principal receptor modeling tool for mathematically analyzing measured air pollutant species concentrations for the purpose of determining their source contributions. In this model the concentration c_i of species i measured at the receptor site is expressed as

$$c_i = \Sigma A_{ij} S_j + e_i \tag{1}$$

where A_{ij} is the abundance of species i in the profile of source j, and S_j is the mass concentration at the receptor site due to source j. This series of equations (one for each fitting species i) is simultaneously solved for each S_j by minimizing the reduced chi square

$$\chi^2 = (N_s - N_c)^{-1} \Sigma (e_i/E_i)^2 \tag{2}$$

where e_i is a residual, N_s is the number of fitting species, N_c is the number of sources, and E_i^2 is an "effective variance" which includes uncertainties in both A_{ij} and c_i.

The U.S. EPA distributes a software module (CMB7) which embodies this approach and has been useful in standardizing CMB applications. It is well-documented (Watson et al., 1990) and contains a number of useful diagnostics. The software was developed for aerosol applications, but is usable for VOC with only modest changes in the input files.

The selection of fitting species is an important part of a CMB calculation. Generally, fitting species should satisfy two selection criteria:

(1) be "unreactive," i.e., atmospheric lifetimes longer than eight hours;

(2) be emitted only by the selected sources, i.e., its measured ambient concentration can be satisfactorily reproduced by those sources included in the calculation.

Figure 1 shows a result of applying CMB7 to ambient VOC species concentrations measured in Atlanta, Georgia, U.S.A., during the 1990 summer (Purdue, 1991). The measurements were obtained on the campus of the Georgia Institute of Technology, located near the urban center of Atlanta. The measurements were made on an hourly basis with an automated gas chromatographic system. The "weekday average" result shown is for an average over five weekdays for which nearly every hourly period was successfully measured (a total of 114 measurement periods).

Sixteen fitting species (of the 47 measured) were used in the CMB procedure. In Figure 1 the "road" source (representing the sum of vehicle tailpipe emissions and running losses) is seen to account for about half of the total measured VOC (defined here as the sum of all chromatographic peaks in the C_2 to C_{10} range). The "gasoline" and "vapor" sources represent two distinguishable kinds of gasoline evaporation, whole gasoline and headspace gasoline vapor, respectively. It is noteworthy that the whole gasoline contribution is 2-3 times greater than the headspace contribution (9.9 % vs. 3.7 %). The relative uncertainties for the source contributions range from 9 - 14 % for the "road," "natural gas," "propane-rich," and "isoprene-rich" sources, 30 - 40 % for the evaporative gasoline sources, and 54 % for the solvent source. The poor accuracy of the solvent source is minimized by its small contribution. The unexplained portion in Figure 1 is likely associated with the cumulative effect of multiple small sources, as well as secondary products of atmospheric photochemical transformation. Additional details and results of this work may be found in Lewis et al. (1993).

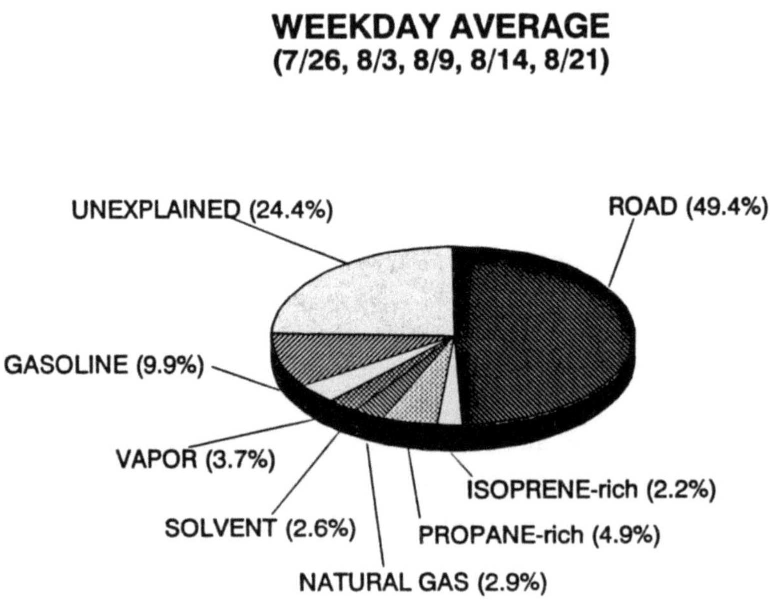

Figure 1. CMB estimate of source contributions to total ambient VOC in Atlanta during 1990 summer (5-day 24-h average).

Extracting Source Profiles from Ambient Data

The most difficult obstacle to applying the CMB method is the frequent unavailability of source profiles which are both accurate and representative of the source emissions in the airshed under consideration. A new procedure has been developed to extract profiles from the ambient data themselves. The procedure combines two graphical/mathematical components, GRACE (Graphical Ratio Analysis for Composition Estimates) and SAFER (Source Apportionment by Factors with Explicit Restrictions).

The basis for the GRACE/SAFER approach is illustrated in Figures 2 and 3. Figure 2 shows a scatterplot of simultaneously measured concentrations of acetylene and ethylene, from the same dataset discussed in the CMB section above. For summertime Atlanta there is good reason to believe that acetylene should be mostly from motor vehicle exhaust emissions (Henry et al, 1994). The approximately linear relationship between ethylene and acetylene then suggests that most of the ethylene has the same source, and that the ethylene/acetylene emissions ratio for the source is nearly unity, as given by the slope of the best-fit line in Figure 2.

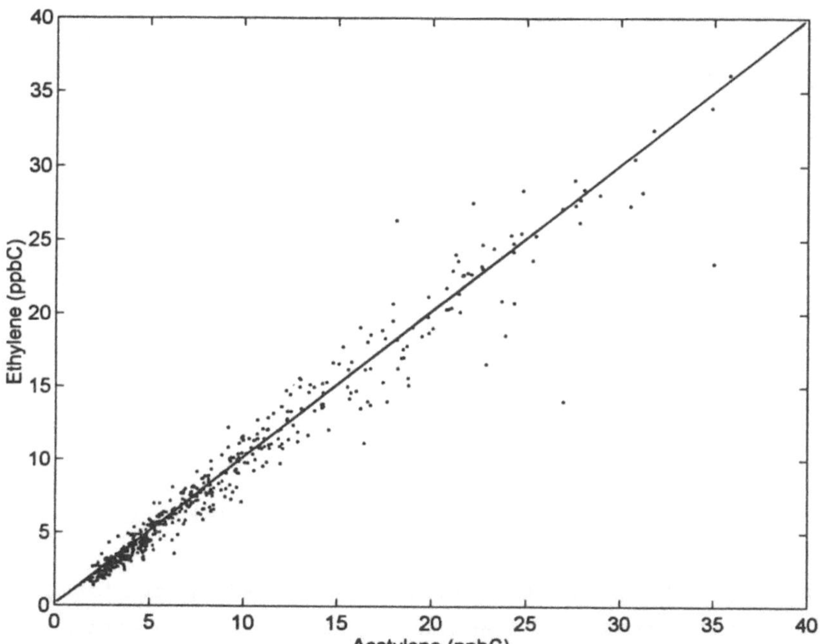

Figure 2. Ethylene vs. acetylene (Atlanta).

Figure 3 shows an analogous scatterplot for toluene and acetylene, but with a very different character from that of Figure 2. The principal feature of Figure 3 is the quite sharp linear lower boundary to an otherwise rather diffuse distribution. The interpretation of this is that acetylene, assumed to be primarily from motor vehicle exhaust, is always accompanied by toluene exhaust emissions at a fixed toluene/acetylene ratio of about 1.8, as given by the slope of the lower boundary. Data points above the lower boundary reflect the additional contribution of other independent sources of toluene.

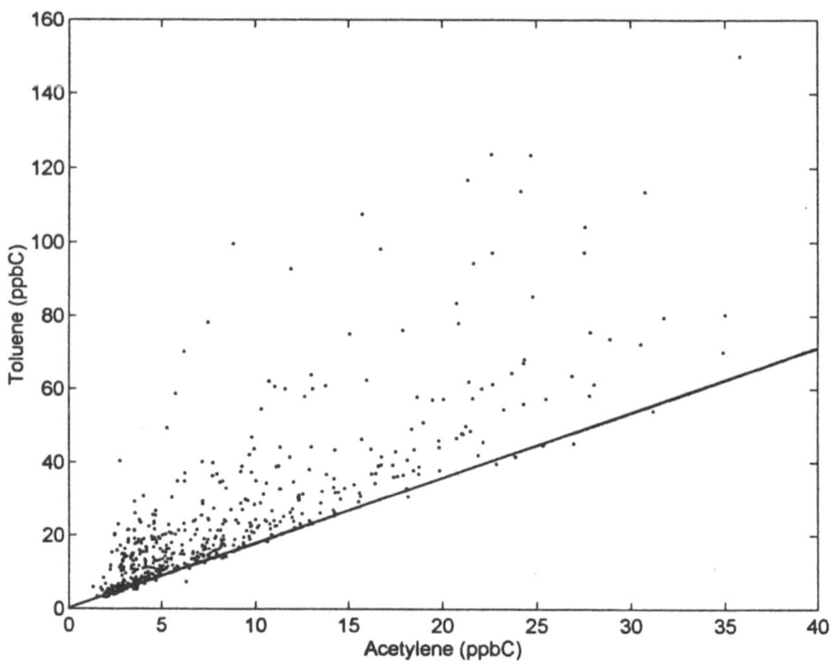

Figure 3. Toluene vs. acetylene (Atlanta).

From such graphical considerations (the GRACE procedure) a set of initial estimates of source profile abundances can be constructed. These are then used as constraints in a combined linear programming-factor analysis step (the SAFER procedure) which generates the final source profiles.

A validation of the GRACE/SAFER methodology has been carried out by comparing the ambient-derived source profiles with profiles directly measured in Atlanta at the same time the ambient data were collected (Conner et al., 1995). For the three vehicle-related profiles -- vehicle exhaust, whole gasoline evaporation and headspace gasoline evaporation -- for which the comparison could be made, the agreement was excellent. Details may be found in Henry et al. (1994).

^{14}C Determination of Biogenic VOC Emissions

A fundamental uncertainty in understanding the origin of urban tropospheric ozone is the lack of reliable knowledge on the size of the biogenic contribution to precursor VOC. Biogenic VOC is known to be very photochemically active and is basically uncontrollable, so that if it constitutes an appreciable fraction of total VOC emissions, control of the remaining anthropogenic VOC may be ineffective (relative to NO_x control) in ozone abatement.

In principle the CMB approach described above is applicable to all source types. In practice however it is not well suited to the quantification of biogenic sources, because of the lack of non-reactive species that are suitable tracers of biogenic emissions. For that reason an independent effort, based on ^{14}C methodology, is being developed. The biogenic fraction f_b of an ambient VOC sample is given by

$$f_b = f_s \;/\; f_{CO2} \tag{3}$$

where f_s and f_{CO2} are the $^{14}C/^{12}C$ ratios for the VOC sample and atmospheric CO_2, respectively. The use of ^{14}C for VOC is an extension of a methodology that has been successfully used to determine the fraction of ambient aerosol resulting from wintertime woodburning (Lewis et al., 1988).

The most critical problems in applying ^{14}C methodology to ambient VOC samples are (1) collecting sufficient sample quantity to allow the ^{14}C measurement to be performed with adequate precision, and (2) insuring that the VOC sample contains minimal, quantified amounts of carbonaceous contaminants that would otherwise render the ^{14}C measurements misleading or uninterpretable.

Problem (1) has been addressed by the use of large-volume (32-L) SUMMA® stainless steel canisters to collect whole air samples. Pressurizing the samples to 30 psig during sample collection results in 0.1 m^3 of air per canister. At the current state of ^{14}C measurement technology the ^{14}C fraction can be routinely measured to a few percent precision on samples containing as little as 50 μg carbon. This amount is achievable for most airsheds of interest by compositing several such canister samples.

Problem (2) was initially addressed by a purely cryogenic method (Klouda et al., 1993) to separate the VOC fraction from those whole air constituents which would otherwise be interferents in the ^{14}C measurement (CH_4, CO and CO_2). While this was effective for CH_4 and CO, the separation of CO_2 from the VOC was not as complete as desired. The solution was to introduce an additional step of passing the whole air sample through a bed of LiOH prior to cryogenic elimination of CH_4 and CO. A 10^4 reduction factor for CO_2 has been achieved, with only modest loss (10-15 % by mass) of VOC.

The full analytical measurement scheme has been demonstrated on canister samples collected during late summer 1992 at the same Atlanta site indicated above. Measurements of the VOC ^{14}C content have been performed on two composited samples, an "AM" composite of mid-morning samples, and a "PM" composite of late evening samples. The AM and PM samples showed biogenic fractions of 12 and 14 %, respectively (but with relative uncertainties of about

50 % that are mostly associated with a large and incompletely characterized processing blank). The results are plausible when compared with the VOC species composition of the samples.

Conclusion

Three projects that are part of a coordinated VOC source apportionment research program at the U.S. EPA have been described. Work is continuing in all three areas. The general goal of emissions inventory validation is being pursued with the CMB approach on additional datasets (Conner et al., 1994). The GRACE/SAFER methodology for source profile extraction is being refined to include estimates of profile uncertainty. Anticipated improvements in the ^{14}C methodology for biogenic estimation include equipment modifications to reduce the blank, and high pressure canister sample collection to increase the amount of sample material.

Acknowledgements

The work summarized in this article represents the collective efforts of a number of individuals and organizations: Robert K. Stevens, Teri L. Conner, William A. Lonneman, Robert L. Seila (U.S. Environmental Protection Agency); Ronald C. Henry, John F. Collins (University of Southern California); George A. Klouda (National Institute of Standards and Technology); Reinhold A. Rasmussen (Oregon Graduate Institute); William D. Ellenson and David C. Stiles (ManTech Environmental Technology, Inc.).

Disclaimer

The information in this document has been funded wholly or in part by the United States Environmental Protection Agency under Cooperative Agreement No. 818410 to the University of Southern California, Interagency Agreement No. DW13935098 to the National

Institute of Standards and Technology, and Contract No. 68-DO-0106 to ManTech Environmental Technology, Inc. It has been subjected to Agency review and approved for publication. Mention of trade names or commercial products does not constitute endorsement or recommendation for use.

References

Conner TL, Lonneman WA, Seila RL (1995) Transportation-related volatile hydrocarbon source profiles measured in Atlanta. J Air Waste Manage Assoc (to be published)

Conner TL, Collins JF, Lonneman WA, Seila RL (1994) Comparison of Atlanta emission inventory with ambient data using chemical mass balance receptor modeling. Proceedings of the A&WMA/EPA Symposium, The emission inventory: application and improvement

Henry RC, Lewis CW, Collins JF (1994) Vehicle-related hydrocarbon source compositions from ambient data: the GRACE/SAFER method. Environ Sci and Technol 28:823-832

Klouda GA, Norris JE, Currie LA, Rhoderick GC, Sams RL, Dorko WD, Lewis CW, Lonneman WA, Seila RL, Stevens RK (1993) A method for separating volatile organic carbon from 0.1 m^3 of air to identify sources of ozone precursors via isotope (^{14}C) measurements. Proceedings of the 1993 A&WMA/EPA symposium, Measurement of toxic and related air pollutants 585-603

Lewis CW, Conner TL, Stevens RK, Collins JF, Henry RC (1993) Receptor modeling of volatile hydrocarbons measured in the 1990 Atlanta ozone precursor study. Paper No. 93-TP-58.04, in Proceedings of the 86th annual meeting of A&WMA

Lewis CW, Baumgardner RE, Stevens RK, Claxton LD, Lewtas J (1988) Contribution of woodsmoke and motor vehicle emissions to ambient aerosol mutagenicity. Environ Sci Technol 22:968-971

Lewis CW, Baumgardner RE, Stevens RK, Russwurm GM (1986) Receptor modeling study of Denver winter haze. Environ Sci Technol 20:1126-1136

Lewtas J, Lewis C, Zweidinger R, Stevens R, Cupitt L (1992) Sources of genotoxicity and cancer risk in ambient air. Pharmacogenetics 2:288-296.

Purdue LJ (1991) Summer 1990 Atlanta ozone precursor study. Paper No. 91-68.8, in Proceedings of the 84th annual meeting of A&WMA

Watson JG, Robinson NF, Chow JC et al (1990) CMB7 user's manual. Report No. EPA-450/4-90-004, US Environmental Protection Agency, Research Triangle Park, NC, USA

Air pollution Characteristics of a Region of Sofia and Data from Field Experiments (1992-1993)

Ekaterina Batchvarova, Dimiter Syrakov, Anna Tzenkova
National Institute of Meteorology and Hydrology
Bulgarian Academy of Sciences
66 Tzarigradsko chaussee
1784 Sofia
Bulgaria

Introduction

The population of 1.2 million inhabitants and the developed industry of all kinds define the high level and great variety of pollutant emissions in the town of Sofia. The climatological and terrain conditions in the valley of Sofia in general are not favourable for pollution dispersion and the combination of both the above factors (economical and natural) leads to significant air pollution in the town. The problem of urban air pollution in Sofia has been studied since the seventies at the National Institute of Meteorology and Hydrology (NIMH). A number of experimental campaigns and model simulations have been performed. Two experimental campaigns covering typical summer and winter conditions, respectively were recently organised within the frame of a Project funded by the Ministry of Science and Education in Bulgaria.

The area chosen for the experiments was a mixed industrial and residential region in the Eastern part of Sofia. In the region there are: a power plant, metal and chemical factories, leather-works and many smaller unorganised sources. The air pollution problem of this area is caused by local industrial sources, by air pollution transport from a big metallurgic complex situated north-east, and by several highways. The relatively low terrain is favourable to the accumulation of pollutants in the area.

NATO ASI Series, Partnership Sub-Series, 2. Environment – Vol. 8
Urban Air Pollution
Edited by Ivo Allegrini and Franco De Santis

Experimental design

The purpose of the experimental work was to study the spatial and time distribution of the concentration fields for typical pollutants in a dense network as well as to provide data for different model (statistical and dispersion) simulations. Based on cluster and factor analysis, the number of stations for the second or winter campaign (November 1993) was reduced by two stations compared to the summer one (June 1992).
Mainly sulphur dioxide (SO_2), nitrogen dioxide (NO_2), phenol (C_6H_5OH), hydrogen sulphide (H_2S) and dust measurements were performed at 9 stations during the summer and at 7 stations during the winter experiment within a region of about 10 sq. km, Fig 1.
The sampling was performed 8 times a day at synoptic observation hours and simultaneous meteorological measurements were performed at all the stations. Additionally to the regular measurements at NIMH a tethered balloon sounding was performed.

Experimental results

The part of the town under investigation is characterised by its own microclimatological conditions, formed by the type of the built-up (density, height, materials), the types of industry and fossil fuels used, the intensity and distribution of transport, etc.
From previous investigations at the National Institute of Meteorology and Hydrology it is known that the heat island of Sofia is characterised by a maximum in the center of the city and a secondary maximum over the north-eastern part - the region of the present investigation. One of the tasks of this new study was to investigate more precisely the urban climate characteristics.
The structure of the heat island observed over the region during the experimental campaigns is not homogeneous but changeable and complicated, depending on the intensity of the factors responsible for its formation - such as radiation balance changes over the area (especially during clear and calm nights), the heat accmulated in buildings, the ground characteristics, the heat generated by human activities and reduced evaporation over the city.

The distribution of air humidity characteristics is also different over and outside the city due to the same factors.

In Fig.1-3 a sequence of maps of air temperature, relative humidity, SO_2 and NO_2 distribution over the area for a chosen calm and clear summer night and the following morning, is presented.

The influence of the town on the local climate and the air pollution is best pronounced under the chosen meteorological

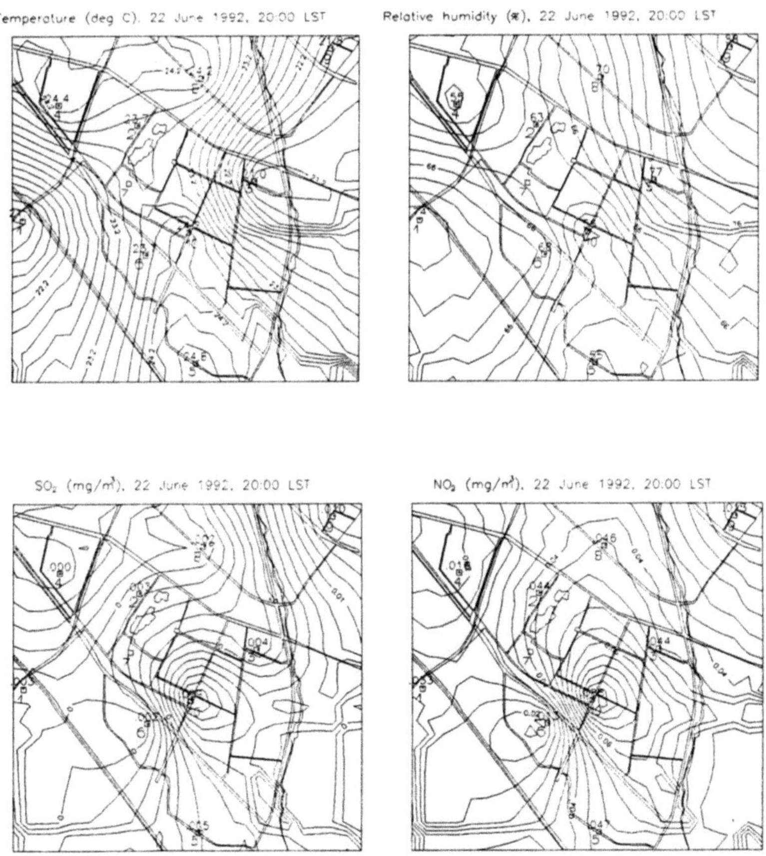

Fig.1. Distribution of air temperature, humidity, sulphur and nitrogen dioxide for 20 LST on 22 June 1992.

conditions. The wind speed is in the range 0 - 1.5 m/s. Calm conditions are observed in the afternoon hours of June 22 and the morning hours of June 23. Quiet winds from the South are observed during the night from 22 to 23 of June, which is a typical local orographic effect - night mountain breeze.

At 8 p.m. on June 22nd the lowest part of the region is warmer than the rural site (9) and the hill (1) with elevation of 50 m. The temperature distribution is homogeneous over the flat part of the region. At station 2, an open place in a small park (grass and bushes) with good ventilation and close to a lake, the air temperature is lower. The maximum difference of 3°C is between the sites 5 and 9. The urban part is drier than the rural with maximum difference of 33% (stations 4 and 9). Station 2, because of the same factors, the relative humidity is a little bit higher (Fig.1). At 11 p.m. the temperature difference between sites 5 and

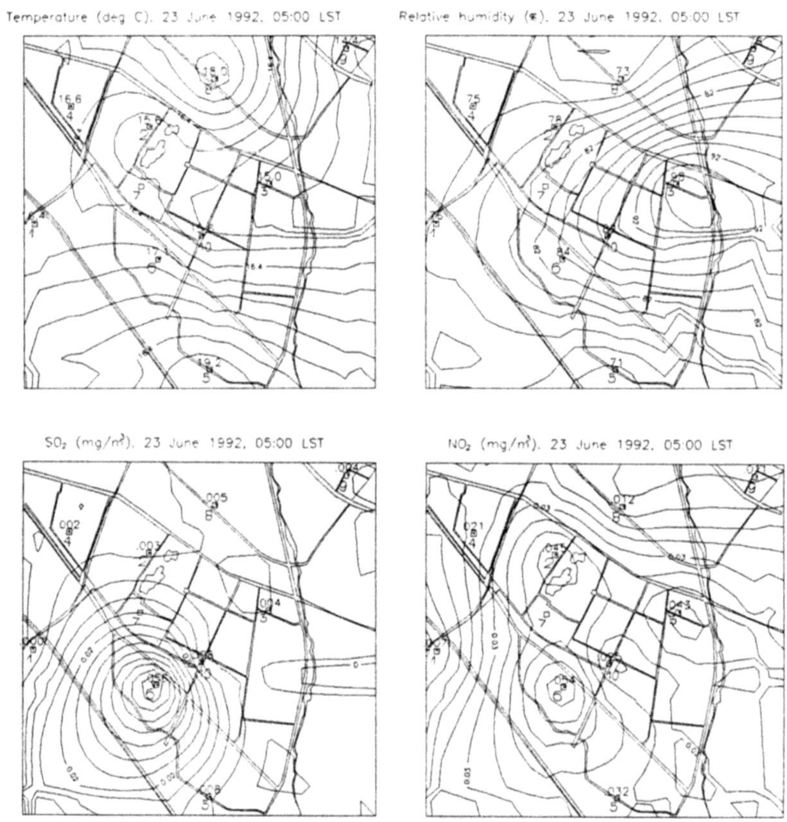

Fig.2. Distribution of air temperature, humidity, sulphur and nitrogen dioxide for 05 LST on 23 June 1992.

9 is still 3°C, but there is a warmer zone to the North at the industrial region (station 8) and the maximum difference is 4.4°C. The town is still about 30% drier. At 2 a.m. the picture is similar but at 5 a.m., before sun rise the relative humidity distribution is significantly changed - no difference between urban and rural sites is observed while the difference in temperature is higher - 4.8°C (st 5- st 9, Fig.2).

At 8 a.m. the temperature difference is decreased to 2.9°C and the relative humidity is higher over the city than over the rural site. At 11 a.m. and 2 p.m. the temperature difference is about 1.4°C, 1.6°C and the relative humidity is still homogeneous, (Fig.3). For the same hours the observations of SO_2 and NO_2 concentrations are presented on the figures. The distributions of

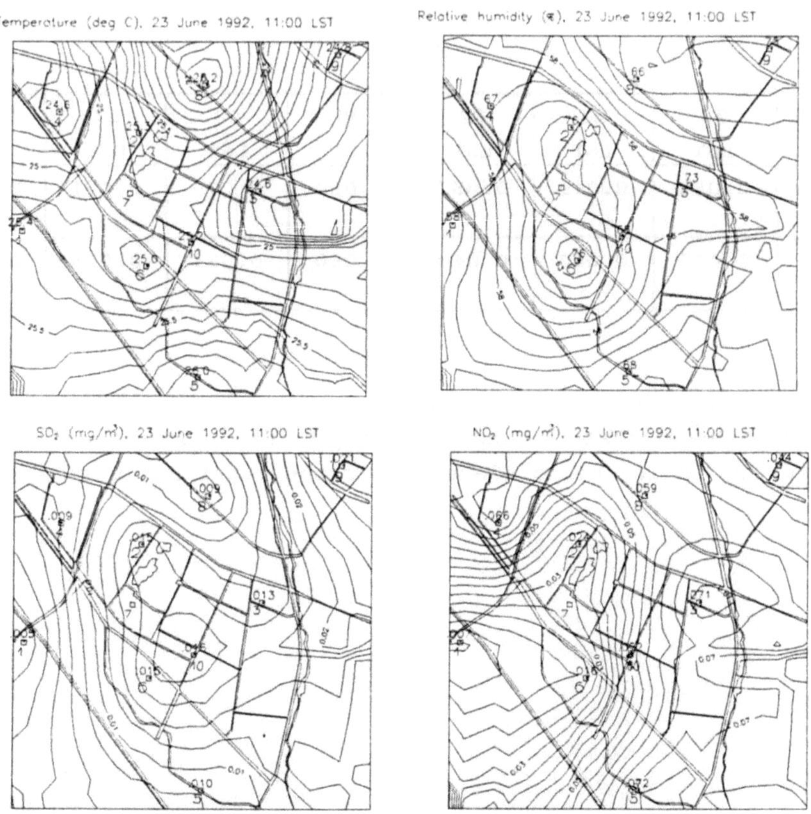

Fig.3. Distribution of air temperature, humidity, sulphur and nitrogen dioxide for 11 LST on 23 June 1992

these concentrations during the night (8 and 11 p.m. and 2 a.m.) is mainly defined by the location of unorganised sources (SO_2 at station 6). As can be seen from the figures, the meteorological and air pollution characteristics are quite changeable in calm and clear weather conditions. Some similarities between the patterns of relative humidity and air pollution was observed for night hours, but this relation needs a more detailed investigation in order to be confirmed.

The maximum concentrations of NO_2 are observed between 5 and 8 p.m. and 8-11 a.m. in coincidence with the rush-hours (Table 1). The SO_2 concentrations are higher during the night. A maximum is observed near a factory for plastic tube covers, where fugitive emissions are expected.

Table 1

	20 LST	23 LST	02 LST	05 LST	08 LST	11 LST	14 LST
NO_2 (mg/m^3)	0.088	0.045	0.057	0.054	0.079	0.071	0.037
SO_2 (mg/m^3)	0.009	0.040	0.040	0.080	0.051	0.016	0.010

The analysis of the mean values for the whole period of the summer experiment shows that during the night the concentrations are twice higher, while this difference for the clear night is 4-8

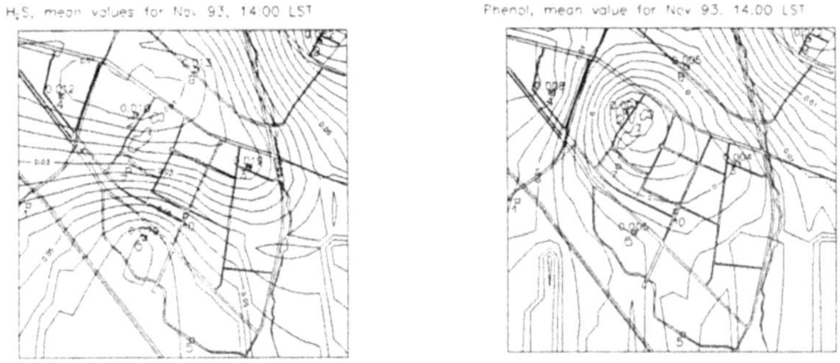

Fig. 4. Phenol and H_2S mean concentrations for November 1992

times for SO_2. The depth of the mixing layer during the day also gives a more homogeneous field of the concentrations. The mean NO_2 concentrations are higher at rush hours and near the highways.
The winter experiment was organised in the same way. The period was cold, with snow covering the area for almost all days. The daily variation of the meteorological and air pollution parameters was not significant.

The mean for the experiment (15-26 November) of phenol and hydrogen sulphide concentrations are homogeneously distributed over the urban area during the day and the night. The values at the rural site (st 9) and the Northern industrial region (st 8) are a little bit higher. The maximum values for the period are about 10 times higher than the mean concentrations, Fig.4 and Fig.5.
Higher values for NO_2 are always observed at stations 2, 4 and 6, situated close to streets with intensive traffic, Fig.6.

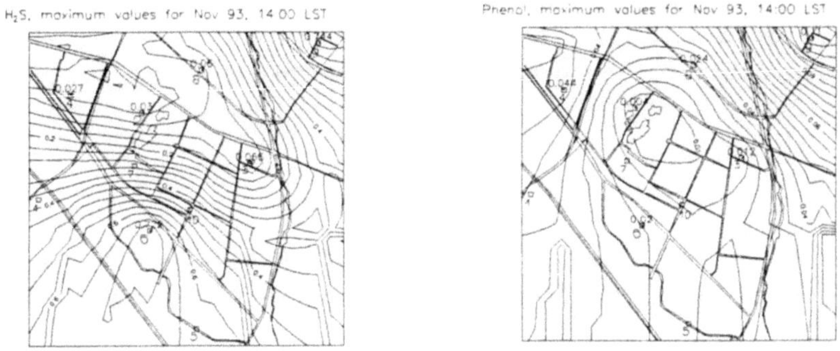

Fig. 5 Phenol and H_2S maximum concentrations for November 1992

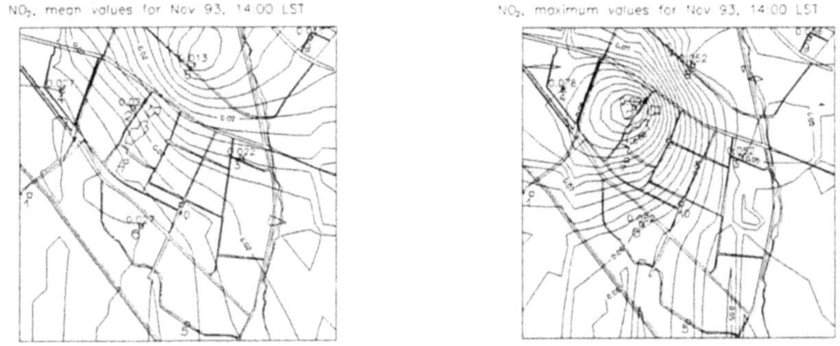

Fig. 6. NO_2 mean and maximum concentrations for November 1992

Conclusions

The main results of the chemical measurements can be summarised as follows:

- The region is significantly polluted with NO_2 in winter as well as during the summer. The main sources for NO_2 are transport and power plants.
- High concentrations of H_2S are also observed during summer and winter conditions.
- The level of SO_2 concentrations is low during the summer in all the region and higher during the winter only at the suburban site.

The data of the spatial distribution of the meteorological parameters show the details of the antropogenic influence on the local climate. The observed characteristics of the heat island are:

- More than 4°C intensity in clear calm summer nights and during the winter.
- Higher temperatures in the city even at noon during summer days.
- Higher gradients at the border town-rural site and more homogeneous distribution over the urban area.

References

Blaskova D et al (1983) Climate and Microclimate of Sofia. Publishing house of the Bulgarian Academy of Sciences, Sofia.

Evaluation of the air quality in Budapest based on the data of a monitoring network and measurement campaigns.

László Bozó and Györgyi Baranka
Institute for Atmospheric Physics of the
Hungarian Meteorological Service
H-1675 Budapest, P.O.Box 39.

Emission of pollutants in Budapest

The present general air pollution conditions in Budapest (more than 2 million inhabitants over 525 km^2) are mostly determined by the emissions from mobile line sources (road traffic). The structure of the emissions has changed significantly over the last 15-20 years in Budapest due to the "natural gas program". As a result of it, the emission of sulfur and nitrogen oxides from industry as well as space and domestic heating has decreased a lot. The annual anthropogenic emissions of CO, VOCS, NO_x, SO_2 and Pb based on a survey referring to 1990 for different source categories are summarized in Table 1.

	CO	VOCs	NOx	SO2	Pb
Mobile sources	111,500	14,000	7,712	1,663	53
Industry and energy production	24,949	2,177	8,705	9,095	10
Space and domestic heating	10,236	2,794	2,516	12,125	4
Total	146,685	18,971	18,933	22,883	67

Table 1. Annual anthropogenic emissions in Budapest (t yr^{-1})

It should be noted, however, that the emission rate of particulate matter also contributes significantly to the air pollution over Budapest. Its estimated annual total value is around 9 thousand tons.

At present, the number of cars is seven times greater than it

NATO ASI Series, Partnership Sub-Series, 2. Environment – Vol. 8
Urban Air Pollution
Edited by Ivo Allegrini and Franco De Santis

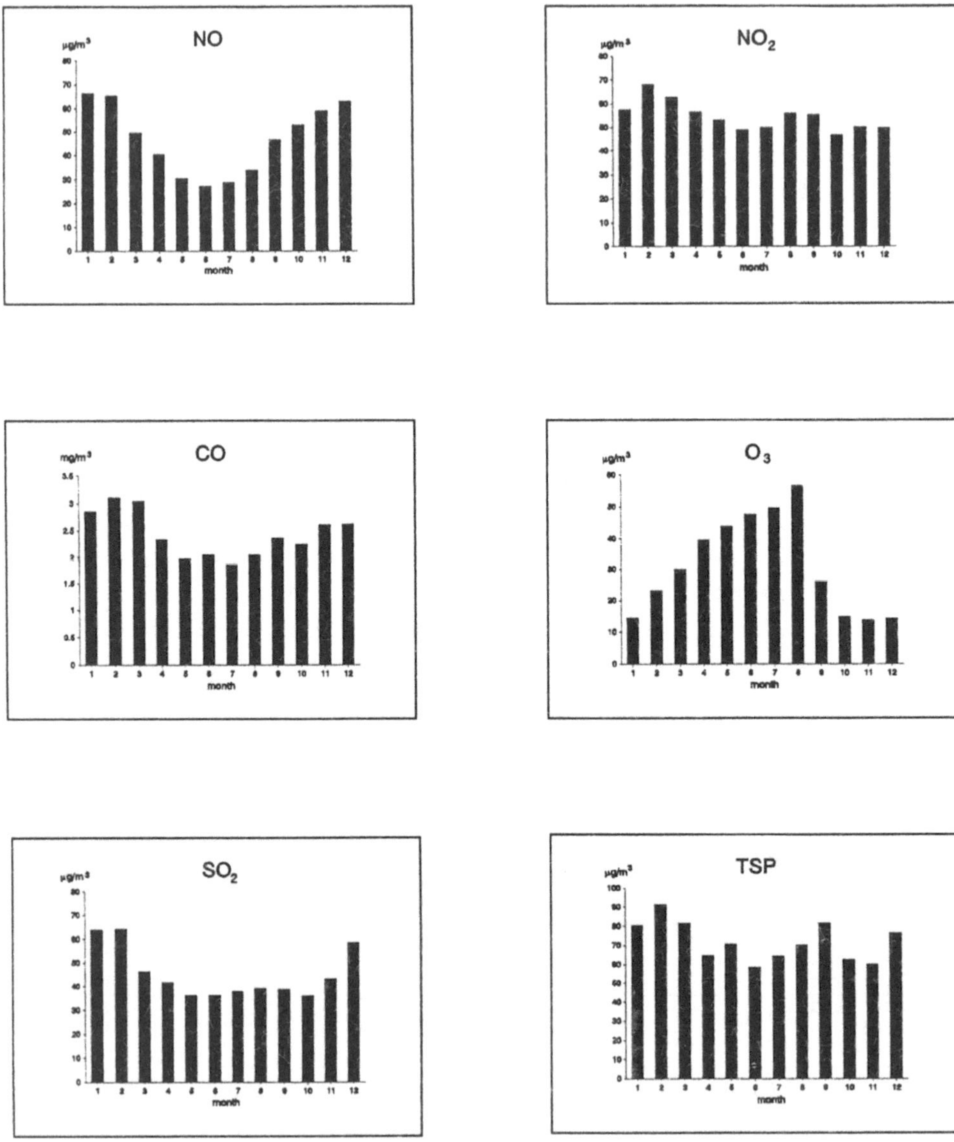

Figure 1. Average annual variation of air pollutants monitored in Budapest during the period of 1991-1993.

was 20 years ago. Most of them are equipped with out-of-date engines *without catalytic converters*. Approximately 68% of the car fleet is more than 6 years old and only 4% are less then 1 year old. The average age of the car fleet is over 10 years old. The improvement of road infrastructure is far behind the rapid development of mobilization: the total extent of public roads is only 50% more than that in the 70's. It can be seen from Table 1 that mobile sources are responsible for the majority of CO (76%), VOCs (74%) and Pb (80%) emissions in Budapest.

Data from monitoring network

The Institute of the Governmental Service for Public Health, Budapest has been operating since 1991 an air pollution monitoring network in Budapest consisting of 8 stations for the measurements of SO_2, NO_x, CO and TSP concentrations. There are also O_3, and TNMHC monitoring at two of these stations. All the substantial meteorological information (wind speed and direction, temperature and humidity) are registered. A central data aquisition system collects and stores the data coming from the stations. There are two public display systems in Budapest at two underground stations to inform the inhabitants on the actual air pollution conditions of the downtown area.
The non-real-time analysis of the data gained by means of the monitoring network is carried out in the Institute for Atmospheric Physics. Daily and seasonal variations of the concentrations as well as the characteristics of the actual meteorological situations are considered altogether. On the basis of our investigations, it can be stated that the average sulfur dioxide concentration is only 25% of the average measured 10-12 years ago. Average annual variations of SO_2, NO_2, NO, CO, TSP and O_3 for the period 1991-1993 (operational period of the new monitoring system in Budapest) are shown in Figure 1. It can be seen that SO_2, NO and CO concentrations have a similar annual variation with summer minima and winter maxima. Due to photochemical formation, the ozone peaks can be detected in summer. It is obvious that all the pollutants have

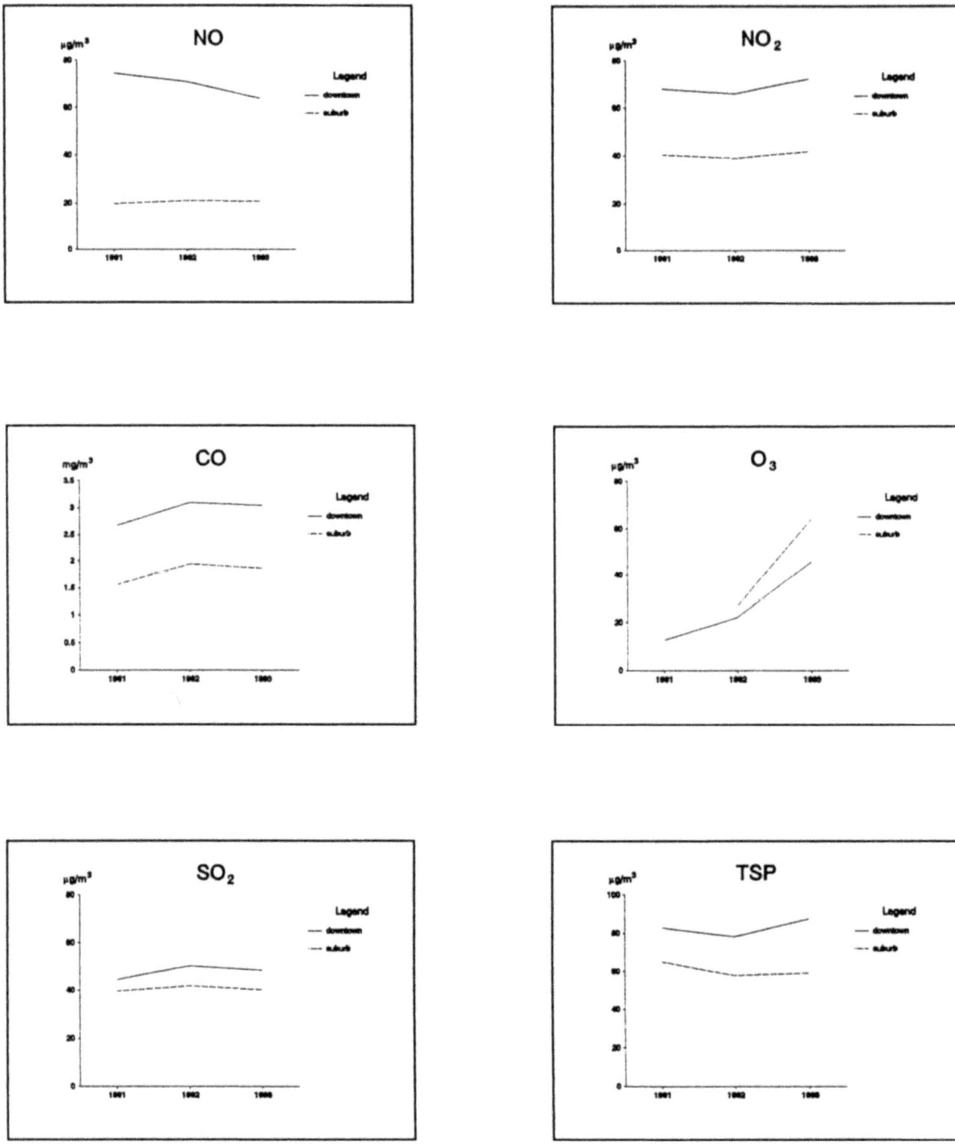

Figure 2. Multiyear variation of air pollutants monitored in Budapest.

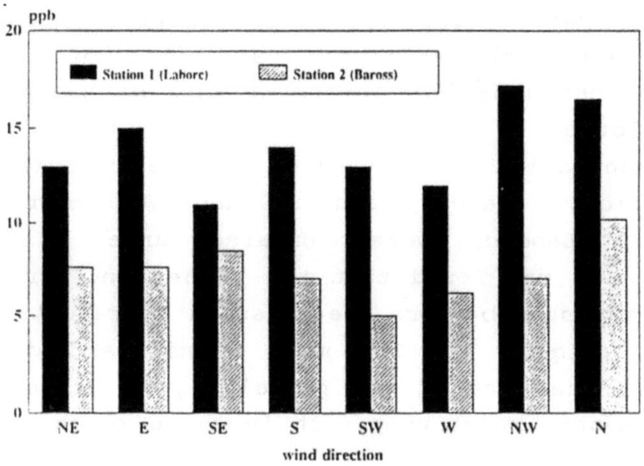

Figure 3. Ozone concentration in suburb and downtown of Budapest

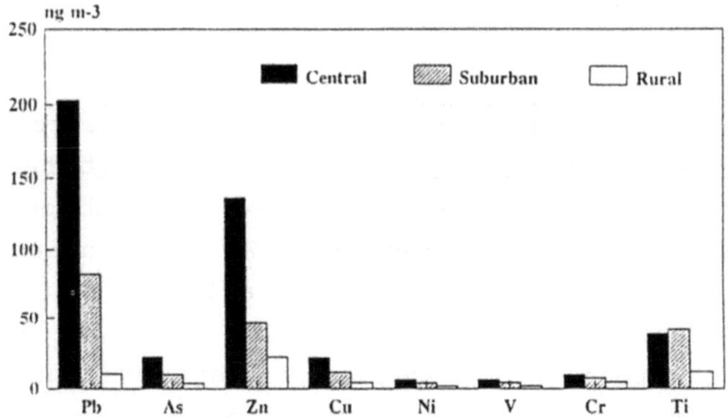

Figure 4. Trace element concentrations at 3 sites in Hungary

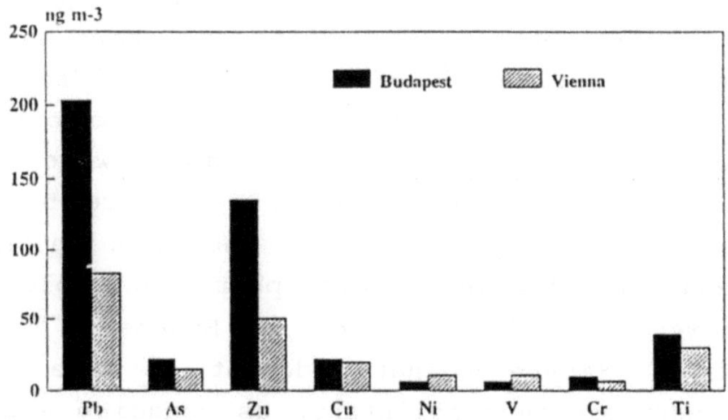

Figure 5. Trace element concentrations in Budapest and Vienna

significant spatial variation in the city. This can be seen in Figure 2, where the multiyear variations of the pollutants measured in the urban monitoring network of Budapest from 1991 are also plotted.

Ozone concentration data measured in the central part of the city *(Station 1 Baross)* and in the NW suburb of Budapest *(Station 2 Laborc)* were compared under different wind directions. It was found that the ozone concentration is much higher in the suburb where the emission of reductive precursors are lower (Figure 3). The most significant difference was detected in the case of the prevailing NW winds when "fresh" rural air masses relatively rich in ozone reach the NW edge of the city.

Data from measurement campaigns

Trace elements. An aerosol sampling and measurement campaign was organized in 1991 (Molnár et al., 1993) at three different locations: Central Budapest, downtown of Budapest and at a rural background station (K-puszta) 70 km downwind of Budapest in order to estimate the spatial variation of trace element (e.g. Pb, As, Zn, Cu, Ni, V, Cr, Ti) concentrations in atmospheric aerosols. The sampling sites were located on a NW-SE axis from Central Budapest down to the background station. The samples were collected by means of absolute Nuclepore polycarbonate filters with a pore size of 0.4 µm. The analyses were done by PIXE-method. The results are shown in Figure 4. It can be seen that a concentration of different elements increases generally from background to downtown. The trace element concentrations measured in Budapest were also compared to those measured in Vienna, Austria (figure 5). Due to the more widespread use of unleaded gasoline in Austria, the average lead concentration in atmospheric aerosols was found to be much lower than that observed in Budapest. Generally, one can see from Figure 5 that Budapest is more polluted than Vienna, except for such pollutants as vanadium and nichel wich are due to either oil combustion or iron and steel industries.

Ozone and precursors in urban plume.

As it can be seen from Figures 1 and 2, the surface ozone concentration is relatively low in Budapest. However, it can be expected that some ten kilometers downwind of Budapest much higher concentrations will occur due to the high emission of precursors from Budapest and photochemical oxidant formation in the urban plume during its transportation from the city towards rural areas. For this reason, the ozone concentrations measured at K-puszta were analysed as a function of the sectors of backward trajectories ending at K-puszta at 1200 and calculated for 850 hPa. Air parcels pick up a significant amount of ozone precursors over Budapest during the early hours in the morning and, due to photochemical reactions under clear weather conditions, elevated ozone concentrations occur at K-puszta station if air masses come from the direction of the city (from NW to SE). The frequency distribution of ozone concentration above 80 ppb in air masses coming from different directions is show in Figure 6. One can see that the largest relative contribution is measured in the case of NW trajectories when the wind speed is relatively low ($<2.5\ ms^{-1}$).

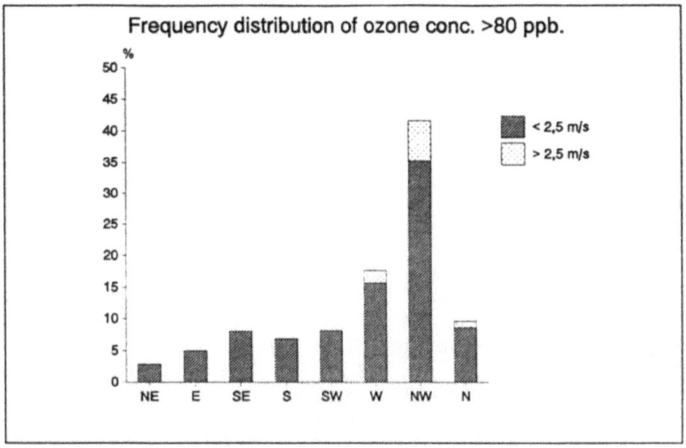

Fig. 6. Frequency distribution of ozone concentration higher than 80 ppb at K-puszta station

The results were also supported by chemical-meteorological model calculations for 28 cases (Bozó and Baranka, 1994) using the CBM-IV mechanism, consisting of 34 species in 82 reactions (Gery et al., 1988) as well as wind and solar radiation data.

Conclusions and further plans

The results presented above may be considered as a rough evaluation of the general air quality in Budapest. In the future, more measurements are needed for ozone and specific hydrocarbons (at the moment only the NMHC is monitored) in the city. It is planned that, in co-operation with EC-countries, a comprehensive chemical-meteorological model will be adapted and developed for the city. This will be done taking into consideration a detailed emission inventory, orography, parameters of atmospheric chemical transformations as well as meteorological fields in order to evaluate the contribution of different pollution categories (mobile and stationary) to the level of air pollution. Based on model computations, environmental impact studies can also be performed to investigate the impact of a new scheme under plan (e.g. a new road or a new industrial plant). This information can be used to determine the acceptability of a scheme or to help modify it to mitigate the impact.

References

Bozó L. and Baranka Gy. 81994): Model simulation of ozone formation in an urban plume over Hungary. Proceedings of EUROTRAC Symposium '94 edited by P.Borrell. SPB Academic Publishing bv. The Hague, The Netherlands.

Gery W.M., Whitten G.Z. and Killus J.P. (1988): Development and testing of the CBM-IV mechanism for urban and regional modeling.USEPA, Research Triangle Park, North Carolina.

Molnár A., Mészáros E., Bozó L., Borbély-Kiss I., Koltay E. and Szabó Gy. (1993): Elemental composition of atmospheric aerosol particles under different conditions in Hungary. Atmospheric Environment 27A, 2457-2461.

Air Pollution Monitoring in the Industrial Centers of Belarus

B.B. Kozeruk, Ivan M. Skuratovich
Main Administration of Hydrometeorology of Belarus
16 Komsomolskaya Street,
220600 Minsk,
Belarus

A monitoring network to estimate air pollution has been set up in 14 industrial areas where 65% of the urban population of the Republic lives. Air pollution is monitored at 47 permanent-operating stations which are fully equipped with devices and apparatuses to produce a running record of the carbon oxide content, to collect air samples for other pollutants and to determine meteorological parameters. Ten of these station are in Minsk and in the other cities there are from two to four. The monitoring program entails taking from three to four daily air samples to estimate the main (dust, sulfur dioxide, carbon oxide, nitrogen dioxide) and specific compounds. The list of specific compounds is based on total emissions of each substance, hazard class and quality standards, dispersion coefficients etc. About 250,000 urban air samples are taken and analyzed annually in Belarus. In some Belarus cities the programme carried out by the stationary-operating stations is backed up by survey and "beneath-the-stack" observations performed by mobile laboratories. Air pollution in large industrial centres is actively monitored by the Ministry of Public Health. There is a clear distinction between the monitoring activities of sanitation-epidemiological stations and the Main Administration of Hydrometeorology. The data produced by the latter, situated outside the sanitation-protection zone is used to obtain statistical characteristics, to compute the information regarding the concentrations and lastly to forecast air pollution whereas the sanitation stations are responsible for carrying out incidental survey and "beneath-the-stack" observations from mobile laboratories in order to obtain the spatial distribution of air pollutants from a given source of emission.

NATO ASI Series, Partnership Sub-Series, 2. Environment - Vol. 8
Urban Air Pollution
Edited by Ivo Allegrini and Franco De Santis

Belarus covers an area of 207,600 sq. km. It is 560 km in length from north to south and 650 km in width from east to west. Belarus borders with Poland in the west, with Russia in the east, with Latvia and Lithuania in the north and with the Ukraine in the south. The population is 10,3 million. people. About 63% of the population live in towns. The Republic is divided into six administrative regions.

Belarus is mainly a plain, where vast hilly terrains alternate with flat valleys, or slightly concave lowlands. The hilly terrains are cut by a dense network of river valleys.

The climate of Belarus is largely determined by its geographical position in the middle latitudes of the northern hemisphere. The Republic lies in that rather wide zone of moderate climatic belt, where maritime climate gives way to a continental one. The domination of west winds blowing from the Atlantic Ocean is a major factor forming the climate.

Complex analysis of meteorological parameters shows that Belarus belongs to a regions where there is a low occurrance of continuous unfavourable weather responsible for the accumulation of pollutants in urban air basins. What is peculiar about this meteorological regime is that there is more air pollution when there is a change of seasons, the longest period being March and August - September. Furthermore, there is likely to be an increase in air pollution at night-time during the warm half of the year. This change of weather results in the pollutive periods not lasting long. However, favourable weather conditions can not ensure complete dispersal of industrial emissions.

In the national economy of the former USSR, the Republic of Belarus played an "assembly line" role. Belarus specialized in such labour consuming and high technological fields as engineering, textiles, wood-working, dairy and meat industries. Besides, Belarus became an important producer of mineral fertilizers, basic chemical and petrochemical products. About 2,100 large-scale and medium enterprises operate in Belarus with 63,000 point emission sources. Urban development and agglomeration tangibly increase adverse affect of mobile sources of pollution, responsible for almost 70% of total emissions in Belarus.

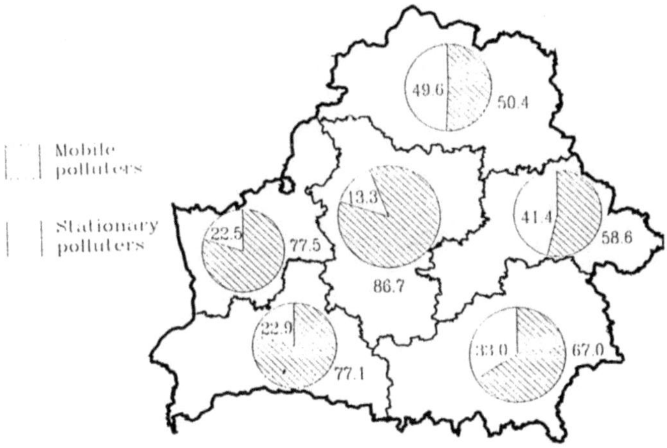

Figure 1. Atmospheric emissions in Belarus.

An increased content of some pollutants is largely associated with the amount of motor vehicles. In some cities air pollution from nitrogen oxides represent an acute problem. In recent years there has been no signs of improvement. Out-dated second-hand cars coming from abroad further aggravate urban air pollution. Noteworthy also are hydrocarbon automotive pollutants, namely, formaldehyde. Its emissions are several orders less than carbon oxide emissions, but produce concentrations exceeding prescribed urban quality standards.

Analysis of atmospheric pollution levels in Belarus centres from different industrial sources testifies that cities with heating, chemical and petrochemical plant sources are the most polluted. Along with these, spatial distribution of pollution in each industrial centre varies with the orographic, microclimatic conditions and the extent of industrial expansion.

Using distribution of background concentrations of nitrogen dioxide in Minsk as an example, it can be shown that the most polluted zone is in the south-eastern and central parts of the city. This is determined by the following factors:

(a) dominating west-quarter winds, which are responsible for carrying the pollutants from the entire city;

(b) general south-east inclination of the area, resulting in the inflow of pollutants from other localities;

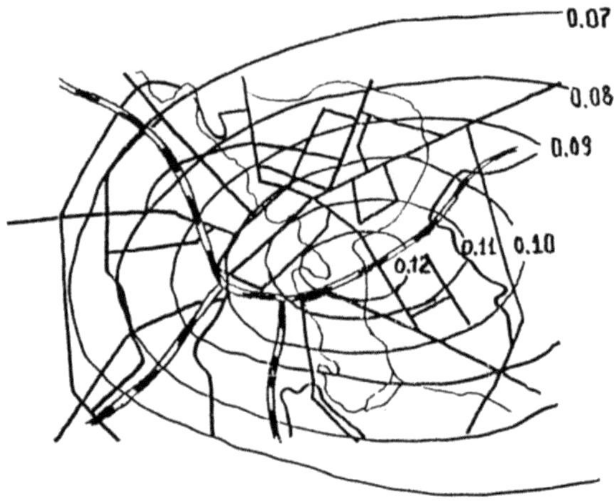

Figure 2. Background concentration field of nitrogen dioxide in Minsk.

(c) high industrial expansion of this part of Minsk;

(d) "Heat islands" are continually formed causing heat and pollution to be drawn in;

(e) intensive traffic.

Air pollution in industrial areas are 2 - 3 times higher than the background one in most of Belarus cities.

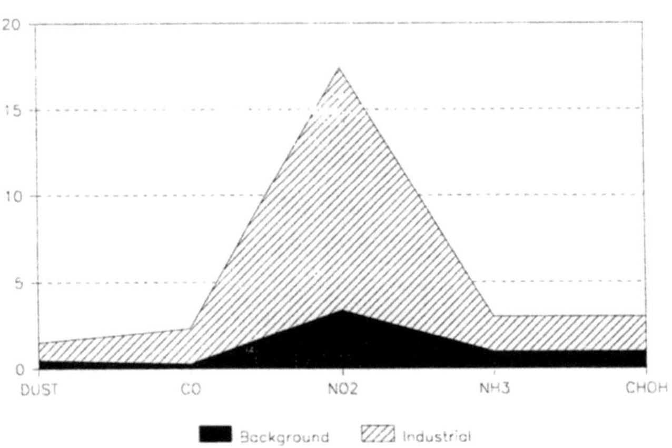

Figure 3. Frequency of exceedance of maximum concentration limit in the air of background and industrial areas of Minsk

Figure 3 illustrates the frequency of air quality standards exceedances both in industrial areas of Minsk and in the outskirts of Minsk, where light anthropogenic load determines lower dust content and where the limit prescribed for all pollutants, namely nitrgen dioxide is exceeded less frequently. In most industrial centres pollutants originate from low level sources. and they poorly dissolve in the troposphere. Therefore, low sources more often than the high ones are responsible for adverse urban ecology. Low emissions are often spontaneous, unaccounted for and neglected. Nevertheless, emissions of road and railway transport remain the major source of urban air pollution, which is considerably higher near railway stations and highways with intensive traffic. Figure 4 illustrates frequency of the exceedances near Vitebsk railway station and round the city. Close to the station dust, nitrogen dioxide, benzapyrene has been found to be from 3 to 5 times higher and increased formaldehyde content is noted.

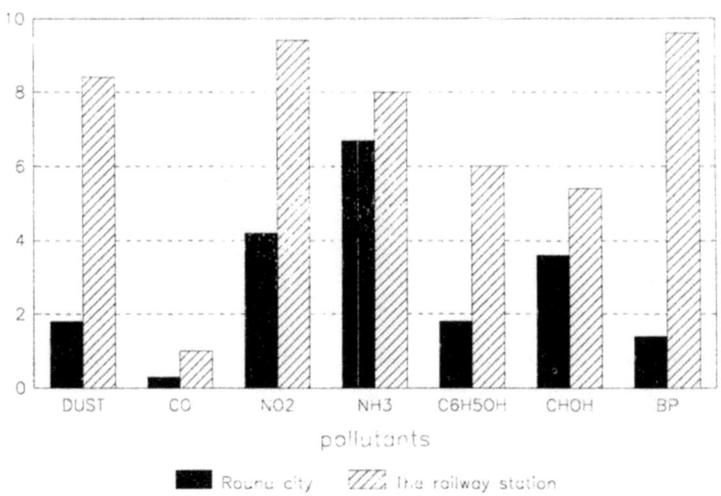

Figure 4. Frequency of excedance of MCL near the railway station and the residential area of Vitebsk

Long-term monitoring revealed that Mogilev is the most polluted city in Belarus. Large-scale chemical factories of the city release a great amount of specific pollutants. Many industries situated high up directly pollute the centre and residential sections on the lee side of the city. The situation is aggravated by the city orography (elevation difference is above 50 m) and its inner planning. The unfavourable conditions are most pronounced in summer time, when the frequency of gentle winds and stagnant conditions are higher (Figure 5). Sharp temperature rises in June - July also increases the carbon bisulphide content in urban air.

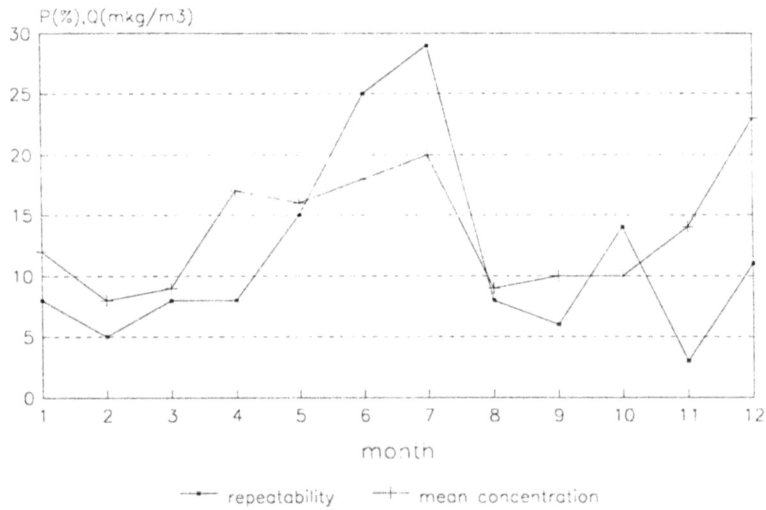

Figure 5. The annual trend of average carbon bisulphide concentrations and frequency of gentle winds in Mogivev

Meteorological conditions rather than the amount of emissions are sometimes more responsible for the average pollution level. Thus, in 1983, 1985-1986 the increase of pollution in Mogilev was determined by dominating southeasterly winds transporting carbon bisulphite from "Khimvolokno" chemical fiber plant, which repeatedly exceeded the acceptable limit. Thresholds for meteorological conditions of 1987 with cyclonic processes and variable rainy and windy weather facilitated surface dispersion of atmospheric pollutants. Gas cleaning system was introduced at the plant in early 1990. This caused a drop in total emissions of carbon bisulphite and related pollution level.

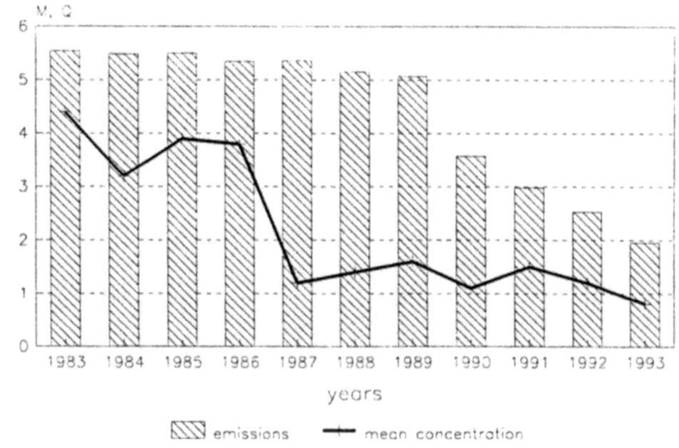

Figure 6 Emissions (Tg/year) and average concentrations (mg/m^3) of carbon bisulphide.

Large-scale chemical and petrochemical plants operate in some Belarus towns. Novopolotsk is one of them. The town was known for high specific emissions density per unit area prior to perestroyka. Petrochemical emissions exerted a pronounced effect on nearby Polotsk urban air. In the last four years the fall of total emissions in Novopolotsk has reduced the average pollution level in the town. Figures 7, 8 illustrate the decrease of average concentrations of phenol and ammonia due to the fall of total atmospheric emissions. The 1991 rise of average concentrations was associated with repeated unfavourable meteorological conditions.

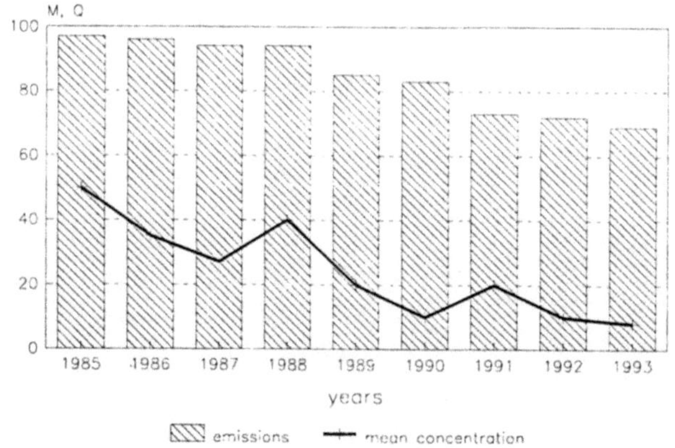

Figure 7. Emissions (Tg/Year) and mean concentrations (mg/m^3) of phenol in Novopolotsk.

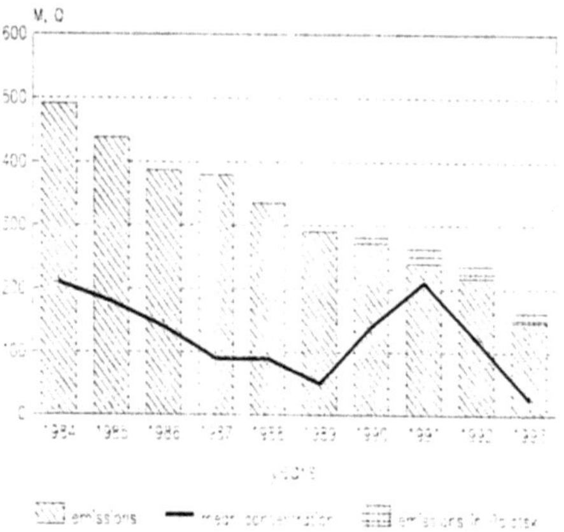

Figure 8 Emission (Tg/Year) and average concentrations (mg/m^3) of ammonia in Novopolotsk.

It is noteworthy that minor plants of Polotsk may also influence Novoplotsk urban air. Thus, the 1990-1991 increase of ammonia average concentrations in Novopolotsk initiated a close inspection of its emissions in Polotsk, which revealed new ammonia sources.

The amount of atmospheric emissions in Belarus has decreased sharply in the last few years basically due to the reduction of fuel supply to the Republic and long-term stoppages of industrial plants (Fig. 9). In 1992 the volume of emissions was 22,4% less than that in 1990. A similar trend is valid at present.

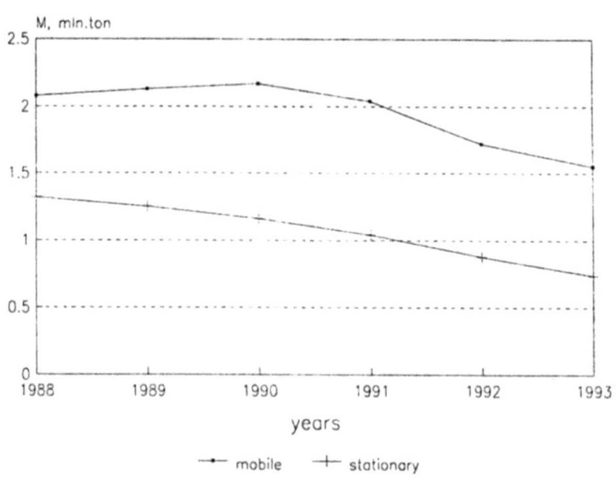

Figure 9. Dynamics of atmospheric emissions from mobile and stationary sources of pollution in Belarus.

To estimate the results of the change of air quality in industrial centres a complex air pollution index (API) is computed in the Main Administration of Hydrometeorology. The calculated API indicates to what atmospheric pollution level (e.g. in units of sulphur dioxide maximum concentration limit) the observed content of urban air pollutants is, i.e. it indicates how many times the total air pollution level exceeds the permissible value. Fig. 10 illustrates the dynamics of the API change over the past ten years (designed for the four most common pollutants). The analysis covered four cities with different industries and included Minsk with its numerous machinery construction, heating power and building materials plants; Brest, where automotive emissions constitute 90% of total pollution; Mogilev, where ferrous metallurgy, chemical and heating power industries have developed, and Novopolotsk, Fig. 10, indicates dissimilar API dynamics.

Figure 10 Dynamics of the air pollution index (API).

For the period of 1988-1993 the general trend for pollution decrease in Mogilev, Novopolotsk and Brest became evident as late as in 1992. In contrast, it increased in Minsk due to the long periods of unfavourable meteorological conditions of 1992. Formaldehyde produced by wood industry and essentially by automobile transport is the major pollutant contributing to the ASI value in Brest. The API growth in 1991 was due to the high intensity of international trasportation.

Input-output modeling data relationship for computer simulation of photochemical oxidants : some relevant examples in urban-industrial areas

J.C. Dechaux, V. Nollet, F. Cazier and F. Volckaert
Laboratoire de Cinétique et Chimie de la Combustion - URA CNRS 876
Université des Sciences et Technologies de Lille
59655 Villeneuve D'ascq Cedex - FRANCE

Introduction

The obsession of most researchers in the modeling of tropospheric chemistry, i.e. the formation of photochemical oxidants, can be put in a verbal form by the colloquial expression: "garbage in, garbage out". Let us understand its meaning. Despite the perfect physical structure of a given model for computer simulation, the poor quality of experimental field input data entails a defective running of even the best up-to-date tool. Consequently, the credibility of the results of computation becomes dubious.

In this paper we look at several categories of important input data: the nature and concentration of initial reactants (i.e. various oxidized nitrogen), the exhaustive (or not) processing of volatile organic compounds, and the best quantification of emissions of ozone precursors from mobile or fixed sources.

On the other hand, besides the question of input data quality the structure of the model is very important. For several years our team has used a lagrangian trajectory single box model; we will describe here some results obtained using this model, and also some preliminary results we recently obtained by attempting to transform the single box model into a polybox one. Finally, we will present very succintly, the direction we are now following which involves running a new eulerian model with close alliance of chemistry and transport.

NATO ASI Series, Partnership Sub-Series, 2. Environment - Vol. 8
Urban Air Pollution
Edited by Ivo Allegrini and Franco De Santis

The so called NO_x analysis

For years the overwhelming majority of worldwide analysts have used the gas phase chemiluminescent reaction of ozone with NO to perform NO titration.This method was developed by Fontijn et al. (1970) and Stuhl and Niki (1970) and applied to the determination of NO_2 according to Brestenbach and Shelef (1973). The low cost and power requirement of the chemiluminescence instrument made it attractive. Its reliability is related to the possibility of interferences. These possibilities were reported by Houlgate and Evens (1989), Bollinger (1983) and Drummond et al. (1985). As far as environmental issues are concerned, positive interferences could be very worrying. This is due to the behaviour of various atmospheric species when using the chemiluminescent detector in the so-called NO_2 mode, when NO_2 is determined as nitric oxide after passage of the sample stream through a thermocatalytic converter which reduces NO_2 into NO. Practically this method is used in all commercial chemiluminescent analyzers for NO_2. Commercial manufacturers have capitalized on this development and now offer a variety of chemiluminescent monitors which are based on the same principle and which differ only in design and configuration whatever their origin is. Numerous researchers worried about the possibility of interferences including nitrogen-containing compounds which may be reduced to NO in the thermocatalytic converter, thereby causing abnormally high signals for NO_x and hence correspondingly fake values for NO_2. From 1974 Winer, Peters, Smith and Pitts indicated a 100 % instrument response for PAN, whereas the HNO_3 response did exist but was non-quantitative. A close survey of the literature (far too long to be detailed in this paper) from that time reveals numerous reports from 65 researchers and concerning studies performed in various experimental conditions e.g. the nature of the converter, the working temperature, and others.

Unfortunately, in conclusion of this survey it appears that the situation is not so clear owing to some results which are more or less at variance with others and the various data to be considered were sometimes obtained by means of restricted studies, or only deduced from the interpretation of field studies performed by using different devices. Thus the state of the art requires more accurate and systematic data i.e. the investigations we describe now.

In the first stage we prepared controlled and calibrated continuous flowing gas streams, allowing ease of operation over a wide range of stable and adjustable concentrations, including very small ones. The response to NO_2 was considered first. This is presented in figure 1, which shows some stability between 200 °C and 450°C converter temperature but a sharp increase above 450 °C. The response to nitric acid is in figure 2 giving the so-called "NO_2 signal" versus the molybdenum converter temperature; one can see that at the manufacturer recommanded temperature (360° C) the response is remarkable. The absolute calibration of the HNO_3 generator (PFE teflon tubing immersed in pure liquid HNO_3, see Bowermaster and Shaw 1981) indicates that the ratio HNO_3 concentration to NO_2 signal is about 70 % at 360°C. Nevertheless we determined that a small quantity of NO_2 was emitted by this device (5 to 7 %); the calibration was made using the colorimetric sodium salicylate method (at 415 nm).

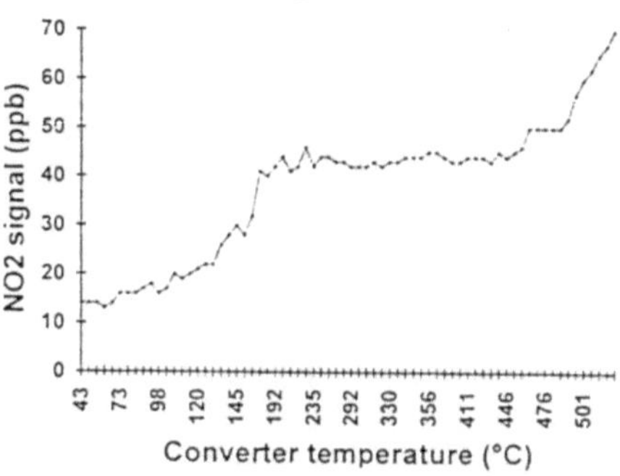

Fig.1: Chemiluminescence device response to NO_2 vs converter temperature

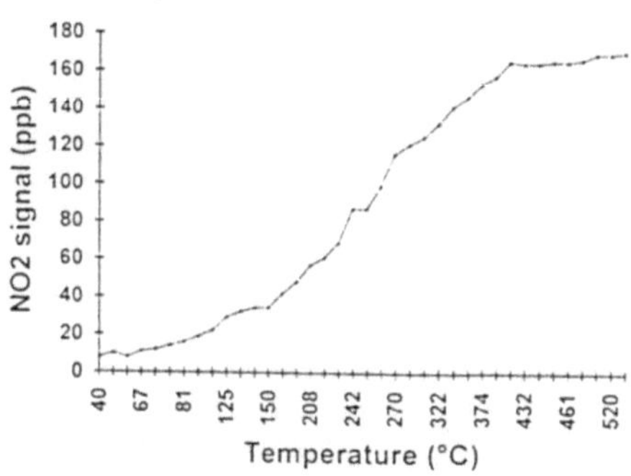

Fig.2: Chemiluminescence device responce to HNO_3 vs converter temperature

For nitrous acid we used the generation device proposed by Braman and de la Cantera (1986) involving the reaction of oxalic acid with solid sodium nitrite, and calibration by the Zambelli colorimetric method. The results are presented in figure 3. The figure shows a slight increase from room temperature to 500° C. The mean interference ratio is 70 % at 360° C. The PAN flow was generated by a zero air microcapillary bubbling into a dodecane solution. In figure 4a one can see a practically constant 100 %

response from 200 to 500° C. The "NO_2 signal" at 360 °C is plotted versus the absolute PAN flow concentration in fig. 4b.

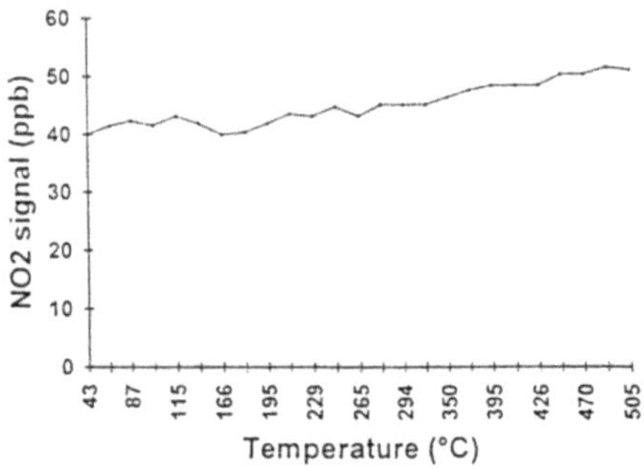

Fig.3: Chemiluminescence device response to HNO_2 vs converter temperature

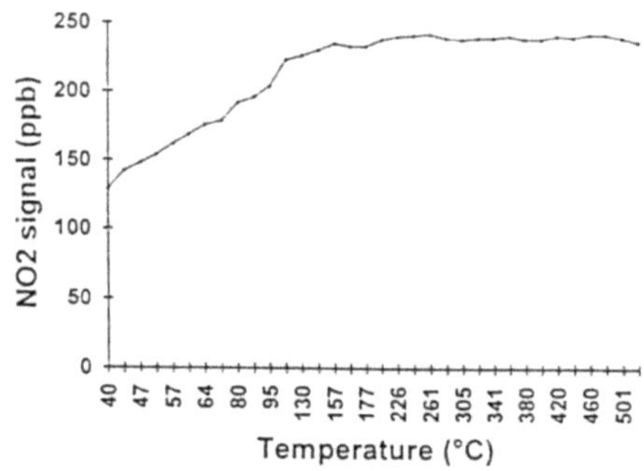

Fig.4a: Chemiluminescence device response to PAN vs converter temperature

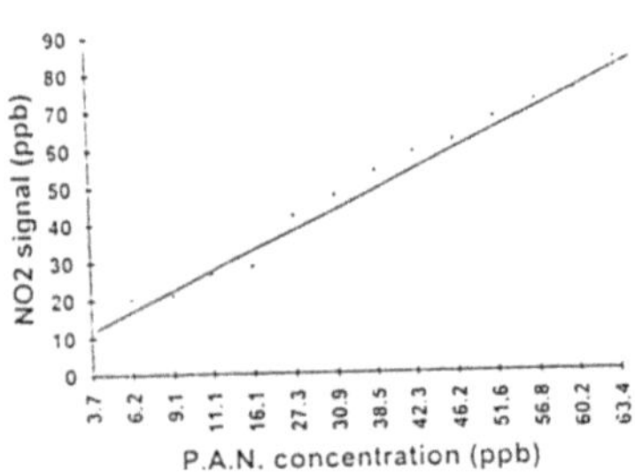

Fig.4b: Variation of the "NO_2 signal" at 360° C vs the absolute PAN concentration

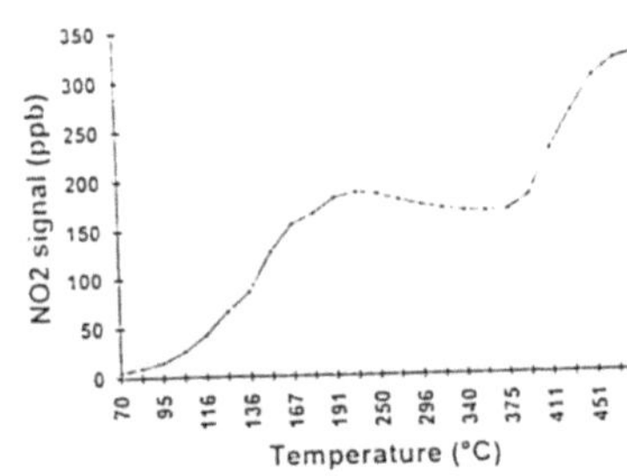

Fig.5: Variation of the chemiluminescence response to NH_4NO_3 vs converter temperature

The pattern of the ammonium nitrate aerosol, generated by using a compressed air nebulizer as described by Mercer et al. (1968), is rather special. In the figure 5 one can see that the "NO_2 signal" is important at 500° C, then passes by a minimum around 375° C, then a new maximum, then a sharp decrease.

The results reported above were obtained by using the French Environment S.A. equipped with a thermocatalytic molybdenum convertor. However, we believe that similar results can be obtained with similar NO_2 chemiluminescence monitors: response about 80 % for HNO_3, 70 % for HNO_2, 100 % for PAN and 60 % for NH_4NO_3. A most important question arises : what is the impact of this uncertainty on the results of modeling? Indeed it is quite usual to use casually the field measured NO and NO_2 concentrations as input data for the tropospheric models, data obtained by using these now diffused chemiluminescence devices which usually assimilate NO_2 to the actual balanced sum NO_2 + HNO_3 + HNO_2 + PAN + NH_4NO_3. According to the papers of Altshuller (1983) and Finlayson - Pitts and Pitts (1986) a rough estimation of current early morning order of magnitude of the nitrogen species considered (i.e. 3 ppb HNO_3, 1 ppb HNO_2, 0.5 ppb of PAN and 0.5 ppb of NH_4NO_3) can be made. When comparing these likely average values with the field data, fake NO_2 were measured during several campaigns in various areas in France and we ascertain that the non-NO_2 nitrogen species correspond to 15 - 20 % of the "NO_2 signal" we had obtained. Thus let us now present one series of a double simulation we made, using as input data for initial nitrogen species, either "fake" NO_2 concentration equal to 26 ppb or "put right" NO_2 which is 21 ppb accompanied by 3 ppb HNO_3, 1 ppb HNO_2, 0.6 ppb PAN and 0.5 ppb NH_4NO_3. The results obtained by using a lagrangian trajectory model fitted with a carbon-bond exp. chemical mechanism (Nollet et al. 1992) are presented in the figures 6a,b,c,d concerning the most important photochemical oxidants :

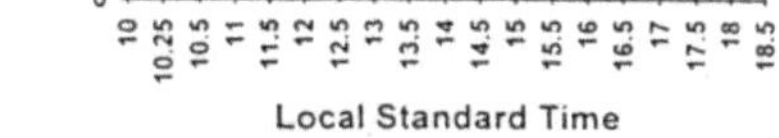

Fig.6a: Calculated time variation of ozone using either "fake" NO_2 or "put right" NO_2

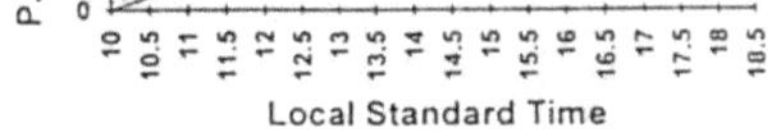

Fig.6b: Calculated time variation of PAN using either "fake" NO_2 or "put right" NO_2

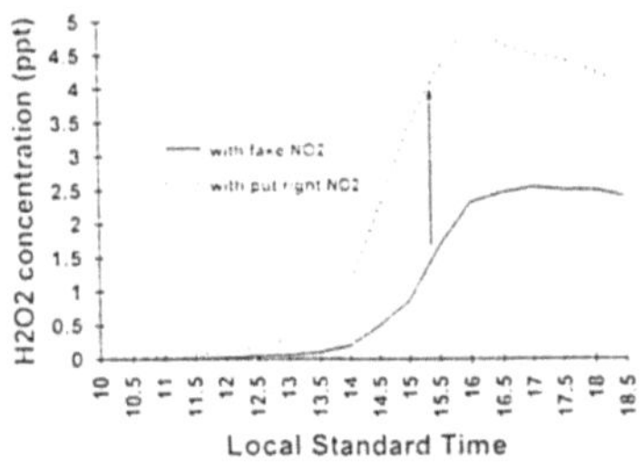

Fig.6c: Calculated time variation of H_2O_2 using for NO_2 input data either "fake" NO_2 or "put right" NO_2

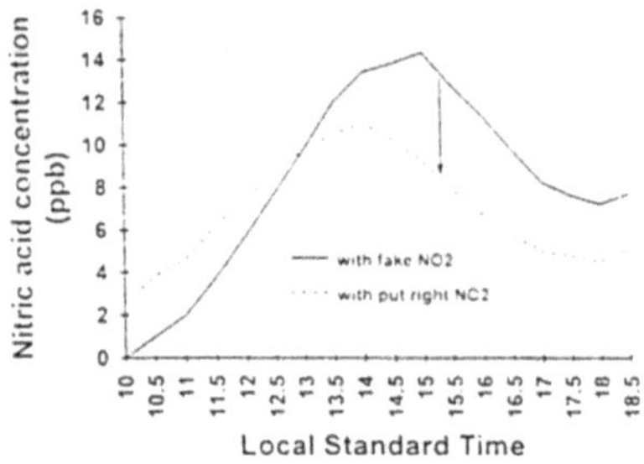

Fig.6d: Calculated time variation of HNO_3 using for NO_2 input data either "fake" NO_2 or "put right" NO_2

As far as ozone is concerned one can ascertain that a 11 % decrease circa is observed when fake NO_2 is used. This decrease is 23 % for PAN and 88 % for H_2O_2. On the contrary an increase of 30 % is observed for HNO_3. This example demonstrates the significant impact of the poor quality of input data when we use the up-to-date chemistry involved in a recent model.

The VOC analysis

Two possibilities may be anticipated: either one uses a F.I.D. global non-methane hydrocarbon apparatus or one uses a more sophisticated system. Everybody knows that the first approach is becoming less and less significant as far as modeling is concerned, due to the increasing complexity of the chemistry involved. The second approach lies in the utilization of a gas chromatography coupled to mass spectrometry analysis, preceded by a suitable sampling method. However, in this second procedure a question arises :i.e whether it necessary to titrate in an exhaustive way all the tropospheric hydrocarbon species or is it sufficient to titrate a restricted number of these species (for example 33) as was proposed three years ago (Kotzias and Hjorth 1991). To answer this question we made two categories of comparative model computer simulations. We limit ourselves here to present one result for which 73 % of the VOC concentration was GC/MS identified. The splitting of the VOC into various structural families of the carbon-bond mechanism we used, is reported in the first line of the table I :

	OLE	PAR	TOL	XYL	FORM	ALD	ETH	NR
all VOC	5.12	56.80	8.95	9.51	0.02	2.54	4.64	12.43
33 VOC	7.12	39.17	14.22	12.12	0.0	1.27	7.6	18.5

On the other hand if we decide to limit the analysis to the 33 species reported in the following list : methane - ethane - ethylene - acetylene - propane - propene - n-butane - i-butane - trans 2 butene - 1-butene - cis 2 butene - 1,3 butadiene - n-pentane - i-pentane - i-pentene - 2-pentene - isoprene - n-hexane - i-hexane - n-heptane - n-octane - i-octane - benzene - toluene - m+p xylene - o-xylene - ethylbenzene - 1,2,4 trimethylbenzene - 1,2,3 trimethylbenzene - 1,3,5 trimethylbenzene - formaldehyde - carbon monoxide, the percentages of the various families are largely different, as can be seen by comparison of lines 1 and line 2 of the table I. We notice that in the last case the ppmC concentration of VOC is reduced to 61 % of the concentration obtained with all the VOC. An example of the impact on the results of modeling is reported in the figures 7.

A sharp difference is observed especially for PAN, the concentration of which is multiplied by about 3.6 when all the VOC are considered, whereas O_3 and HNO_3 are multiplied by 1.85 and 1.3 (Fig. 7a,b,c) respectively.

Consequently, this is a serious decision to reach when choosing or not to titrate all the VOC species. This result led us to think that it is of interest to calculate the ozone percentage which can be attributed to various hydrocarbon precursors. We used a procedure connected to the one previously proposed by Hough (1986) Thus we performed computer simulations consisting of repeating calculations with the amount of one particular structural family decreased by a small fraction, a method applied

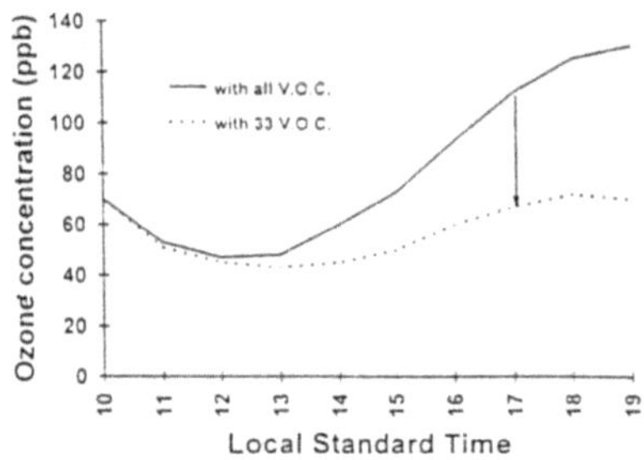

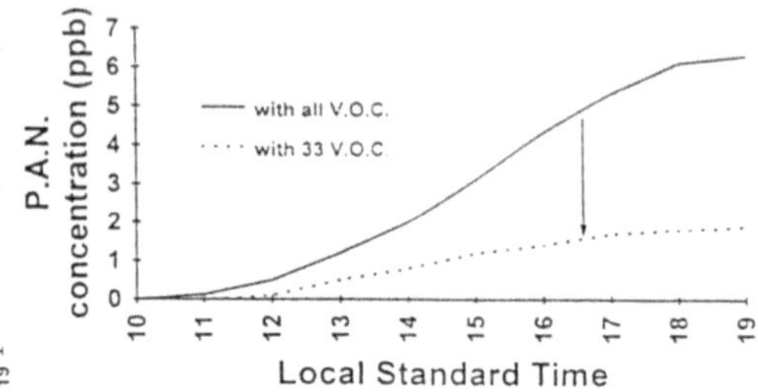

Fig.7a: Calculated time variation of ozone either with detailed measurement of all VOC or with only 33 VOC

Fig.7b: Calculated time variation of PAN either with detailed measurement of all VOC or only 33 VOC

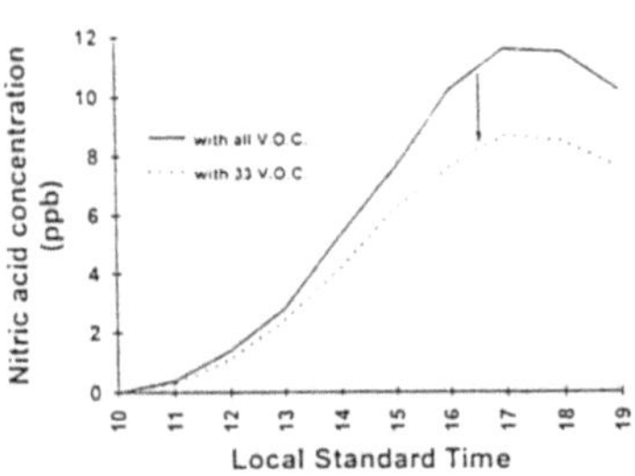

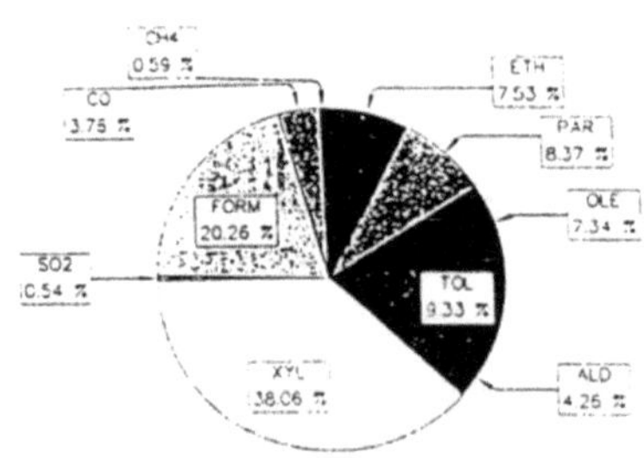

Fig.7c: Calculated time variation of HNO_3 either with detailed measurement of VOC or with only 33 VOC

Fig.8: The percentage of O_3 present in the air mass which can be attributed to the various VOC

to both the initial concentrations and to VOC emissions and then comparing the ozone production all day long in each model run. The average of the results for three campaigns in France is presented in the figure 8 which gives the 18.00 percentage of ozone present in a representative city plume, which can be attributed to the various structural reactivity families of the more recent carbon-bond mechanism. This dependence upon the whole families, which consider all the molecular species, is a further convincing argument in favour of the absolute necessity for those involved in the conception of control strategies of photochemical oxidants, to titrate all the VOC without limitations.

The quantification of ozone precursors emission

It is on purpose that we use the term quantification and not estimation. At the moment what do we know about emissions of NO_x and VOC from fixed and mobile sources? Till today this work is more often, than not, made from estimations,the emissions inventories being generally obtained by using emission factors, for lack of better experimental data. However, modeling is demanding and should be rigorous. The recent literature contains recommendations concerning this issue : the report of the French "Programme Environnement" of the CNRS (1992) recommends to determine massic emissions on an hourly basis for mobile and fixed sources. The CORINAIR inventory project (1992) of the European Commission is aware of the limitations of the results of the work they performed themselves and strongly encourage any initiative tending to develop knowledge of the emissions. The US EPA OAQPS report (1992) gave recommendations as follows: emissions must be provided for each individual cell of a grid system, specified as hourly rather than annual or daily rates, disaggregated into several classes of VOC, NO and NO_2. The report of the UK Photochemical Oxidants Review Group (Ozone in the United Kingdom (1993)) recommends to emphasize the development of refined inventories based upon better emissions data together with improved spatial dissagregation of source strengths and trend estimates to beyond the turn of the century. This lack is a major limitation for the development of reliable model calculations of photochemical oxidants concentrations. In consideration of these reports we began with a preliminary computer modeling test. In the following figure 9 are presented calculated time variations of ozone obtained either by using the most recent admitted standard French emissions or the results obtained if these emissions are multiplied by 2 for VOC, or for NO_x, or for both.

One can see that the influence of emissions has to be fully considered notwithstanding the moderate factor 2 we chosen in this test. Thus our team decided two years ago to go in the field and perform direct, at the source, measurements of emissions of ozone precursors in various industries in the area of Dunkerque, a

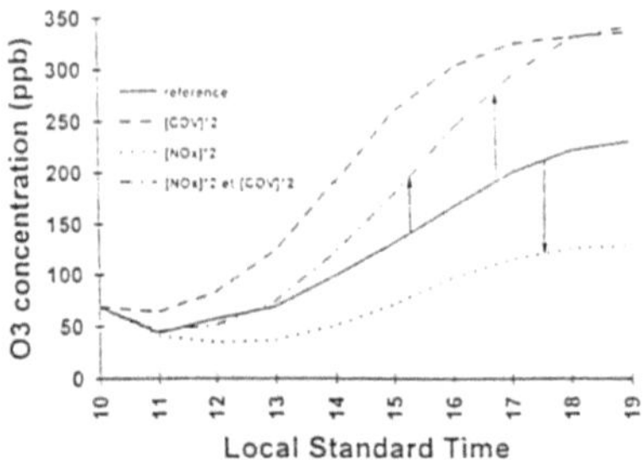

Fig.9: Calculated time variation of ozone according to the emissions of VOC and NO_x

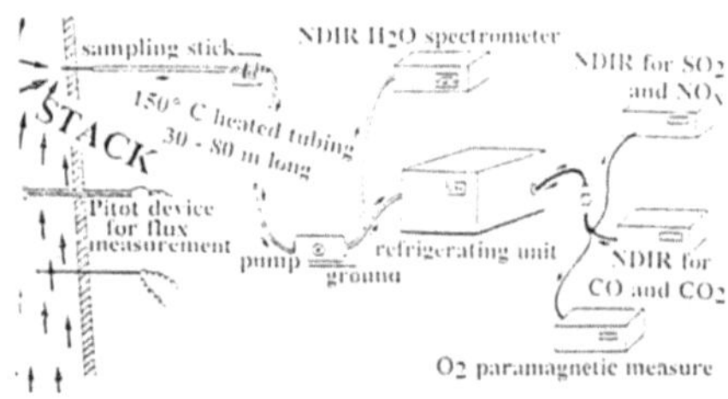

Fig.10: Schematic view of the device for sampling and titration of emissions

French northern industrial and urban zone near the Belgian border and on the north sea side, one of the most active among the industrial French areas. This field work, rather tiring and sometimes hazardous due to the high elevation of some stacks bound us to set up a mobile laboratory with sampling instrumentation lines and analytical devices, using a method the principle of which is presented in the figure 10. We limit ourselves here to present two examples of daily concentrations and flow variations of emitted pollutants concerning a waste incinerator and a power plant.

We compared our experimental measurements with expected values calculated from accepted standard equations used by Industry to declare their annual emissions in 12 different plants. We found that a multiplicative factor representing the differences between the two categories of method, lies between 0.05 and 2.5. with occasional good agreement between our experimental results and their official statement. Nevertheless these more or less important discrepancies are sufficient to entitle the scientific community to require actual emissions measurements support from representatives of research authorities, public administration and private entreprises.

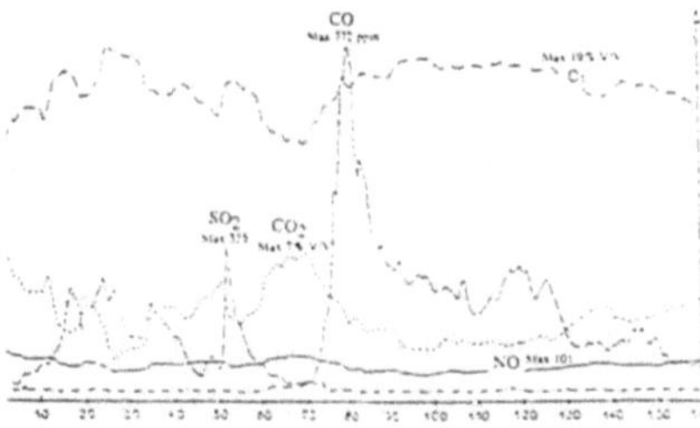

Fig.11a: Time variation of emissions from a waste incinerator (mn)

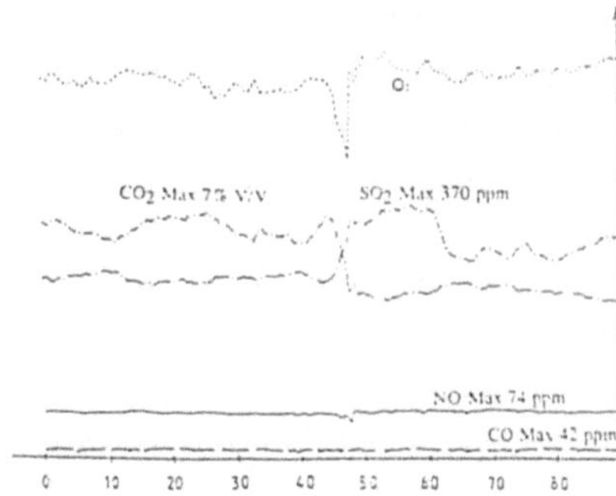

Fig.11b: Time variation of emissions from a thermic power plant (mn)

This discussion should be completed by debating with regard to the emissions issued from the traffic, which have a reputation to be at least as important as those from fixed sources. Those in the scientific community who have already tried to be rigorously successfull in this task, know how much effort has to be made to produce realistic emissions data due to the great complexity of the urban situations. Our team performed a study concerning Mexico City and the results obtained allowed us to define VOC, NO_x and CO hourly emission grids over this town. This was done with the aim of using the data for the modeling treatment of the formation of photochemical oxidants in the air of this city. It would take far too long to present this almost completed work in this paper.

The physico-chemical structure of the model

Our main purpose in this paper was to convince the Scientific Community involved in atmospheric research that it is essential to supply the troposphere models with as excellent as possible input data. But of course the other aspect of the problem has to be considered, we mean the specific quality of the modeling tool, i.e., its representativity dealing with the physical and chemical reality. For a long time our team has utilized a lagrangian single box model. As an evolution of this simple model, we tried to perform a polybox methodology as follows : at the very beginning of computation (i.e. in the early morning) we admitted that in the vertical direction a single box did exist, as far as temperature is concerned, because most often the mixing height is low (order of magnitude 50 to 100 m). Thus, the temperature difference between the ground level and the top of the box is equal to about one degree, a difference which could be neglected when comparing with the afternoon vertical temperature profile when the mixing height is at a maximum. Consequently our trial was as follows : at 9 a.m. only one box is considered and the computation begins. We stop it at 10 a.m. and divide the global box into several boxes according to the measured vertical temperature profile, and then separate computer runs are made in each box during one hour, till 11 a.m. We stop again at this time, and considering that efficient mixing occurs at that time, we begin again the computation in more boxes with identical initial concentrations but different temperature till 12 a.m. We stop again at this time, mix again the boxes, and so on all day long, according to the vertical temperature data coming from radio-soundings. By using this tentative approach we were convinced of the opportunity to use a polybox i.e. an eulerian sophisticated approach. We present in the figure 12 a,b,c,d, the compared calculated time variations of O_3, PAN, HNO_3 and H_2O_2 when considering or not the polybox approach. One can see that even though the difference is moderate as far as ozone is concerned (6 %), and a little bit more important for PAN (20 %), it is largely significant for HNO_3 and H_2O_2 : about 80 %. Considering these significant differences our team is at the moment continuing our research by running a new eulerian multibox model involving close alliance of chemistry and transport.

Conclusion

We expect that the Scientific Community involved in monitoring and control strategies for urban-industrial air pollution agree with us when we state that there is a great deal of wisdom in the colloquial expression "garbage in, garbage out" which means we should feel more interest in order to obtain input data of the best quality for modeling. This is of course dependent on analytical chemistry achievements. Indeed, it is generally recognized that the control strategies for urban and suburban photochemical smog must rely on the support of rational modeling.

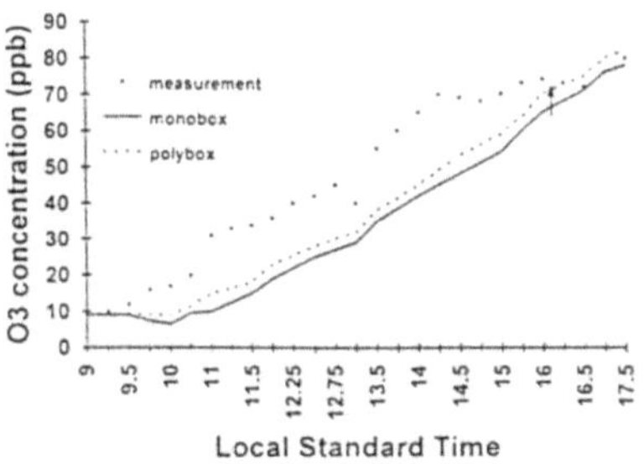

Fig.12a: Calculated time variation of ozone according to a mono or a polybox approach

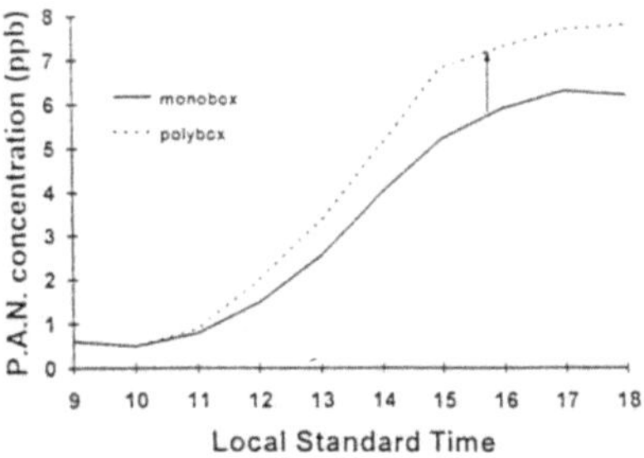

Fig.12b: Calculated time variation of PAN according to a mono or a polybox approach

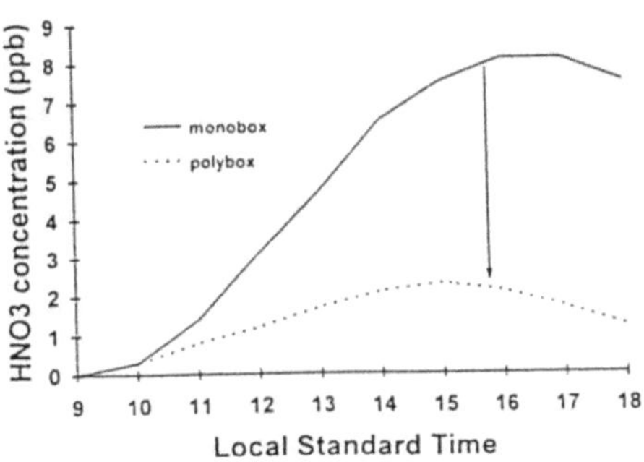

Fig.12c: Calculated time variation of HNO_3 according to a mono or a polybox approach

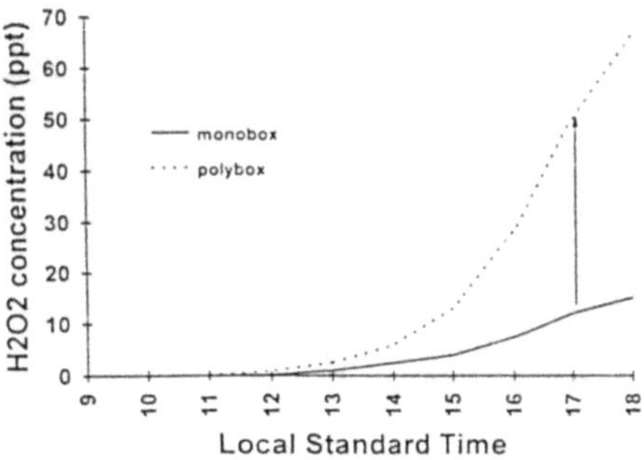

Fig.12d: Calculated time variation of H_2O_2 according to a mono or a polybox approach

If this tool is largely validated it will make a positive contribution to define the best, and realistic, reductions in the emissions of ozone precursors to meet air quality guidelines defined by the European Community directive, or the WHO, or the US EPA.

Moreover, we must add that the modeling tools must meet high standards concerning their physical and chemical structures. It is only the combination of these two constituants: input data quality and model quality which will allow atmospheric science to progress.

References

Altshuller A.P., 1983, Measurements of the products of atmospheric photochemical reactions, Atmos. Envir., 17, 2383 - 2427

Bollinger M.J., Sievers R.E., Fahey D.W. and Fehsenfeld F.C, 1983, Conversion of nitrogen dioxide, nitric acid and n-propylnitrate to nitric oxide by gold catalysed reduction with carbon monoxide, Anal. Chem., 55, 1980 - 1986

Bowermaster J. and Shaw R.W. Jr., 1981, A source of gaseous nitric acid and its transmission efficiency through various materials, J. Air Poll. Control. Assoc., 31, 782 - 787

Braman R.S. and de la Cantera M.A., 1986, Sublimation sources for HNO_2 and other nitrogen compounds, Anal. Chem., 58, 1533 - 1537

Brestenbach L.P. and Shelef M., 1973, Development of a method for the analysis of NO_2 and NH_3 by NO measuring instruments, J. Air Poll. Control. Assoc., 23, 128 - 131

Corinair Inventory Report, 1992, Default emission factors, Commission of the European Community

Drummond J.W., Volz A. and Ehhalt D.H., 1985, An optimized chemiluminescence detector for tropospheric measurements, J. Atmos. Chem., 2, 287 - 306

Finlayson-Pitts B.J. and Pitts J.N. Jr, 1986, Atmospheric chemistry, fundamentals and experimental techniques, p. 42, John Wiley and Sons Ed.

Fontijn A., Sabadell A.A. and Ronco R.J., 1970, Homogeneous chemiluminescence measurement of nitric oxide with ozone, Anal. Chem., 42, 575 - 579

Hough A.M., 1986, The production of photochemical pollution in Southern England and the effects of vehicle exhaust emission control strategies, Aere Report R-12069 - Harwell

Houlgate P.R. and Evans W.H., 1989, Artifactual chemiluminescence in the determination of NO in the vapour phase, Analyst, 114, 71 - 75

Kotzias D. and Hjorth J., 1991, Working Group VOC measurement techniques, final proposal on sampling and analysis of selected VOC, Commission of the European Community, Ispra

Mercer T.T., Tillery M.I. and Chaw H.Y., 1968, Operating characteristics of some compressed air nebulizers, Amer. Indus. Hygiene Assoc. Journal

Nollet V., Déchaux J.C. and Volckaert F., 1992, Simulation numérique de l'impact de la nature des COV sur l'ozone troposphérique - un nouveau type de stratégie de contrôle, Critical issues in the global environment, Atmospheric Chemistry, vol. 2, IUAPPA and Air and Waste Management Assoc. Ed.

Ozone in the United Kingdom, 1993, Third report of the UK Photochemical oxidants review Goup, Air Quality Division, Department of the Environment

Programme Environnement du CNRS, 1992, Rapport final du groupe d'experts, Lettre Environnement n° 8, CNRS Paris

Stuhl F. and Niki H., 1970, An optical detection method for NO in the range 10^{-2} to 10^{-3} ppm by the chemiluminescent reaction of NO with O_3, Ford Motor Company Dpt, Dearborn, Mich.

US EPA Office of Air Quality Planning and Standards, Emission Inventory Branch, 1992, Emission inventory requirements for photochemical air quality simulation models, 454/R- 92-026

Winer A.M., Peters J.W., Smith J.P. and Pitts J.N. Jr, 1974, Response of commercial chemiluminescent NO - NO_2 analysers to other nitrogen compounds, Envir. Sci. Technol., 8, 1118 - 1121

Advanced observation methodologies of urban atmospheric boundary layers

A. Longhetto[*o] - C. Giraud[o]

* - Istituto di Fisica Generale dell'Universita` di Torino - Via P. Giuria 1 - 10125 Torino - Italy

o - Istituto di Cosmogeofisica - C.N. R. - C.so Fiume 1 - 10133 Torino - Italy

1 - Introduction

Before entering upon the problem of observation methodologies of the urban atmospheric boundary layers, it is opportune to trace back a few outstanding features of natural PBLs, developing at the Earth's surface in rural regions.

The Earth's PBL, resulting from a balance between turbulent inertial and Coriolis terms (Ekman layer), is almost constantly evolving in response to the diurnal heating cycle and to the change of synoptic conditions. As a result, its depth and structure can vary a lot over space and time.

The usual classification of PBL states is based on buoyancy effects, connected to different vertical temperature profiles. These effects allow us to define three main stability classes, called unstable, stable and neutral respectively.

A very simplified view of the vertical structure of a convective, unstable, PBL makes three distinct regions appear. The surface layer, near ground, has strong gradients in the wind, temperature and moisture; it is typically a few tens of meters deep, and the stronger the convection is, the thinner and more intense it becomes. Above it, extending nearly to the inversion base at $z_i \cong u_* / f$ (the scale height of the neutral Ekman PBL), or lower, is

NATO ASI Series, Partnership Sub-Series, 2. Environment – Vol. 8
Urban Air Pollution
Edited by Ivo Allegrini and Franco De Santis

the mixed layer, within which the mean gradients are usually smaller. This layer grows during the day due to solar heating, and a good estimate of the time development of its height has been given by Carson (1973): $h(t)=\left[2\int_{t_o}^{t} H_T dt \Big/ \rho c_p(\gamma_a-\gamma)\right]^{1/2}$, where $\int_{t_o}^{t} H_T dt$ is the integral of the sensible heat flux H_T from dawn (t_o) to time t, and g is the stable lapse rate at dawn. Above this, is the inversion layer, where the turbulence is damped by stable stratification and the profiles approach the condition of the free atmosphere, where there is geostrophic balance.

When the solar radiation flux is very high and the wind is very low, the mixed layer can be considered as made up of two layers: the free convection layer, covering the height range $-L<z<0.1z_i$, and the proper mixed layer, extending to z_i. In the free convection layer, the turbulence is no longer influenced by u_*, and a new velocity scale has to be introduced; this is the convective mixed layer velocity $w_*=\left[\frac{g}{T_o}H_T z_i\right]^{1/3}$, by which the turbulence statistics has to be scaled. In the mixed layer, neither u_* nor z influence the turbulence any more and the only height scale is z_i. Beneath the surface layer, a very thin (order of $\cong 5\nu/u_* \cong 10^{-1}$ mm) viscous sub-layer exists (mainly over near calm water or very smooth surfaces), within which the molecular viscosity n prevails over the eddy viscosity ($\cong u_*$).

When the PBL has an adiabatic temperature vertical profile (corresponding to a lapse rate of $10^{-2}\,\mathrm{K\,m^{-1}}$, or potential temperature constant throughout), which can happen during windy and overcast weather conditions, the turbulence is due entirely to the wind shear. This situation gives rise to a neutral PBL, which establishes its depth $H \cong u_*/f$, as we have already seen.

Finally, when radiative heat losses of the Earth's surface during night-time bring about positive temperature vertical gradients up to a height $z_i'=a\sqrt{t}$, where t is the time elapsed from sunset (hours), we have the so- called stable boundary layer. z_i' is also called the depth of the ground-based nocturnal inversion layer.

The depth z_i' of the stable boundary layer can be divided in three sub-layers; the first one, ranging from the surface to H_s, is the proper turbulent layer. Above this, is a region with very low, intermittent turbulence (sporadic layer) and higher up is another turbulent region where dynamic instability is brought about by wind shear at the height z_i (residual layer). This vertical structure is today well detected by remote sensing instruments, like Doppler Sodar.

Any progress in the knowledge of the statistics and stability conditions of the atmospheric boundary layer means a progress in the modelling of pollutant dispersion. Limiting ourselves to the simple gaussian model, some improvement can be obtained by introducing in it more realistic dispersion coefficients, derived from relationships between turbulence statistics and pollutant dispersion. Furthermore, if atmospheric stability can be assessed in a satisfactory way, it is possible to evaluate the distribution of wind and diffusion coefficients with height, in the different layers of the PBL, and then to improve the dispersion models.

2 - The Urban Boundary Layer (UBL)

After this short introduction to the natural atmospheric PBL, we turn our attention to the modifications induced in this layer by large urban areas.

Just as the antropic effects on climate are the result of interferences in the operation of natural systems, so the urban atmospheric environment is the result of three dimensional adjustment to alterations, induced in the natural surface and atmosphere by massive urban structures, which sometimes exhibit the signs of inadvertent climate modifications. This adjustment, which involves the transformation of natural aerodynamic, radiative, thermal and moisture regimes of the atmospheric environment of a region, appears under manyfold aspects, The best documented aspect is the so-called "urban heat island", while the less known is the vertical structure of the PBL in this heat island.

The nocturnal boundary layer, which is characterised by the existence of a definite elevated temperature inversion above the urban area, together whith urban heat island play important roles in urban pollution. This occurs mainly during clear night-time, when the lower surface loses energy, and the vertical diffusion is limited, leading to high ground-level concentrations .

The main causes of alteration of the urban atmospheric environment at the surface include (Oke, 1982): reduced evaporation from more impervious urban surfaces, increased heat storage in urban material, anthropogenic heat and water input associated with combustion and urban water supply, increased atmospheric I.R. radiation from polluted urban atmosphere, and enhanced short-wave radiative absorption along with reduced net long-wave radiative losses resulting from complicated urban geometry. Other causes of alteration of UBL come from short-wave radiative absorption from atmospheric pollution, sensible heat input from the warm area below, and increased sensible heat entrainment from above, as a result of more developed urban turbulence.

Quantitative estimates of urban effects would require, ideally, extensive sets of *pre-urban* measurements of the regional climate, to be compared with the present, *urban*, situation. Instead, what could be done is to just compare climatic data from the downtown of an urban area with those from its rural surrounding areas. Even if this is the best we can do, nevertheless in most cases it is only an approximation of the actual urban modifications. Particular attention must then be paid when selecting the rural or sub-urban stations. In fact, they should *not be influenced* by extraneous effects due to local topography features, water bodies and downwind advection from the city.

Just as an urban settlement brings about alterations on the local pre-existing climate and atmospheric environment, which allow one to speak of urban climate, so changes in mesoscale weather conditions, or in local climate, can induce changes in the turbulent structure of the UBL, independently of surface conditions in the urban site (Yoshikado and Kondo, 1989; Westcott, 1989). This point is very important, because it opens a challenge to the design of modern cities against inadvertent climate changes, which can alter the local wind regimes and air stability conditions and, as a consequence, the turbulent structure of the UBL and its depth and, in doing so, the mechanical and thermal energy exchanges through it.

The *urban boundary layer* is a local to meso-scale phenomenon whose characteristics are governed by the nature of the general urban "surface". Its climate is an ensemble of micro-climates each of which is dominated by the characteristics of its immediate surroundings (Oke, 1987).

The most distinctive feature which makes the main differences between the urban and the rural BLs is the particular structure of the roughness elements of the lower boundary of the former, The ULC gives rise to a new layer, the Urban Canopy Layer (UCL). This UCL extends vertically between the level of zero net heat flux in the ground up to an upper level, beneath which all structures of the surface contributing to the energy storage of the canopy layer lie (Kerschgens and Kraus, 1990). As a matter of fact, the top of the UCL is situated above roof level, like the rural canopy layer is situated above the crop, or the grass or the trees.

The existence of the UCL involves the need to carefully assess the energy storage in it and the energy exchanges with the UBL which, under fair weather conditions, extends from the top of the canopy layer up to the base of the subsidence inversion.

As it appears from the above general consideration, the main areas of study to implement in order to build up predictive and diagnostic models of UBL structure have to be addressed toward the assessment of thermodynamic energy budget equations at its surface and inside its canopy layer, characterized by the so called "urban canyons", and toward observations of its vertical structure. This report will address the second aspect of the problem, by considering the capability of advanced remote sensing techniques in obtaining the required information.

3 - Modification induced by buildings on the surrounding environment.

Any placement of a building on the landscape gives rise to radiative, thermal, moisture and aerodynamic modification of the surrounding environment. The most important radiative effects are local decreases and increases in the solar radiation input due to the following: shadowed areas, reflection from sunny walls, and reduction of net long-wave cooling from surfaces near the building caused both by a decrease in upward I.R. flux (due to the reduced sky view factor) and to an increase in downward I.R. flux (due to warm contructions). Also, the wind shelter provided by buildings contributes to raised soil and air temperatures in the UCL.

The water balance around a building is perturbed because of spatial variability in soil drainage, evaporation and precipitation collection due to differences in interception and wind shelter. Most important, however, are the perturbations of the airstream, produced by the building acting as an obstacle to the wind: the net effects of this perturbation are a decrease in average wind speed, an increase of turbulence, and a distorsion of vertical wind profile at the top of the building.

An exception to this behaviour is exhibited by tall buildings: the large momentum brought downward by the impinging air flow makes the wind speed increase at ground level, with values up to 3 times as high as those which could have been observed in the absence of the tall obstacle: in this case, both average wind speed and turbulence increase at the surface (Oke, 1987).

Moving toward interactions of more than one building, different flow patterns appear, connected with buildings of equal height H with different spacings W, allowing or preventing the cavity circulations from exchanging air with the overlying flows. An additional effect of tall buildings is their downwind suction zone, producing important downwash effects and local cavity circulations, of particular relevance as far as atmospheric pollutant dispersion is concerned, and whose pattern strongly depend on the H/W ratio.

In order to assess the energy and water balance in urban environments, we have to define a "building-air volume". This volume is vertically limited by the urban boundary layer depth and the balance of energy and water is assessed at a horizontal surface at about roof-level.

The energy and water balance equations now contain terms which are directly controlled by human decision. For example, the already mentioned heat and water sources in the city associated with combustion, urban water supply piped in the city from rivers or reservoirs and runoff which can upset the other, natural, terms of the balance and their seasonal trends.

We have mentioned before that some important modifications which an urban environment can induce on the radiation budget are represented by short-wave radiation attenuation, due to scattering and absorption brought about by urban air pollution, and by reduced albedo of the surface (about 0.15). As a consequence of the partial compensation of these competing effects, urban/rural net short-wave radiation differences are not large (except for snow-covered conditions).

With regard to the long-wave radiation budged in a city, as the urban area is usually warmer than its environs at night (urban heat island effect), it emits more long-wave radiation to the atmosphere. The return long-wave flux from the atmosphere is also greater in the urban area because of the overlying pollution layer. As with the short-wave terms, the changes in the long-wave radiation fluxes tend to oppose each other, therefore urban/rural differences of the absorbed radiation are small. At night the loss is likely to be greater from the city.

4 - Urban heat island and urban boundary layer climate

In a previous section, the *urban heat island effect* has been related to a particular kind of inadvertent climate modification. Even if this subtle and important subject-matter is outside the scope of this report, nevertheless we shall give a short review of the main peculiarities of urban climatology.

The characteristic horizontal trend of air temperature from the countryside to the downtown of an urban area exhibits a steep gradient (as great as 4 K km^{-1}), or *cliff,* at the rural/urban boundary, followed by a *plateau* of warm air with weaker horizontal gradient of increasing temperature towards the center of the city, where a maximum temperature is found. The difference between this value and the background rural temperature defines the *urban heat island intensity* ΔT_{u-r}.

Urban heat islands can extend their effects even in the nearest surrounding countryside and, under stationary weather conditions, their intensity can vary in a peculiar way through the day, with a reduced cooling downtown in the late afternoon and evening, resulting in a higher nocturnal minimum temperature in the city. The urban area also shows slower warming-up after sunrise, followed by a more rapid growth of ΔT_{u-r} just after sunset (reaching a maximum 3 to 5 hours later). Usually the intensity relaxes then through the rest of the night and is rapidly dispersed after sunrise. In some cities ΔT_{u-r} may even be negative in the midday period, i.e. urban centre can be cooler than the countryside.

This simplified diurnal picture is considerably modified by changes in weather conditions. At night, when the heat island is best developed, ΔT_{u-r} is inversely related to wind speed $(\bar{u})$ and cloud cover, being greatest under the conditions of weak winds and cloudless skies. The heat island intensity is most sensitive to wind speed, probably due to turbulent and advective activity.

With reference to the urban *structure*, represented by an index given by the already discussed ratio H/W, a strong relation has been found between this index and the maximum heat island intensity, given by: $\Delta T_{(u-r)\max} = 7.54 + 3.97 \ln(H/W)$.

Other interesting considerations, concerning the vertical structure of a UBL, can be suggested by vertical profiles of potential temperature observed early in the morning at different along-wind distance in large urban areas. In addition to the warming effects coming from anthropogenic heat emanating from the UCL, buildings roofs, chimneys and industrial stacks, and transported upwards by turbulent diffusion, it also seems that some heat can enter the UBL from above: in fact, both by day and by night, the urban layer is capped by an inversion and hence there is warmer air aloft.

Thermal modifications of the UBL in presence of weak horizontal wind advection from the surrounding countryside give rise to heat island thermal breeze systems and converging motions with uplift in the downtown region.

Downwind of the city, the lower roughness of rural surfaces can favour some subsidence, but an elevated *plume* of rising air may persist for tens of kilometres.

During the day, the influence of a large city may extend up to 0.6-1.5 km, virtually extending throughout the whole PBL, while at night the urban heat island contracts to a depth of only 0.1-0.3 km, because the bulk of the PBL is stable and this suppresses vertical transfer.

As a result of the combination of urban warmth and increased forced convection, the urban atmosphere, unlike rural areas, does not experience a strong diurnal change in stability. Both by day and by night the urban layer is well mixed, and this tends to destroy vertical temperature gradients and therefore both strong instability and stability are absent. Another climatic feature peculiar of UCLs is the canopy layer distribution of water vapour pressure during a summer night at about midnight with almost ideal conditions. The night-time humidity excess in the city, especially in summer, though rather small with respect to the rural values, can exhibit a *moisture island* similar to that of temperature: the humidity island also has a *cliff* which corresponds well with the urban-rural boundary, while in the daytime (when the city air is drier) the spatial pattern is more amorphous.

5 - Doppler radar evaluation of vertical flux of turbulent kinetic energy

We now come to examine some recent experimental approaches to the problem of measuring the most relevant physical quantities which characterize the structure of the UBL. Due to its importance in prognostic models of boundary layer structure, we start with the flux, or transport, of turbulent kinetic energy (TKE) in the urban boundary layer. This is difficult to measure, not only because it requires fast response instruments (ultrasonic anemometers) installed on a mast (whose height is forcedly limited) or on an aircraft, but also because it requires calculation of third-order moments of turbulence over long averaging times (Wyngaard, J. C., 1973), which means significant losses in the kinetic energy budget (Wyngaard, J. C., 1984).

An interesting solution to this problem has been suggested by Frisch et al., 1989, who showed the capability of a NOAA-WPL X-band Doppler radar, with suitable scanning and analysis technique (VAD, Velocity Azimuth Display), to provide vertical fluxes of TKE at a fixed height over averaging times long enough to satisfy the homogeneity and stationary assumptions. This represents an extension of Doppler radar capability to derive boundary - layer profiles of turbulent wind stress, variances and mean wind speed and direction (Lhermitte, 1968; Kropfli, 1986).

The operation principle of the Doppler radar is the usual one, based on measurement of the Doppler shift of diffused radio-echoes from naturally-occurring scatterers (like seeds, insects and other millimetre-sized particles) which are faithful tracers of instantaneous air motions.

The VAD technique can be summarised as follows. When the radial velocities (VR), derived from the above mentioned Doppler shifts, and their variances $\left\{V_{ar}\left[V_R\right]=\overline{\left(V_R-\overline{V_R}\right)^2}\right\}$, are measured by scanning the radar at a constant elevation angle (J) above the ground (conical scan around a circle), it is possible to related them (Browning and Wexler, 1968) to the speed and direction of the wind at a fixed height above the ground level and to its second moments $\overline{u'^2}, \overline{v'^2}, \overline{w'^2}, \overline{u'v'}, \overline{v'w'}, \overline{u'^2+v'^2}$.

The condition of horizontal homogeneity of statistical quantities, which is necessary for the validity of the above relationships, can be attained by averaging several scans, taken during periods of 20-30 minutes (Kropfli, 1986).

It has been shown (Frisch et al., 1989) that the VAD technique for obtaining first and second moments of radial wind velocities (and the related first and second moments of the components of the wind field) can be expanded to third-order moment calculations (turbulence transport terms).

Under the assumption of horizontal homogeneity of turbulence, the equation expressing the relationships between $\overline{\left(V_R-\overline{V_R}\right)^3}$ and the third moments of the components of the wind field can be integrated with respect to azimuth b, from 0 to 2p (conical scan), yielding:

$$\int_0^{2\pi}\overline{\left(V_R-\overline{V_R}\right)^3}\,d\beta=3\pi\left[\overline{u'^2w'+v'^2w'}\right]sin(\theta)\cos^2(\theta)+2\pi\overline{w'^3}sin^3(\theta)$$

One can immediatly see that, for J = 50.8°, the coefficients on the right side are equal and can be factored out. This means that by scanning the radar antenna at a zenith angle J = 50.8°, the vertical flux of TKE can be directly calculated from the above equation as:

$$\frac{\overline{w'\left(u'^2+v'^2+w'^2\right)}}{2} \equiv \overline{w'q'} = (1.86\pi)^{-1}\int_0^{2\pi}\overline{\left(V_R-\overline{V_R}\right)^3}\,d\beta$$

A comparison of the vertical profiles of TKE flux (obtained by operating an X-band Radar in a scanning sequence of circles at elevation angles of 35.3^o and 50.8^o every 380s, at a pulse length of 112.5 m, with range gate spacing of 70m from 0.20 to 4.05 km and with a beam width of 0.8^o) with aircraft measurements made by other investigators indicated that the radar technique gives reasonable flux values (Frisch et al., 1989).

Another interesting result obtained by this comparison is the possibility to assess the inversion height, which turns out to correspond to the maximum height where the scatters can be found in the PBL. As a matter of fact, the vertical flux of TKE measured by radar reaches a minimum near the inversion (z/z_i = 1), and this is in good agreement with the condition that the vertical flux of TKE must go to zero near the inversion height, unless entrainment processes or gravity waves are active at that level.

6 - Doppler Sodar evaluation of mean wind and turbulent fluxes in the UBL

The Doppler Sodar is a development and improvement of the classical acoustic radar, in which the analysis of the Doppler shift of the backscattered echo makes it possible to obtain both the vertical profile of average horizontal wind speed and direction and the statistical moments of the vertical component of the wind velocity even under calm or very low wind conditions. This prevents a satisfactory structure function from being formed.

Doppler Sodar measurements still deserve great attention; in fact, information they can provide on the whole turbulent and average vertical structure of stationary and non-stationary PBLs allows not only an improvement of their diagnostic and prognostic models, but also an increase of reliability of pollution and heat exchange models, through a more realistic and effective calibration of their input parameters. For this reason, comparisons between Doppler Sodar and fast response, fixed point instruments are still crucial and up-to-date.

Recently, Keder et al. (1989) have presented the results of such an intercomparison between a Sensitron Doppler Sodar (2300 Hz sounding frequency), a Kaijo-Denki DAT 300 acoustic anemometer, and a 20 m platinum-wire fast thermometer.

Starting from the mean horizontal wind at a fixed height (80 m), comparisons of 30-min averages Sodar and acoustic anemometer wind speed data showed that Sodar underestimated the wind speed by 20 % with respect to the acoustic anemometer, and that this underestimation could be even larger for wind speeds exceeding 10 ms^{-1} (due to the noise).

As far as the standard deviation of the vertical wind velocity was concerned, 30-min averages of Sodar data showed that the latter overestimated the acoustic anemometer measurement below $\sigma_W = 0.8 ms^{-1}$, and underestimated them above this threshold.

However, when 10-min averaged data were compared, the agreement between Doppler Sodar and Acoustic Anemometer appeared to be better, but with a poorer correlation between the two series of data (r = 0.68, instead of r = 0.82 for the 30-min averaged data).

With reference to the thermal structure of the PBL, the comparison between 30 min averaged Doppler Sodar echo amplitudes <E> and the temperature structure parameter $C_T^2 = <[T(x) - T(x+r)]^2> / r^{2/3}$ - this last obtained by turbulence measurements provided by a fast thermometer - confirmed the existence of a linear regression between C_T^2 / T^2 and the quantity $<E>^2 e^{2\alpha z}$, $\left(C_T^2 / T^2 = k <E>^2 \; e^{2\alpha z}\right)$ suggested by the Sodar equation (Little, 1969), where T is the average absolute temperature at the height z of the backscatter volume, a the coefficient of sound absorption and k a proportionality factor = $9.48 \cdot 10^{-14}$.

This result allows the use of Sodar data for estimating sensible heat fluxes in the PBL through the relationship between <w'T'> and C_T^2 suggested by Coulter and Wesely in 1980: $< W'T' > = \gamma \left(\frac{1}{2.46} C_T^2 \left(\frac{g}{T} \right)^{2/3} z^{4/3} \right)^{3/4}$, where $\gamma = (1 + 0.07/\beta)^{-1}$ is a correction factor which takes into account the humidity influence, and b is the Bowen ratio (when $\beta > 0.4,\ \gamma \cong 0.9$).

As a matter of fact, the comparison between Sodar flux estimates through the above equations and flux measurements by the acoustic anemometer at 80 meters showed a diurnal trend with parallel development.

The Doppler Sodar observations can also be used to compute some estimates of the "low frequency" friction velocity u_* (Beyrich and Kotroni, 1993) from the covariance's between single measurements of the three velocity components.

This parameter, according to the diagnostic equation for the Ekman layer $\left(h = c \frac{u_*}{f} \right)$ (where c is a constant and f is the Coriolis parameter), is very important under near neutral and slightly stable conditions for a parametrization of the boundary layer depth h.

This last equation can be confidently used only after making sure that the friction velocities estimated from Doppler Sodar data are consistent with h, which represents a bulk boundary layer parameter: this becomes particularly relevant in stable conditions, when the applicability of the Ekman layer equation can become more questionable.

A correlation analysis made by Beyrich and Kotroni (1993) among the u_* from Doppler Sodar data, the u_* from local, fast-response instruments, and the SBL depth h determined by the Doppler Sodar, showed that a better agreement existed between h and u_* obtained from Sodar than between h and u_* obtained from local measurements. This situation was reversed during daytime.

These facts have been explained by considering that the size of the measuring volume for u_* deduced from Sodar observations (length scale of about 30 m) is comparable to that for the depth of the SBL estimated by the vertical antenna, differing from typical values of h ($\cong 100 \div 200 m$) by less than one order of magnitude (Beyrich and Kotroni, 1993) . On the contrary, the u_* measured by *in situ* instruments are local measurements, whose measuring volumes are more than two order of magnitude less than those of Doppler Sodar. Also the cut-off of the high frequency part of the turbulence spectrum does not seem to bring about any limitation, because the SBL depth is better correlated with some characteristics of the largest eddies, which in turn are associated with the lower frequencies.

In conclusion, it seems that the two reasons that (in principle) would make one expect poorer reliability of Doppler Sodar estimates u_* in assessing the depth of SBL, do not intervene in this sense but, on the contrary, play a beneficial role in allowing this technique to be successfully applied to the assessment of the time trend of the SBL height during night-time.

7 - Temperature vertical profiles measured by the Radio-Acoustic Sounding System (RASS)

The RASS is one of the latest developments of advanced remote sensors of the lower troposphere; its operation principles are based on both acoustic and radio scattering properties of the atmospheric medium, but differ from sodar and radar operation by a most peculiar fact.

While sodars and radars, when interacting with the atmosphere, undergo scattering from the already existing turbulent fluctuations in the natural atmospheric temperature and velocity (the so called "atmospheric structure"), the RASS, on the contrary, uses the sound pulse emitted by its acoustic transmitter not to probe the atmospheric structure, but just to create, by compression, its own well defined typical perturbation pattern in the atmospheric refractive structure function.

This perturbation pattern will propagate into the atmosphere along the transmitting antenna axis, with a speed proportional to the local temperature of air, and will be tracked by the continuous radio wave emitted by the radio transmitting antenna of the RASS. Doppler analysis of the backscattered echo of the latter will allow the vertical distribution of air temperature in the PBL to be assessd from surface up to 2500 m.

Essentially, the radio-acoustic sounder consists of a powerful acoustic source, beaming a short burst of sinusoidal waves toward the zenith. The upward speed of this pulse is proportional, at every height, to the square-root of the local temperature. The pulse speed is continuously measured by means of a Doppler radar at ground. The record of the measured sound speed (which is proportional to the Doppler shift of frequency of radio wave return) as a function of the delay leads to the acquisition of the temperature vertical profile.

The formula which gives the temperature as a function of the measured parameters is the following: $T^{1/2} = 4.7\ 10^{-2} f_D \Big/ \left\{ 1 + 4.714\ 10^{-5} (H-30) \cos\left[\operatorname{arctg}\left(\frac{d}{h} \right) \right] \right\}$, where f_D is the Doppler frequency shift, H the relative humidity (%), d half the distance between the two radar antennas $(\cong 10m)$ and h the height of observation. The term $\cos\left(\operatorname{arctg}\left(\frac{d}{h} \right) \right)$ is used to convert radial velocities into vertical velocities.

The time required for the measurement of one vertical profile is less than 2 minutes, during which the RASS performs about twenty instantaneous soundings, at a rate of one about every 5 seconds.

The typical range of a *metric* RASS (*metric* means that the wavelength of its acoustic pulse is of the order of 1 m) is about 1000 meters, with extreme soundings arriving at an altitude of more than 2000 m (Bonino et al., 1981). The space resolution is

20 m and a reliable vertical temperature profile can be obtained every 15 minutes as an average of a sequence of about 15 soundings.

Two outstanding properties of the RASS are represented by its ability to provide real-time information on the upper thermal structure of the atmosphere and to operate during very thick and deep fog layers, when other systems cannot be used (also for security reasons). This last property can be essential to air-quality control under severe environmental conditions.

Temperature profiles, measured by RASS during deep fog episodes (Bonino at al., 1985), showed the ability of RASS to provide reliable and continuous observations both inside the fog layers and at their tops: the inversion layers capping the fogs corresponded very well to the regions where independent Doppler Sodar measurements showed sharp reductions in the turbulence intensity (given by σ_W) .

These results explain the great interest of boundary layer scientists in the RASS, also on account of the possibility of acquiring from it automatic and continuous data concerning the vertical stability of the bulk of the atmospheric boundary layer in such a form to be directly processed by computers-

References

Beyrich, F. and Kotroni, V. (1993) - "Estimation of surface stress over a forest from sodar measurements and its use to parameterize the stable Boundary-Layer height" - *Boundary Layer Meteorol.* **66** pp. 93-103.

Bonino, G. , Longhetto, A. et al. (1981) - "Radio-acoustic measurements of fog-capping thermal inversions" *Nature,* **Vol 290 n. 5802**, pp. 121-123.

Bonino, G., Longhetto, A. et al. (1985) - "Remote sensing of stability conditions during fog episodes" - in *C. De wispelaere (ed.) "Air Pollutions Modelling and its Applications"* - **vol. 7**, Plenum Press.

Browning, K. A. and Wexler, R. (1968) - "The determination of Kinematic Properties of a Wind Field Using a Doppler Radar", *J. Appl Meteorol.* **7**, pp. 105-113.

Carson, D. J. (1973) - "The Development of a Dry Inversion-capped Convectively Unstable Boundary Layer" - *Q. J. R. Meteorol. Soc.,* **99,** pp. 450-467

Clarke,R.H. (1970)-"Observational studies in the Atmospheric Boundary Layer" - *Quart. J. Roy. Meteorol. Soc.,* **96,** 91-114.

Coulter, R. L. and Wesely, M. L. (1980) - "Estimate of Surface Heat Flux from Sodar and Laser Scintillation Measurements in the Unstable Boundary Layer" - *J. Appl. Meteorol. ,* **49,** pp.1209-1222.

Frisch, A. S. et al. (1989) - "Measurements of the Vertical Flux of Turbulent Kinetic Energy with a Single Doppler Radar" - *Boundary Layer Meteorol.,* **49,** pp. 331-337.

Keder, J., Foken, Th., Gerstmann, W. and Schindler V. (1989) "Measurement of Wind Parameters and Heat Flux with the Sensitron Doppler Sodar" - *Boundary Layer Meteorol.,* **46,** pp. 195-204.

Kerschgens, M. J. and Kraus, H. (1990) -"On the Energetics of the Urban Canopy Layer" - *Atmospheric Environment,* **24B,** pp. 321-328 -

Kropfli, R. A. (1986) - "Single Doppler Radar Measurement of Turbulence Profiles in the Convective Boundary Layer" - *J. Atmos. and Oceanic Tecnology,* **3,** pp. 305-314.

Lhermitte, R. M. (1968), "Turbulent Air Motion as observed by Doppler Radar", *Proc. 13th, Conference on radar Meteorology,* Montreal, Amer. Meteor. Soc.'Boston, M.A., pp. 489-503.

Little, C. G. (1969) - "Acoustic Methods for the Remote Probing of the Lower Atmosphere" - *Proc. IEEE,* **57,** pp. 571-578.

Oke, T.R. (1982) - 'The energetic basis of the urban heat island"- *Quart. J. Royal Meteorol. Soc.,* **108,** pp. 1-24.

Oke, T.R. (1987) - "Boundary Layer Climates" - *Second Edition - Methuen, London and New York,* **Sect. 8,** pp. 262-303

Oke, T.R. (1988) - "The Urban Energy Balance", - *Prog. Phys.*Geogr. **12,**pp. 471-508.

Westcott, N. (1989) - Influence of Mesoscale Winds on the Turbulent Structure of the Urban Boundary Layer over St.Louis" -*Boundary Layer Meteorol.* **48.** pp. 283-292.

Wingaard, J C. (1973) - " On the Surface Layer Turbulence", - *in D. A. Haugen (ed), Wokshop on Micrometeorology, Amer. Meteor. Soc. Boston, M.A.,* pp. 332.

Wingaard, J. C. (1984) - "Atmospheric Turbulence and Air Pollution Modelling" - *in F.T.M. Nieuwstadt and H. Van Dop (eds).- Boundary Layer Modelling, D. Reidel Pub. Co.. Boston M. A.,* pp. 358.

Yoshikado, H. and Kondo, H. (1989) - "In Land Penetration of the Sea Breeze Over the Suburban Area of Tokyo" - *Boundary Layer Meteorol.,* **48,** pp. 389-407.

Use of proper variables to describe some aspects of urban pollution.

A. Febo, C. Perrino, C. Giliberti and I. Allegrini
CNR Istituto Inquinamento Atmosferico
Via Salaria Km 29.300-C.P. 10
00016 Monterotondo Stazione (Roma) Italy

The description of the time evolution of urban pollution is a very difficult task. This is due to the simultaneous presence of emissions and physico-chemical transformation processes, coupled with turbulent diffusion and advective processes.
A vital key to decribe such a complex system is the use of '*proper variables*', that is to say a series of parameters which, when appropriately integrated into general physico-chemical relationships, facilitates the description of this system, thus giving us a deeper insight into urban pollution processes.
The measurement of atmospheric pollutants is based upon observations which are mainly carried out at ground level. This offers practical advantages, but it causes many problems when interpreting experimental data because atmospheric processes are dependent upon the dynamic evolution of the boundary layer. Therefore, identification of parameters which are able to describe the behaviour of the boundary layer and its time evolution is of paramount importance.
Conseguently, meteorological observations are a significant part of any monitoring network, although classical meteorological observations, which are normally carried out, only provide rough information about the atmosphere, while finer information is strongly required. This is especially true in urban areas, where complex terrain and thermal disturbance, due to antropogenic activity, do not permit a simple and clear description of these processes. For example, the temporal trend of the ground concentration of non-reactive pollutants is fully known if the dynamic properties of the boundary layer are well-defined and, obviously, if the characteristics of the sources are known.

NATO ASI Series, Partnership Sub-Series, 2. Environment – Vol. 8
Urban Air Pollution
Edited by Ivo Allegrini and Franco De Santis

To clarify the relevance of this phenomenon, we can consider how the diurnal evolution of the planetary boundary layer influences ground level concentration of primary and secondary pollutants in a completely different way. A good example of this process is the daily cycle of temporal evolution of the boundary layer when meteorological conditions are dominated by high pressure and low wind. In this case there is a transition from a Stable Nocturnal Boundary Layer to a Thermal Convective Layer. During the day the boundary layer is completely developed and the concentration levels are spatially quite homogeneous. After sunset, the turbulence drops to very low levels and a new stable layer is developed. This process decouples the residual layer from the surface layer, which becomes stable. In this phase, the concentration levels of primary pollutants rises, notably due to the occurring of stability conditions, whereas, with regard to the secondary pollutants, chemical process and dry deposition processes, become dominant (for example ozone reacting with nitrogen monoxide). It must be stressed that during stability conditions at night, vertical mixing of the atmosphere is very slow, causing a large time for vertical mixing of the pollutants (Daewon et al. 1995). Consequently, the concentration levels of primary and some secondary pollutants exhibit high spatial gradients. During the night, in the residual layer, chemical processes will proceed but, obviously, they cannot be observed by ground based measurements. However, the following day, when convective conditions return after sunrise, they reappear on the surface.

This cycle is a good example of the temporal evolution of the boundary layer when meteorological synoptic conditions are dominated by high pressure and low winds.

Meteorological measurements carried out with radio-acoustic-sounding-tecniques (RASS, SODAR) may add much information to the air pollutants monitoring networks, but their use is limited by cost and complexity. In addition, they provide information which need further elaborations to be effectively used for the interpretation of pollution data. A very important approach to the description of the pollutants evolution may be given by measurements of Radon concentration (Allegrini et al.

1994, Fontan et al. 1979, Lopez et al. 1974, Fontan et al. 1976, Porstendorfer et al. 1991, Mahrt et al. 1982). In fact, Radon emission may be considered almost steady source, at least along the observation time scale of the measurements. In addition, it is spacially homogeneus over some kilometers. Therefore, the temporal evolution of Radon concentration and of its partial derivative on time are dependent upon the dynamics of the boundary layer; it can be formally defined by the following relationship:

$$\frac{\partial C_R}{\partial t} \cong \alpha\ \Phi_R - \beta\ \{C_R\} + Adv \qquad (1)$$

where α is a parameter which links the mixing properties of the boundary layer with the source intensity, $\beta\{C_R\}$ is a term which takes into account the mixing properties of the atmosphere, Φ_R is the emission rate for Radon and *Adv* is the advective term.

Generally, the term *Adv* is negligible because the Radon emission may be regarded as a constant, at least on the spatial scale of several kilometers.

From the shape of the Radon time derivative trend, it is possible to recognise either high stability and instability conditions, or phases of transition. In fact, high stability conditions at ground level maximise the contribution of the term $\alpha\,\Phi_R$ ($\beta\{C_R\}$ term is negligible), while transitions to unstable conditions maximise the contribution of the term $\beta\{C_R\}$.

Figure 1 shows temporal trend of Radon together with the concentration of two primary non reactive pollutants: carbon monoxide (CO) in gas phase and Poli Aromatic Hydrocarbons (P.A.H.), on particles.* It can be seen that the trend of these two primary pollutant concentrations are very similar and that they have the same modulation as Radon concentration. From the quantitative point of view, obviously, the trends of primary pollutans reflect the temporal variability of source intensity.

*The experimental data refer to a field campaign, carried out in the city of Milan during the winter of 1994 within the framework of a CNR project concerning the monitoring of pollution in urban areas.

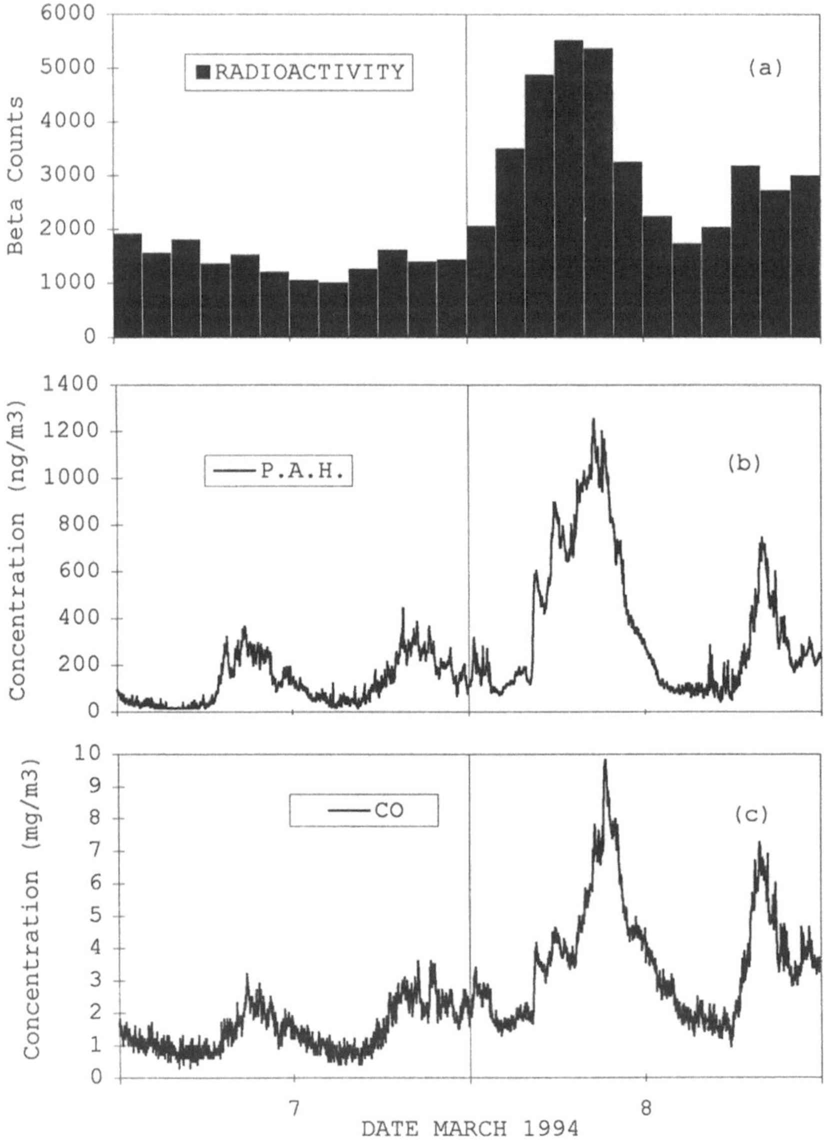

Figure 1. Temporal trend of Radon (a), Poli Aromatic Hydrocarbons (b) and Carbon Monoxide (c) for the period 7-8 March 1994.

Time evolution of carbon monoxide (CO) is then formally described by the following relationship:

$$\frac{\partial C_{CO}}{\partial t} \cong \alpha\ \Phi(t) - \beta\ \{C\} + Adv \qquad (2)$$

and for P.A.H.:

$$\frac{\partial C_{P.A.H.}}{\partial t} \cong \alpha\ \Phi(t) - \beta\ \{C\} + Adv - L_s \qquad (3)$$

The relationships (2) and (3) differ for the removal term L_S, which takes into account the deposition losses affecting the concentration of Poly Aromatic Hydrocarbons. In fact, such compounds are associated to particulate matter which is characterised by deposition velocity which is larger than that of CO; for this reason, the removal term for P.A.H. cannot be neglected, especially during unstable periods, in which turbulent deposition is very effective.
Then the temporal trend of the two primary pollutant concentrations ratio (*P.A.H./CO*) is not a constant but it shows a well defined pattern, depending on the relative importance of the losses. If we compare the derivative of Radon concentration with *P.A.H./CO* trend (figure 2), it appears that high values of the ratio occur when stability conditions are present, while minimum values occur during unstable periods.
Another example of application of this approach is the investigation of the main sources of HONO in urban atmosphere. Nitrous acid plays an important role in urban area because it is precursor of $OH^{\bullet}$ radicals. Concerning the source apportionment of this specie, the main cause for the discrepancies on the interpretation of field measurements is the difficulty to identify appropriate variables capable of characterising the lower part of the boundary layer during night-time (Notholt et al. 1992, Calvert et al. 1994, Winer et al. 1994). The use of Radon as a tracer of the mixing properties of the boundary layer can be useful for the description of the processes concerning the time evolution of ground level HONO mixing ratios (Febo et al. 1995).

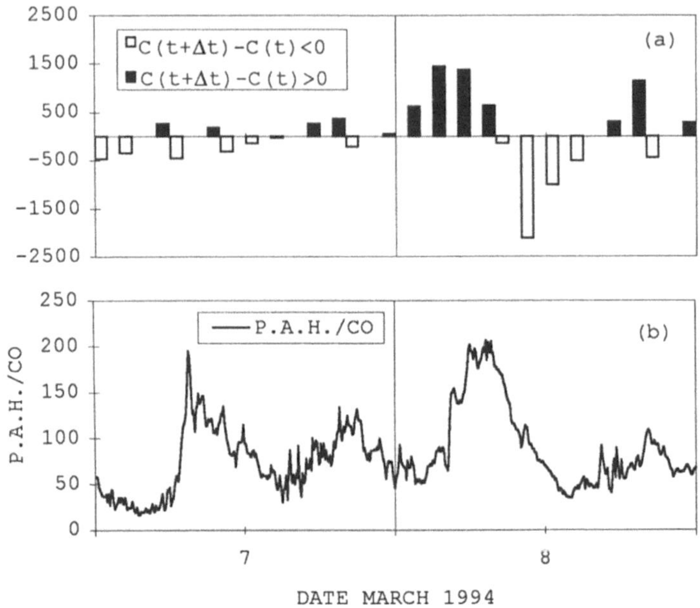

Figure 2. Temporal trend of derivative of Radon concentration (a), and of P.A.H./CO ratio (b) for the period 7-8 March 1994.

It is known that high amounts of HONO can be measured in surface layer during stability conditions. Figure 3 shows the temporal trend of HONO, radioactivity and CO for the period March the 7-12th 1994. From the observation of the trends, it appears that HONO temporal pattern is similar to that of primary pollutants, such as carbon monoxide, and much more to Radon concentration; it confirms the hypothesis that direct emission and the heterogeneus reactions on urban surfaces, in particular those involving nitrogen dioxide, may represent the main sources of HONO in urban ares. This result is in agreement with laboratory studies.

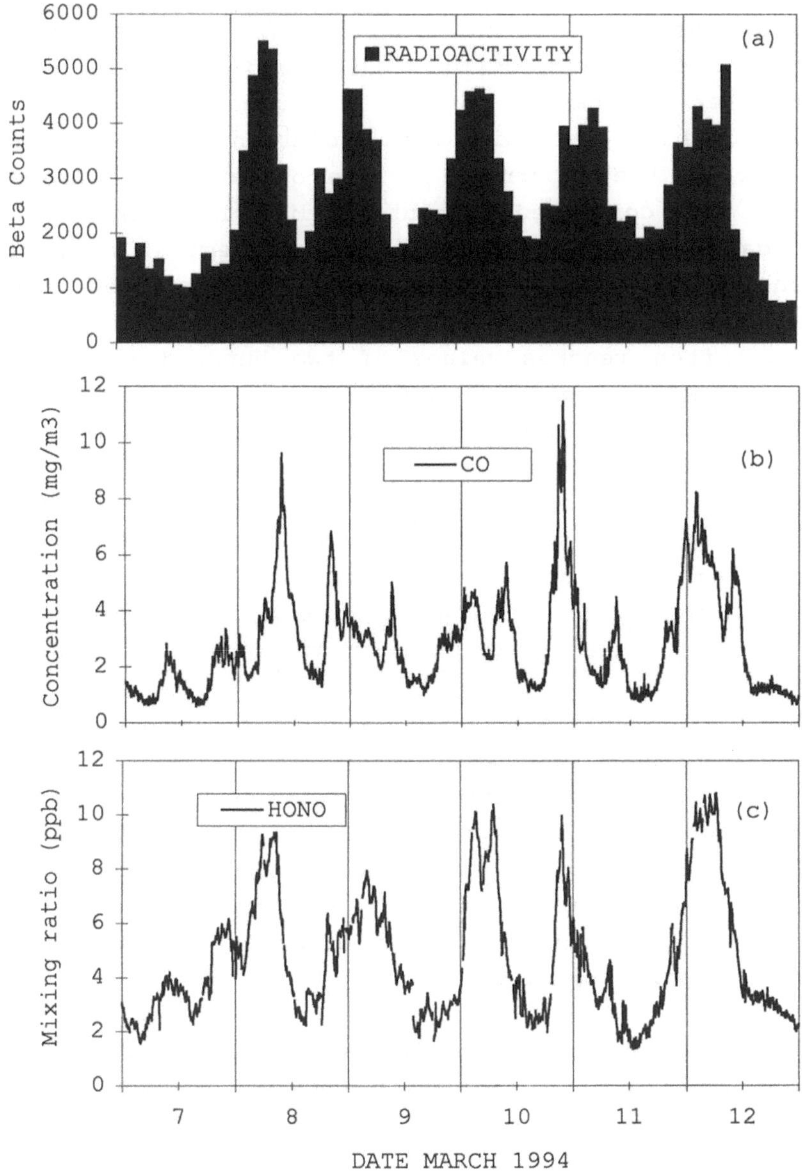

Figure 3. Temporal trend of Radioactivity (a), Carbon Monoxide (b) and Nitrous acid for the period 7-12 March 1994.

It has been stressed that a good correlation between HONO and primary non-reactive pollutants should be expected because, in both cases, we are dealing with ground-based sources: thus the parameter α has a strong influence in both functional relationships describing time evolution of the concentration.
As far as the occurrence of the oxidative processes in urban areas are concerned, recent studies have shown that strong radicalic oxidative processes are present in many urban areas even during winter and when ozone concentration keeps to very low levels. For example, during winter in Milan, NO_2 mixing ratio often reaches values of two hundred ppb while nitrate particulate exceeds 100 $\mu g/m^3$. The presence of these intense oxidative phenomena is moreover confirmed by high concentrations of nitric acid and P.A.N.
In order to understand the processes which are responsible for the formation of these compounds, the knowledge of photochemical processes evolution is required. However, this important step is somewhat limited by two negative aspects of the monitoring activity usually implemented in urban areas:

A) The pollutants monitored are limited in number, so a complex picture is hard to obtain;
B) Usually the measurements are carried out at a given location and height near the ground level; thus they are critically influenced by the dispersion properties of the boundary layer and then they cannot be used, as such, for kinetic analysis. Probably, this is the major drawback in the management of monitoring networks.

It is now opportune to discuss several metodologic aspects which concern the choice of "*proper variables*" capable of giving a good description of the evolution of oxidation processes.
If a complete picture of the oxidation processes in urban areas needs to be obtained, a clear reference to NO_2 and O_3 should be made; in fact, these compounds are, at the same time, main products of the oxidation processes and main precursors of OH· radicals, because they are easily and rapidly photolyzed.

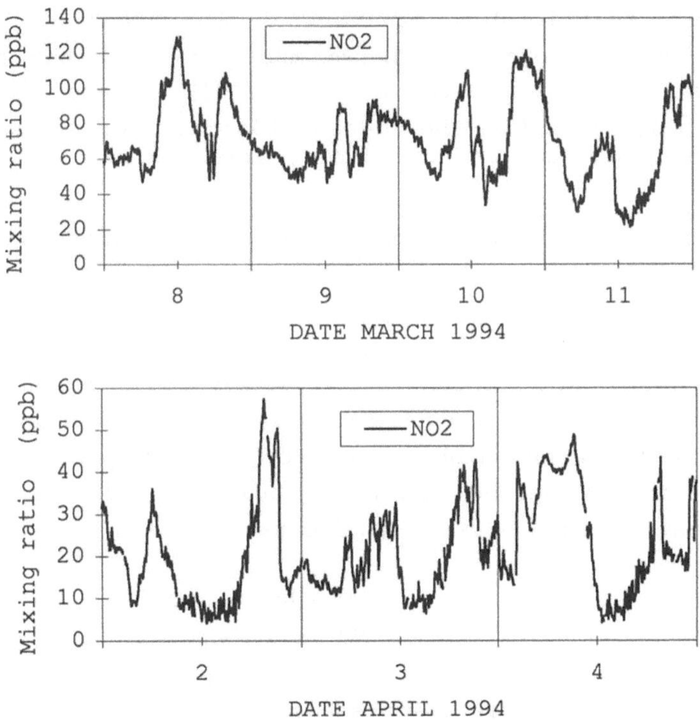

Figure 4. Typical temporal trend of NO_2 concentration during the periods 8-11 March 1994 and 2-4 April 1994.

Typical temporal trends of NO_2 concentration are shown in figure 4. It can be seen that the latter show a very complex temporal behaviour, which is due to a high number of factors which affect the concentration value at ground level.

Time evolution of NO_2 concentration at ground level can be formally described in the following way:

$$\frac{\partial\{C\}_{NO_2}}{\partial t} = \alpha\ \Phi_{NO_2} - \beta\ \{C\}_{NO_2} + Adv + \sum P_k - \sum L_v - L_s \tag{4}$$

The upper relationship is very complicated; in fact, to describe NO_2 temporal trend, we must consider the production processes ($\sum P_k$), the removal term ($\sum L_v$) and the surface losses (L_s), besides the meteorological terms, α, β, the advection term (*Adv*) and the direct emission rate (Φ_{NO_2}).

Let's consider the most important NO_2 production processes ($\sum P_k$):

-*Tritation*: $O_3 + NO \rightarrow NO_2$

In this case, NO_2 concentration would only depend upon the amount of ozone and nitrogen monoxide. It is remarkable that in urban areas, when high stability conditions occur, this reaction causes strong spatial gradients and, in particular after sunset, high vertical gradients.

-*NO radicalic oxidation*: $NO + RO_2^{\cdot} \rightarrow NO_2$

$NO + HO_2^{\cdot} \rightarrow NO_2$

-*Direct emission*: it is responsable for a small fraction of NO_2 concentration in air, because $\Phi(NO_2) << \Phi(NO)$.

-*Three-body reaction*: $2NO + O_2 \rightarrow 2NO_2$ which is negligible.

Among NO_2 removal processes ($\sum L_v$) we have:

-*Photolysis*: $NO_2 + h\nu \rightarrow NO + O$

-*Radicalic removal*: $NO_2 + OH^{\cdot} \rightarrow HNO_3$

-*Surface removal*: $NO_2 \xrightarrow{surfaces} HNO_2 + HNO_3$

-*Night-time chemistry*: $NO_2 + O_3 \rightarrow NO_3^{\cdot} + NO_2$

$NO_3^{\cdot} + NO_2 \rightarrow N_2O_5$

Let's suppose that radical processes and direct emission are negligible ($P_r=0$, $\Phi_{NO_2}=0$). In this case, for NO_2, the main production and removal processes directly involve ozone. The same considerations can be applied also for ozone, whose main products and removal processes directly involve NO_2.

Then time evolution of NO_2 and O_3 concentration at ground level can be described in the following way:

$$\frac{\partial NO_2}{\partial t} \cong -\beta_1 \{NO_2\} + Adv + P_{NO_2}(NO; O_3) - L_{NO_2}(Phot.) - L_{night}(NO_2) - L_s(NO_2) \quad (5)$$

$$\frac{\partial O_3}{\partial t} = -\beta_2 \{O_3\} + Adv + P_{O_3}(O^{\cdot}; O_2) - L_{O_3}(NO; O_3) - L_{night}(O_3) - L_s(O_3) \quad (6)$$

But $P_{O_3} \cong L_{NO_2}(Phot.)$ because of NO_2 photolysis, and $P_{NO_2} \cong L_{O_3}$ because of the tritation.

Then if we consider the variable *S(t)*, which is the sum of NO_2 mixing ratio and O_3 mixing ratio, we obtain:

$$\frac{\partial S}{\partial t} \cong -\beta\{S\} + Adv - L_s' - L_{night}' \quad (7)$$

where L_S' term is related to NO_2 and O_3 surface removal processes and L_{night}' is related to NO_2 and O_3 night-time chemistry.

To solve this equation is sufficient to know the boundary conditions.

With the help of the above descriptive model, it is possible to interpret experimental data in a urban area in strong advective conditions.

In this case, the ozone outside the box which defines urban area coincides with the ozone available from the planetary boundary layer, where it is spatially homogeneous and where its value never exceeds 50-60 ppb, at least in the European southern regions. In this meteorological condition, the most important mass exchange is from the box which defines urban area to the mixing layer outside the urban area: then the *Adv* term becomes dominant, and $\frac{\partial S}{\partial t}$ can be considered approximately zero. Consequently *S(t)* is a constant ($S(t) \cong K$), and its value coincides with that of ozone background. Then, the resulting nitrogen dioxide will be equal or less the value of ozone background concentration.

Figure 5 (a) shows experimental data in strong advective conditions: Radon trend is not structured and concentration values are very low.

In figure 5 (b) it is shown the temporal trend of *K* variable; its value is near fifty ppb.

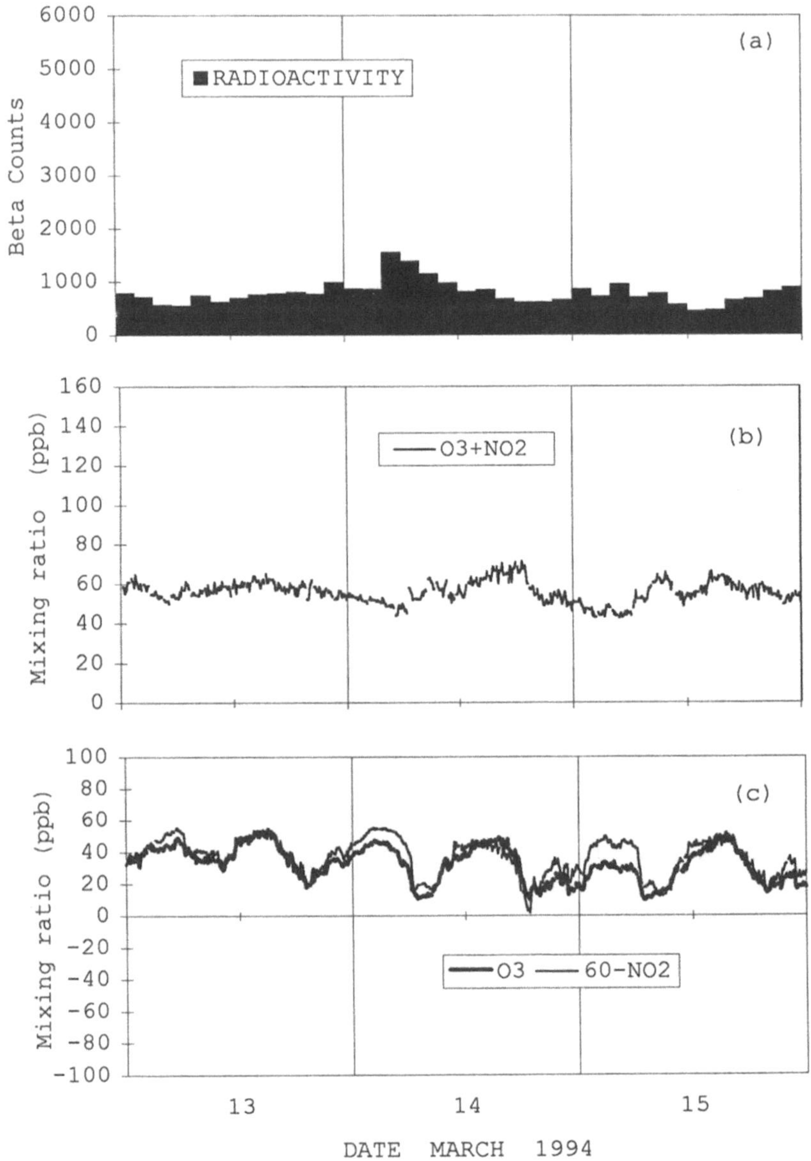

Figure 5. Temporal trend of Radon concentration (a), S(t) mixing ratio (b) and of ozone with the complement to 60 of the concentration of nitrogen dioxide in strong advective conditions.

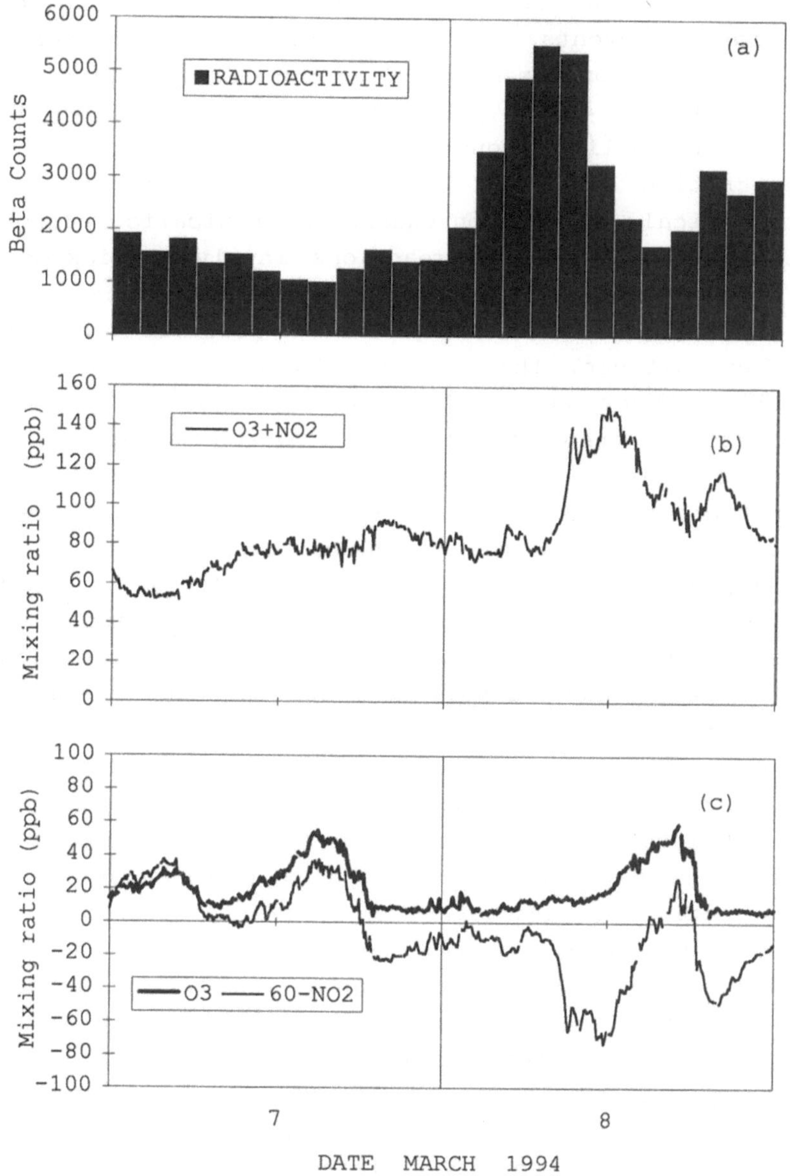

Figure 6. Temporal trend of Radon concentration (a), S(t) mixing ratio (b) and of ozone with the complement to 60 of the concentration of nitrogen dioxide (c) in high pressure conditions.

Figure 5 (c) shows an effective way to represent the occurrence of oxidative events, that is the plot of the concentration of ozone and the complement to 60 of the concentration of nitrogen dioxide ($[O_3] \cong K - [NO_2]$); they are perfectly in phase. In fact, instability conditions cause a decrease of the concentration of the compounds which are precursors of photochemical smog; consequently, radicalic processes are neglegible and then the reactions involving dioxide nitrogen and ozone, which are complementary, become dominant.
Let's study now what happens in high pressure conditions; this is, for instance, the case of figure 6 (a) which reports a stability event well defined by a relative maximum in the radioactivity trend on the night of March the 8th.
The plot showing the ozone and the complement to 60 of the concentration of nitrogen dioxide indicates that the pattern, characterized by waves in phase, typical of the advection periods, disappears during stability conditions (figure 6 (c)).
The *S(t)* trend in high pressure conditions and low wind is reported in figure 6 (b). In such meteorological conditions, this variable shows a well definite pattern, characterized by the following features:
a) during the night, in high stability conditions, the time derivative of *S(t)* is approximately zero. This is due to the fast titolation process involving ozone and nitrogen monoxide. In these conditions, *S(t)* at ground level coincides with NO_2 mixing ratio. Consequently, the removal processes of *S(t)* near ground level are strongly reduced, because the deposition process of NO_2 is slow and the other removal processes, such as those related to N_2O_5 formation, are negligible (O_3 mixing ratio in surface layer is very low).
For this reason, it should be expected that the night-time vertical profile of *S(t)* is quite homogeneous.
This behaviour is strongly different from that of *S(t)* in rural areas, where *S(t)* coincides more or less with O_3 mixing ratio.
b) after sunrise, *S(t)* increases rapidly and it reaches its maximum value at ~10:00. In figure 6 (c), it can be seen that, in the relative early morning, there is a quick increase of NO_2 mixing ratio, while ozone mixing ratio is at very low value;
c) S(t) shows a relative minimum in the afternoon;
d) S(t) shows another relative maximum after sunset.

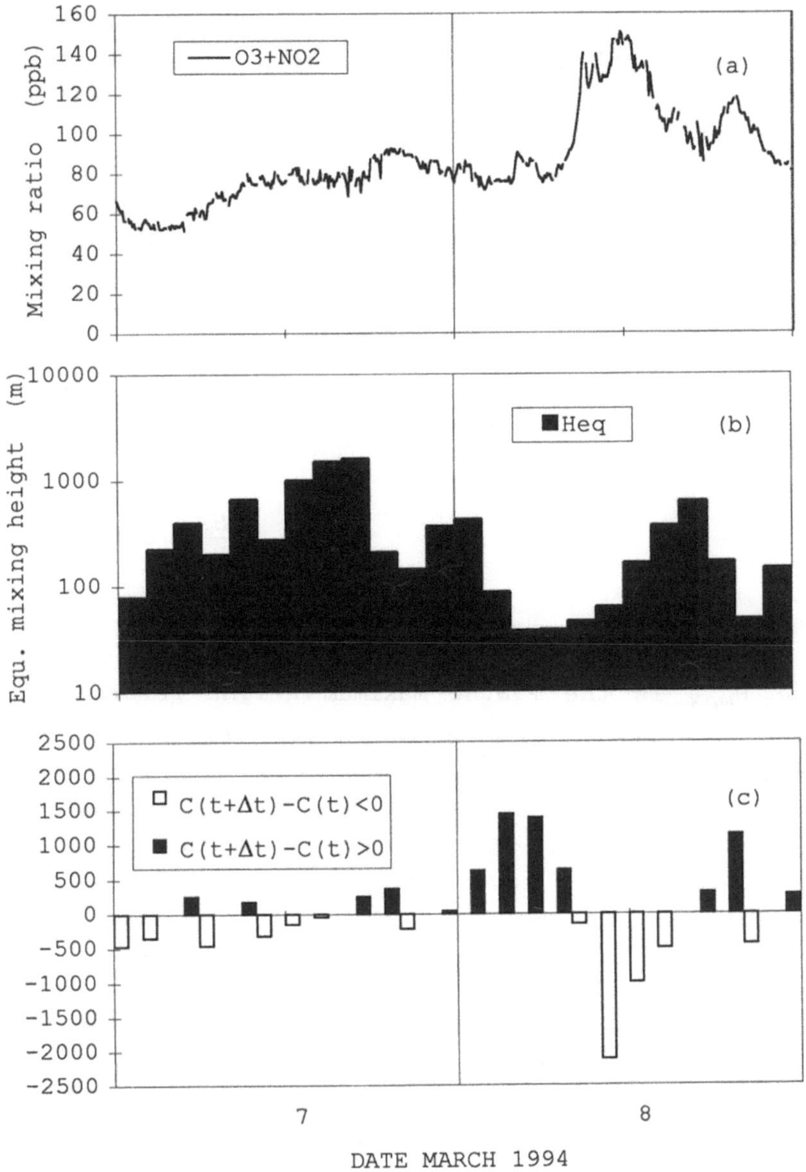

Figure 7. Temporal trend of S(t) mixing ratio (a), the equivalent mixing height (b) and the time derivative of Radon concentration (c), for the period 7-8 March 1994.

Such behaviour of $S(t)$ can be explained with the occurrence of strong radicalic oxidative processes. The presence of high concentration of secondary pollutants, such as nitric acid, *P.A.N.* and ammonium nitrate, in such meteorological conditions (wintertime, high stability) confirms the occurrence of oxidative processes.

In order to explain the temporal structure of $S(t)$, we must take into account the dynamic properties of the lower part of the boundary layer. On this subject, the daily $S(t)$ temporal trend can be formally described, when advective processes are negleglible, in the following way:

$$\frac{\partial S}{\partial t} = P_r - L_s - \frac{[S_G - S_{RL}]}{H_e}\frac{\partial H_e}{\partial t} \qquad \frac{\partial H_e}{\partial t} > 0 \qquad (8a)$$

$$\frac{\partial S}{\partial t} = P_r - L_s \qquad \frac{\partial H_e}{\partial t} < 0 \qquad (8b)$$

$$H_e \leq H_{max}$$

where H_{max} is the P.B.L. maximum height related to the day before; S_G is the mixing ratio of $S(t)$ at the measurement location; S_{RL} is the mixing ratio of $S(t)$ in the residual layer; H_e (equivalent mixing height) is a parameter with the size of a length, which is able to describe the mixing properties of the lower part of the boundary layer. This parameter can be deduced applying the relationships (8) to Radon concentration data.

In figure 7 the trends of $S(t)$ mixing ratio, time derivative of Radon concentration and the equivalent mixing height for the period 7-8 March 1994 are reported.

We can observe that, during the late night of March the 7-8th, strong stability conditions appear (high time derivative of Radon concentration and then low value for H_e) and $S(t)$ is approximately constant until sunrise. In fact, in this period the equation 8b is valid ($\frac{\partial H_e}{\partial t} < 0$); furthermore, radicalic

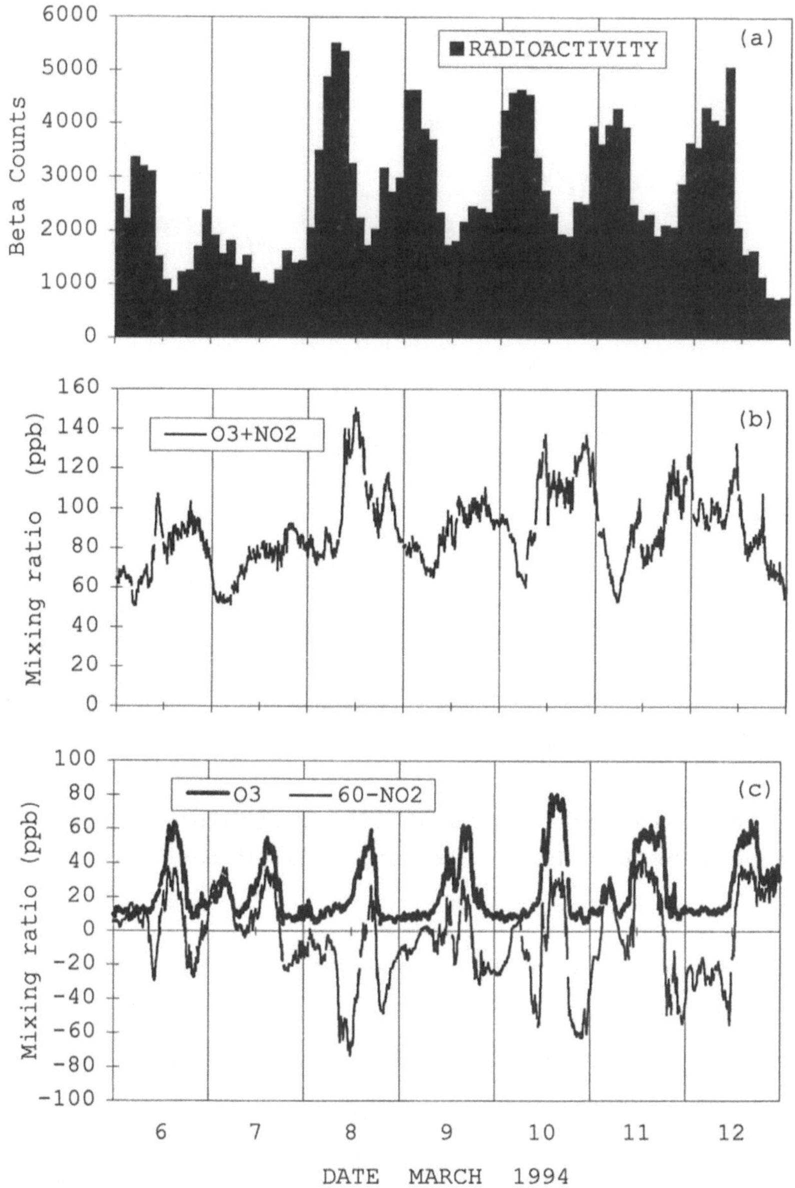

Figure 8. Temporal trend of Radon concentration (a), S(t) mixing ratio (b) and of ozone with the complement to 60 of the concentration of nitrogen dioxide (c) for a period characterised by high pressure conditions.

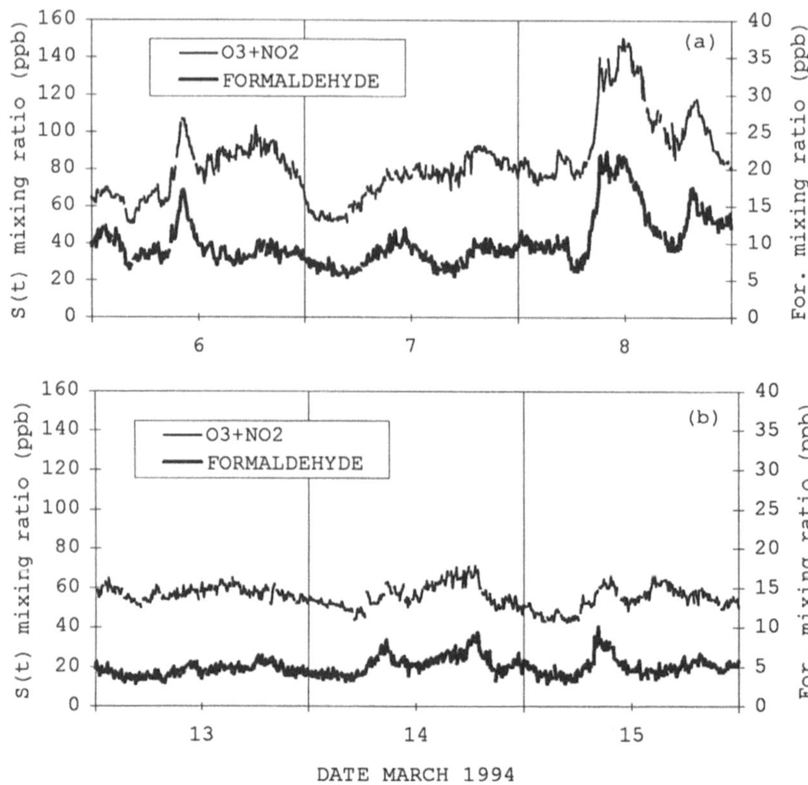

Figure 9. Temporal trend of S(t) variable and Formaldehyde for the periods 6-8 March 1994 and 11-13 March 1994.

activity and removal processes are negligible and then $\frac{\partial S_G}{\partial t} \cong 0$.

After sunrise $\frac{\partial S_G}{\partial t} >> 0$ ($\frac{\partial NO_2}{\partial t} > 0$); at this time, the term $\frac{[S_G - S_{RL}]}{H_e}\frac{\partial H_e}{\partial t}$ is very low ($\left|\frac{\partial H_e}{\partial t}\right| \cong 0$: the vertical mixing is still weak and $[S_G - S_{RL}] \cong 0$: the vertical profile is quite homogeneous), then if $P_R >> 0$ we justify the high positive time derivative of $S_G(t)$.

Subsequently, when the equivalent mixing layer height increases, in the equation (8a) the term $\frac{[S_G - S_{RL}]}{H_e}\frac{\partial H_e}{\partial t}$ is in

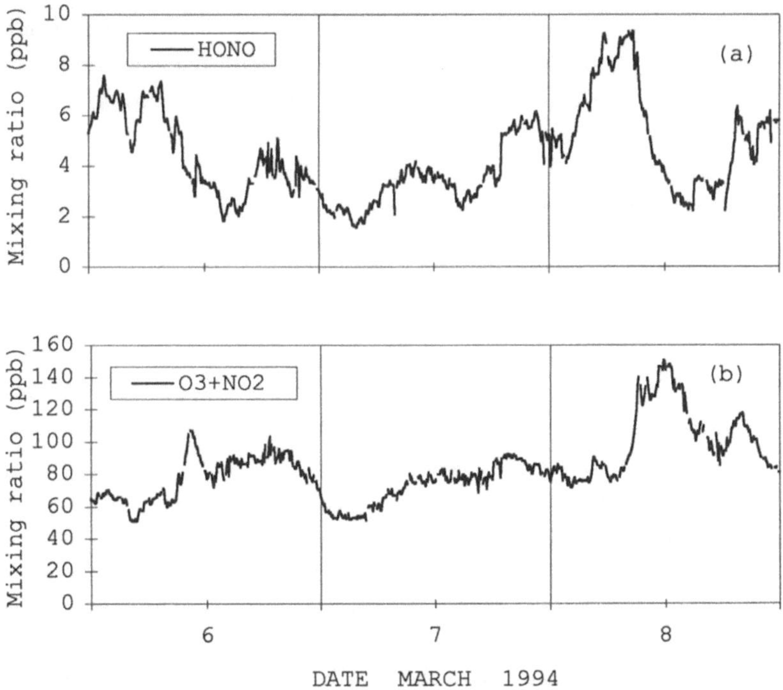

Figure 10. Temporal trend of Nitrous acid concentration (a) and of S(t) mixing ratio (b) for the period 6-8 March 1994.

competition with the term P_r; this explains the morning maximum. In the afternoon, the mixing layer is completely developed and then $S_G(t)$ reaches its relative minimum.

After sunset, the mixing layer height decreases (figure 7b), while the residual radicalic activity causes initially an increase for $S(t)$ ($P_r >> L_S$ equation 8b) but subsequently, in late night, P_r approaches zero; it explains the $S(t)$ relative maximum in the evening.

In figure 8 the temporal trend of the above mentioned variables for a period characterised by high pressure conditions is shown. It is clear that their daily temporal trends are well described by the equation (8).

Now the problem is to find the main sources of OH· radicals in the early morning during wintertime. The main candidates are, a

priori, ozone, formaldehyde and nitrous acid. But the oxidative phenomena start in the early morning, when ozone mixing ratio is very low and when the photolysis rate is weak.
The same considerations are true for formaldehyde.
On this subject, it is interesting to compare *S(t)* and formaldehyde temporal trends, which are reported in figure 9; it can be seen that they are tuned. In particular, there is a strong temporal coherence at the beginning of the oxidative processes. This means that in Milan, in wintertime, the processes responsible for any increase in the *S(t)* variable, in the early morning, are also responsible for the increase in the concentration of formaldehyde. Furthermore, the data show that, for formaldehyde, direct emission is a negligible source.
The above discussion suggests that, in this case, the main source of OH· radicals is nitrous acid. This hypotesis is further supported by the high values that HONO reaches at the occurrence of meteorological stability conditions. It can be seen that, in any case, there is a coincidence between HONO high levels during the night and the strong increase of *S(t)* variable in the next morning (figure 10).

CONCLUSIONS

The choise of a passive scalar, such as the Radon concentration, as a tracer of the mixing properties of the lower boundary layer, and the choise of S(t), (sum of ozone and nitrogen dioxide mixing ratios), as a variable capable to describe some aspects of the oxidative processes in urban areas, seems to give us a useful help either for the interpretation of pollution phenomena or to validate models.

References

Allegrini I, Febo A., Pasini A, Schiarini S (1994) Monitoring of the nocturnal mixed layer by means of particulate radon progeny measurements. J. Geophys. Res. 99: 18,765-18,777.

Byun DW, Dennis R (1995) Design artifacts in eulerian air quality models: evaluation of the effects of layer thickness and vertical profile correction on surface ozone concentrations. Atmos. Environ. 29: 105-126.

Calvert JG, Yarwood G and Dunker AM (1994) An evaluation of the mechanism of nitrous acid formation in the urban atmosphere. Res. Chem. Intermed. 20: 463-502.

Febo A, Perrino C and Allegrini I (1995) Measurements of nitrous acid in urban areas by DOAS and diffusion denuders. (submitted to Atmos. Environ.).

Fontan J, Druilhet A, Lopez A, Bernis NJ and Cautenet G (1976) Surveillance de la stabilité verticale dans un site urbain. Pollut. Atmos. 71: 231-240.

Fontan J, Guedalia D, Druilhet A and Lopez A (1979) Une methode de mesure de la stabilité vertcale de l'atmosphere pres du sol. Boundary Layer Meteorol. 71: 3-14.

Lopez A, Guedalia D, Servant J and Fontan J (1974) Advantages of the use of radiactive tracers 222 Rn and 212 Pb for the study of Aitken nuclei within the lower troposphere. J. Geophys. Res. 79, 1243-1252.

Mahrt L, André JC and Heald RC (1982) On the depht of the nocturnal boundary layer J. Appl. Meteorol. 21: 90-92.

Notholt J., Hjorth J and Raes F (1992) Formation of HNO_2 on aerosol surfaces during foggy periods in the presence of NO and NO_2. Atmos. Environ. 26A: 211-217.

Porstendorfer J, Butterweck G and Reineking A (1991) Diurnal variation of the concentrations of radon and its short-lived daughters in the atmosphere near the ground. Atmos. Environ. 25(A): 709-713.

Winer AM and Biermann HW (1994) Long pathlength differential optical absorption spectroscopy (DOAS) measurements of gaseous HONO, NO_2 and HCHO in the California South Coast Air Basin. Res. Chem. Intermed. 20: 423-445.

Air Pollution Modelling and its Interactions with Experimental Studies - The Situation in Southern Europe

N. Moussiopoulos
Laboratory of Heat Transfer and
Environmental Engineering
Aristotle University Thessaloniki
GR-54006 Thessaloniki, Greece

Introduction

The mechanisms of air pollutant transport and transformation have been studied for decades in Western Europe, in North America and in Japan. Nowadays physical and chemical processes leading to high air pollution levels and, especially, the formation of photochemical smog are to a large extent understood. Nevertheless, air pollution still represents a serious problem for many countries. Specifically, photosmog levels increase dramatically in numerous urbanized and/or industrialized areas in the world. Some of these areas are located in Southern Europe and Northern Africa, where climatological parameters strongly favour the occurence of smog episodes.

Previous studies have revealed the rarity of air pollution data from Mediterranean countries. This shortcoming is associated with both the inadequacy of monitoring networks and the lack of coordinated activities to better understand smog characteristics in heavily polluted areas in Southern Europe.

From the information available there is enough evidence that air pollution in Southern European countries is primarily attributed to local emission sources, though long-range transport may significantly affect background concentration levels. Air pollution episodes represent usually local problems and should therefore be studied at an appropriate scale (horizontal resolution of about 1 km). A requirement for such

NATO ASI Series, Partnership Sub-Series, 2. Environment – Vol. 8
Urban Air Pollution
Edited by Ivo Allegrini and Franco De Santis

studies is the availability of a comprehensive emissions inventory including both anthropogenic and biogenic emissions.

Another prerequisite for comprehensive air pollution studies is the knowledge of wind flow in the area of interest. Given that most of the study cases in Southern Europe are characterized by complex topographical and meteorological conditions, the description of (usually thermally driven) local circulation systems is not an easy task. In view of the already mentioned inadequacy of the existing monitoring networks, the only way to describe mesoscale flow characteristics in Southern European areas of interest is by applying a suitable mathematical model.

In the past there have been some attempts to explain physicochemical processes related to air pollutant transport and transformation in Southern European countries, both at regional and local scale. Unfortunately, most of these attempts were of only moderate success from the scientific point of view because they were lacking in sufficient cooperation between individual scientific teams and operators of monitoring networks. Moreover, so far modelling studies sometimes failed to reach their goals because of insufficient knowledge of the emission sources or because of insufficient observations to allow model validation.

Air Pollution Modelling and Environmental Management

The rapid hardware development in the last years enabled the formulation of comprehensive air pollution model systems. From the scientific point of view, such model systems are important tools for the description of air pollutant transport and transformation, especially if - as it is the case in most Southern European areas of interest - available observations are too scarce to allow drawing conclusions on the air pollution situation. The same systems are, however, at the same time extremely valuable for environmental planning and management, as they may be used to study complex source/ receptor relationships (*Figure 1*). In this sense, air pollution model systems may be used:

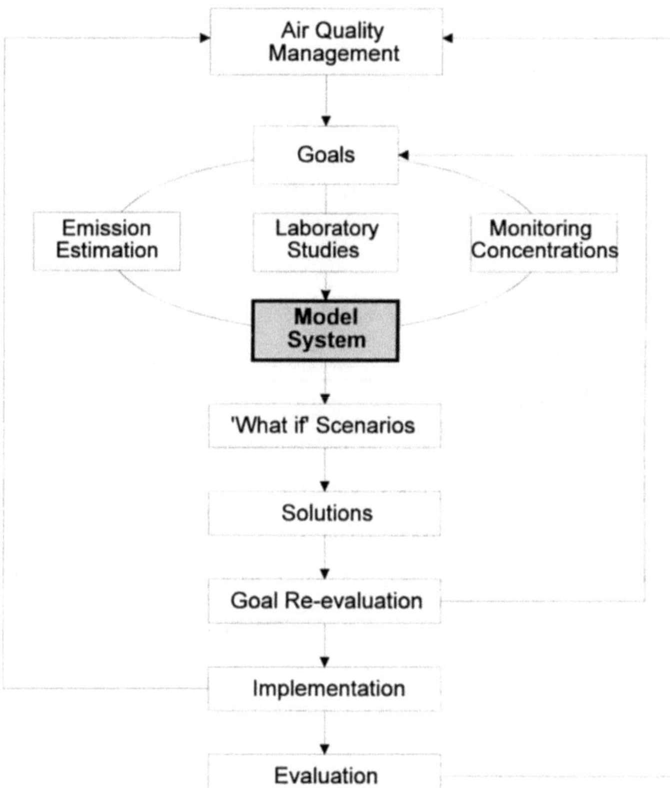

Figure 1. *The role of an air pollution model system in air quality management (adapted from Longhurst et al., 1994).*

- for the optimization of air pollution abatement strategies either locally or in a larger region;
- for short-term predictions of smog episodes in highly polluted areas.

Apparently, model validation with available measurements is indispensable in order that environmental managers can quantify the degree of reliability and accuracy inherent in air pollution model systems.

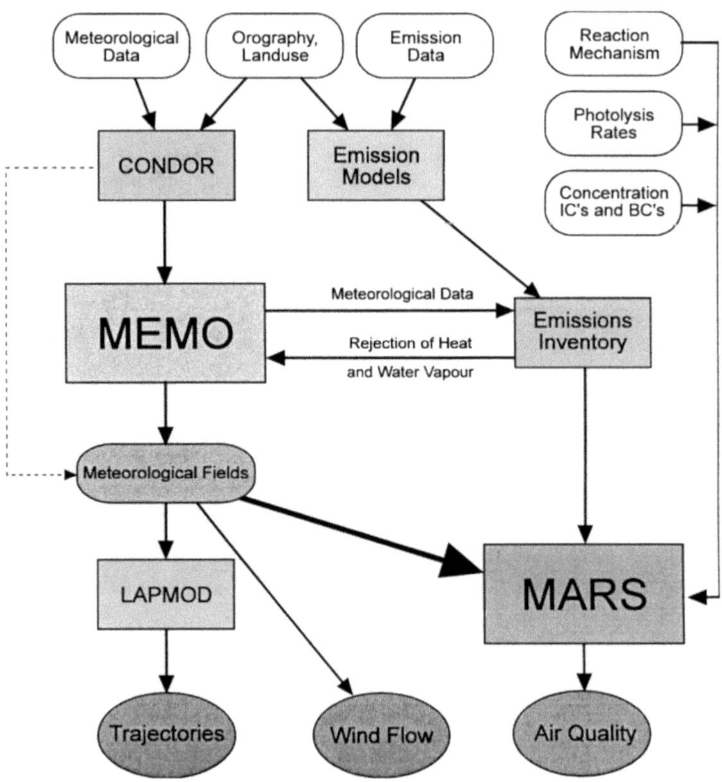

Figure 2. *Structure of the EUMAC Zooming Model (EZM).*

The EUMAC Zooming Model System (EZM)

An example for a contemporary air pollution model system is the EUMAC Zooming Model (EZM; Moussiopoulos, 1994) designed for simulations of wind flow and air pollutant transport and transformation in subregions of the EUMAC domain. Most probably, it represents the most widely tested European air pollution model for urban applications. Although initially designed as a tool to be used in conjunction with a continental scale model, the EZM evolved to be now a stand-alone model system which may be driven directly with measured data. Core models of the EZM are the nonhydrostatic prognostic mesoscale model MEMO (Flassak, 1993; Moussiopoulos et al., 1993) and the photochemical dispersion model MARS (Moussiopoulos, 1989; Moussiopoulos, 1993a). The structure of the EZM is illustrated in *Figure 2*.

The EZM is being utilized by several research institutions all over Europe for analyses of atmospheric transport and transformation phenomena in the local-to-regional scale. In a number of cases EZM results have already supported environmental policy decision making. In the remainder of this paper results of selected case studies with the EZM are briefly presented. More details on these cases may be found elsewhere (Moussiopoulos, 1994).

Athens Case Study

The capital of Greece is a good example for a large city continuously suffering under extremely high air pollution levels. The Athens case is very challenging also from the scientific point of view because it combines elevated anthropogenic emission levels and complex topographical and meteorological conditions. For this reason, within EUMAC the model intercomparison activity APSIS was initiated (APSIS: Athenian Photochemical Smog, Intercomparison of Simulations; Moussiopoulos, 1993b).

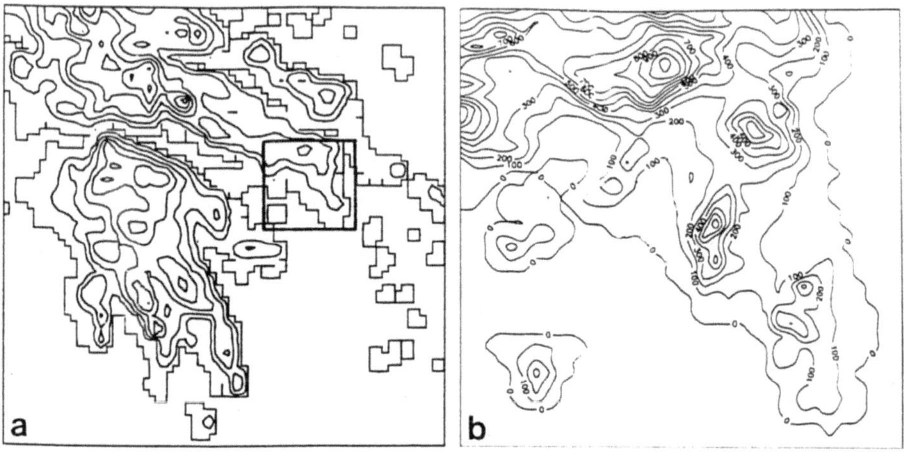

Figure 3. *Coarse grid (a) and fine grid (b) for the calculation of wind flow and the dispersion of chemically reacting pollutants in the Greater Athens Area. The frame in (a) corresponds to the position of the fine grid.*

The Athens Basin is surrounded by mountains at three sides and is open to the sea to the SW. This results in a frequent occurence of sea breeze circulation events, primarily during summery anticyclonic situations. Most smog episodes in Athens take place under such meteorological conditions. The highest air pollutant concentrations in Athens occur, however, in the case of a weak synoptic forcing acting against the development of a sea breeze. Such conditions, which may easily lead to air mass stagnation above the city, were prevailing on 25 May 1990, i.e. the day selected for most of the model simulations within the APSIS activity.

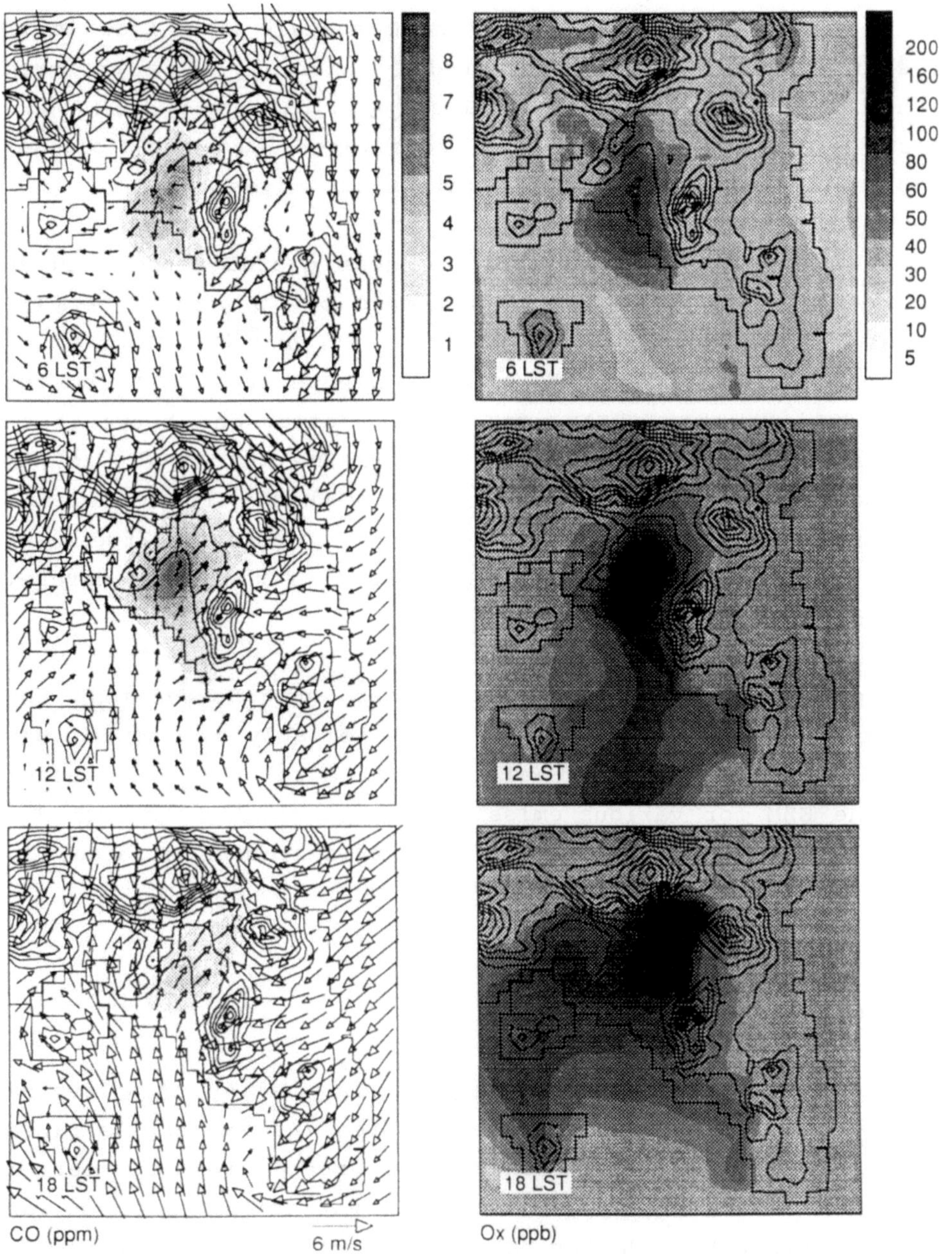

Figure 4. *Results of the EZM for the surface level pattern of the horizontal wind velocity and the concentrations of CO and* O_x *(=* NO_2 *+ ozone) in the Greater Athens Area at 0600, 1200 and 1800 LST on May 25, 1990.*

For an adequate description of the mesoscale wind flow in the Greater Athens Area, nested grid simulations were performed with the EZM: Results of a first application of MEMO on a coarse grid were used to formulate improved boundary conditions for the fine grid simulations (*Figure 3*).

Figure 4 shows EZM results for the near ground wind field and the near ground concentration patterns of CO and O_x (= NO_2 + ozone) in the Greater Athens Area at 6,12 and 18 LST on 25 May 1990. As already indicated, the sea breeze appears to be very weak during the whole day. The interface between the circulation cell in the Athens Basin and that in the neighbouring Mesogia Plain coincides with the East border of the basin. Correspondingly, air mass transport between the Athens Basin and the surrounding areas is found to be severely restricted during that day.

Detailed comparisons of model results and observations show that the observed situation is very reliably reproduced with the EZM (Moussiopoulos et al., 1994). Results of simulations with the EZM for various emission scenarios were considered at the optimization of the air pollution abatement strategy for Athens (Moussiopoulos et al., 1992).

Barcelona Case Study

The EZM was applied to investigate air quality on a typical summer day in the Greater Barcelona Area (*Figure 5*). As far as the wind field is concerned, the model results largely correspond to the expected diurnal variation, i.e. a land breeze combined with katabatic air motion at night and a deeply penetrating sea breeze during the day. According to the results of the dispersion simulation, in the course of the night large amounts of polluted air are transported onto the sea, the area there serving as a reservoir in the hours until sunrise. Driven by the sea breeze, this air mass is transported back to the urban area in the course of the day. During this transport ozone is formed, the corresponding photochemical reactions being augmented by freshly emitted primary pollutants. By this mechanism considerable amounts of photooxidants may be transported far inland.

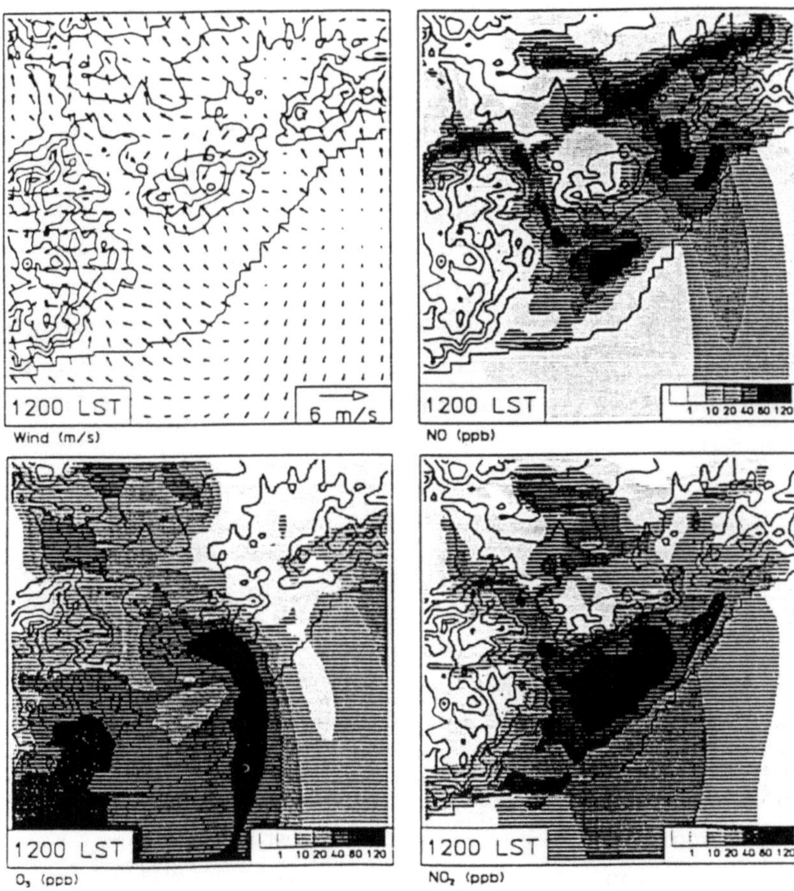

Figure 5. *Results of the EZM for the wind flow and the concentrations of NO, NO_2 and ozone near the ground in the Greater Barcelona Area at noon of 5 August 1990.*

The EZM results are in good agreement with the observed diurnal variation of pollutant concentrations during typical smog episodes in Barcelona (Coutinho et al., 1994). Results of a series of simulations with the EZM were used to support the decisions taken with regard to traffic regulations in Barcelona during the Olympic Games of 1992 (Baldasano et al., 1993).

Concluding Remarks

Air pollution model systems are being increasingly used to support environmental planning and management. As one of the contemporary air pollution model systems, the EUMAC Zooming Model (EZM) has been successfully applied for air quality studies in several Southern European areas including Athens and Barcelona. In these applications, the results of the EZM agree fairly well with available observations. Nevertheless, the lack of sufficient data from observations did not yet allow a conclusive validation of the EZM for applications in Southern Europe.

In the light of the above, it is mandated to perform a concerted activity at the proper location(s) in Southern Europe with the main aims (i) to create an adequate experimental database on photochemical air pollution over complex terrain and (ii) to evaluate state-of-the-art air pollution model systems. To achieve these goals, the above activity should include

- a coordinated experimental campaign in a pre-selected area maintaining the necessary characteristics and,
- the application of several air pollution model systems for predictions of local-to-regional scale wind flow and the combined dispersion and transformation of air pollutants in an attempt to evaluate them by the aid of the aforementioned experimental database.

Acknowledgement

The author is grateful to Dr. J.-M. Baldasano for the permission that results of his EZM application be incorporated in the present paper.

References

Baldasano J.M., Costa M., Cremades L., Flassak Th. and Wortmann-Vierthaler M. (1993), Air quality in the Barcelona Geographical Area for a typical summer day. Comparison with the situation during the Olympic Games '92, Report PB193, ITEMA-UPC, Terrassa (Barcelona).

Coutinho M., Baldasano J. and Calbo J., (1994), Simulation of the typical meso-meteorological circulations occuring during winter in the Barcelona region, in **Computer Simulation** (J.M. Baldasano, C.A. Brebbia, H. Power and P. Zannetti, eds), Computational Mechanics Publications, 23-30.

Flassak Th. (1990), **Ein nichthydrostatisches mesoskaliges Modell der planetaren Grenzschicht**, Fortschr.-Ber. VDI, Reihe 15, Nr. 74, 203 pp.

Longhurst J.W.S., Lindley S.J. and Conlan D.E. (1994), Towards local air quality management in the United Kingdom, in **Pollution Control and Monitoring** (J.M. Baldasano, C.A. Brebbia, H. Power and P. Zannetti, eds), Computational Mechanics Publications, 527-532.

Moussiopoulos N. (1989), **Mathematische Modellierung mesoskaliger Ausbreitung in der Atmosphaere**, Fortschr.-Ber. VDI, Reihe 15, Nr. 64, 307 pp.

Moussiopoulos N., ed. (1994), **The EUMAC Zooming Model (EZM): Model Structure and Applications**, EUROTRAC Report, 266 pp.

Moussiopoulos N., Flassak Th., Sahm P. and Berlowitz D. (1993), Simulations of the wind field in Athens with the nonhydrostatic mesoscale model MEMO, *Environmental Software* **8**, 29-42.

Moussiopoulos N. (1993a), **The Athens Experience**, Tutorial Notes, Air Pollution '93 Conference, Monterrey, Mexico, 128 pp.

Moussiopoulos N. (1993b), Athenian photochemical smog: intercomparison of simulations (APSIS), background and objectives, *Environmental Software* **8**, 3-8.

Moussiopoulos N., Flassak Th. and Knittel G. (1988), A refined diagnostic wind model, *Environmental Software* **3**, 85-94.

Moussiopoulos N., Pattas K. and Samaras Z. (1992), Measures for reduction of road traffic emissions and consequences to ozone formation in Athens, in **Traffic Induced Air Pollution**

(R. Pischinger, ed.), Mitteilungen des Institutes fuer Verbrennungskraftmaschinen und Thermodynamik, 351-362.

Moussiopoulos N., Sahm P., Kessler Ch. and Kunz R. (1994), Simulation of mesoscale wind flow and photosmog formation in the Greater Athens area, in **The EUMAC Zooming Model (EZM): Model Structure and Applications** (N. Moussiopoulos, ed.), EUROTRAC Report, 67-86.

Neural Network Modelling: Perspectives of Application for Monitoring and Forecasting Physical-Chemical Variables in the Boundary Layer

Antonello Pasini, Sergio Potestà
Servizio Meteorologico dell'Aeronautica
2° CMR - Aeroporto "De Bernardi"
Via di Pratica di Mare
I-00040 Pratica di Mare (Roma)
ITALY

Introduction

The aim of this paper is to present environmental applications of a technique of neural network modelling, which demonstrates its usefulness in the simulation of real systems that are complex, time-changing and characterised by many feed-back processes, such as the atmospheric pollution environment.

The first part of this article is devoted to a brief introduction to the topic of neural networks and to the features of our model. We then present some results about case studies of meteorological forecasting (concerning visibility) by means of neural computations. Finally, perspectives of a more specific use of neural networks for monitoring and forecasting physical or chemical quantities in the boundary layer are envisaged.

The structure of a feed-forward neural network and our model

Usually, when one talks about artificial neural networks, he imagines the natural neural network in our brain. In this article we shall not deal with the parallelism between

NATO ASI Series, Partnership Sub-Series, 2. Environment – Vol. 8
Urban Air Pollution
Edited by Ivo Allegrini and Franco De Santis

artificial and natural neural networks. Today, artificial neural networks are generally very simple models which try to simulate the working of our brain. Of course, the biological parallelism is not complete and the complexity of these networks is far-away from the complexity of the human brain.

Anyway, our approach to neural networks and to the simulation of a real system is very simple, if seen from the system theory point of view (see fig.1). In the simulation activity of a real system, the dynamical equations of a model (for monitoring, controlling or forecasting) link inputs and outputs (for example, causes and effects), thus determining the behaviour of the model itself and the correct simulation of the answer of the real system to certain inputs. Our idea is to replace the dynamical equations of a model with the computational structure of a neural network linking inputs and outputs.

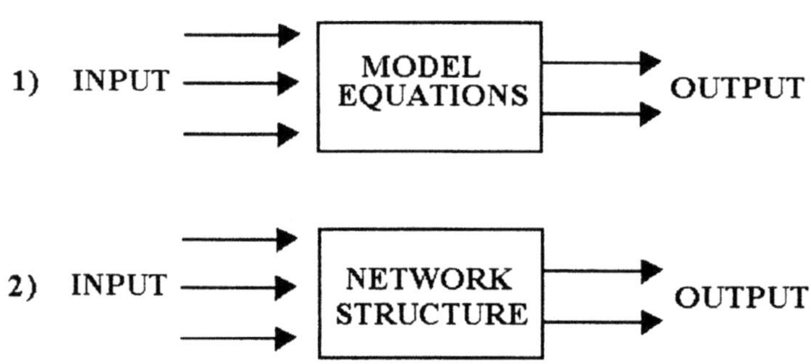

Fig.1. Approach to the simulation of a real system: (1) dynamical models; (2) neural network models.

In fig.2 the structure of a feed-forward neural network is shown; its fundamental elements are the connections, with associated weights, and the neurons, i.e. the single computational units. Each neuron is connected to all the neurons of the previous and the following layer; in this topology there are no connections between neurons of the same layer.

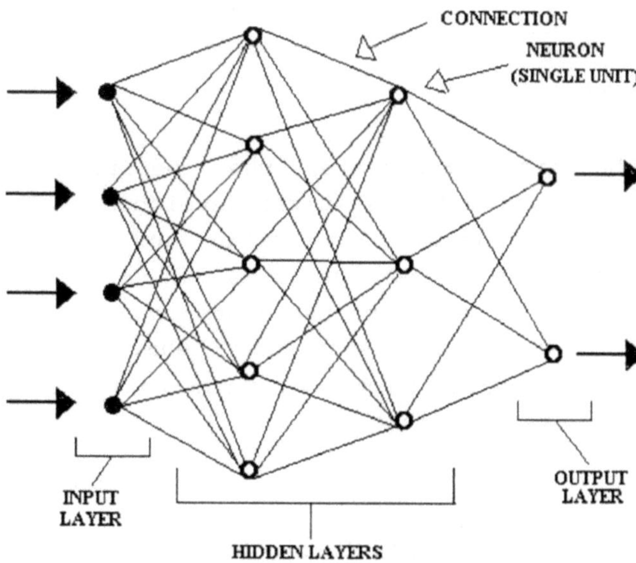

Fig.2. The structure of a feed-forward neural network.

Weighted sums of inputs converge to every hidden and output neuron. Then, at the single neuron a strongly non-linear function is usually calculated. In our model we adopt a network with only 1 hidden layer and use normalised sigmoid functions, so that the outputs O_i have the following form:

$$O_i = g_i\ (h_i) = g_i\ (\sum_j W_{ij} V_j) = g_i\ (\sum_j W_{ij}\, g_j\ (\sum_k w_{jk} I_k)) \quad , \tag{1}$$

where $I_k =$ inputs,

$w_{jk} =$ connection weights between input and hidden layers,

$W_{ij} =$ connection weights between hidden and output layers,

V_j = output of the hidden neuron N_j = input element at the output neuron N_i,

$g_j\,(h_j) = \dfrac{1}{1+\exp(-A\,h_j)}$, $A = (N_c)^{-1/2}$, N_c = number of input connections.

After the first forward step, when the initial weights are randomly chosen, our software (first developed by E. Setti from the Polytechnic University in Milan and then adapted to meteorological simulations) performs an iterative learning procedure in order to minimise the distance between outputs and targets (T_i , which represent the behaviour of the real system). In more precise terms, we minimise the following cost function E for every pattern μ:

$$E = \frac{1}{2}\sum_{\mu,i}\left(T_i^\mu - O_i^\mu\right)^2 = \frac{1}{2}\sum_{\mu,i}\left[T_i^\mu - g_i\left(\sum_j W_{ij}\,g_j\left(\sum_k w_{jk}\,I_k^\mu\right)\right)\right]^2 \quad , \tag{2}$$

by means of the so-called error backpropagation procedure for the updating of the weights, using a generalised Widrow-Hoff rule (for all these topics and for a more technical introduction to the neural computation see Hertz *et al.*, 1991). The learning rate and the momentum coefficient are fixed at 0.9. The iterative procedure goes on till

$$\frac{\sum_{\mu,i=1}^{M,N}\left(O_i^\mu - T_i^\mu\right)^2}{M\cdot N} < MSS \quad , \tag{3}$$

where MSS is a previously fixed coefficient.

We can test the generalisation phenomenon by choosing a statistical sample of input - output pairs in a training set. We try to teach the most general relationship between all these pairs to the network in the training procedure. Then, we test its generalisation performance on a disjoint set, during the test procedure.

At the end of the training procedure by means of the error backpropagation method, the weights of the network are fixed at their final values. The test procedure consists of a forward step, inserting new inputs and propagating them through connection weights and functions g to new outputs. Finally, the comparison between outputs and targets gives the generalisation performance on this new case.

After testing many new cases we can define a generalisation coefficient GC as follows :

$$GC = 1-\sqrt{\frac{\sum_i\left[(O_i - T_i)-\overline{(O_i - T_i)}\right]^2}{N}} \tag{4}$$

Because the input data are normalised between 0 and 1, then $0 \leq GC \leq 1$. If GC is close to 1, the network generalisation is very good.

We perform the training of the network in a way which is adaptive and useful for the successful treatment of quick time-changing data.

We choose a time window of our data archives, that becomes our initial training set of input-output cases. The training procedure fixes the connection weights. Then we test on the case which immediately follows. Now, we put aside the oldest input-output case in the training set and include the new case (previously tested) in the training set, then we perform a new training procedure and a new test on the immediately following case. ... And so on.

This procedure is completely automatic in our development environment. We stress that this window training is useful to simulate a forecast operational activity.

CASE STUDY
ROMA CIAMPINO (VISIBILITY)

DATA ARCHIVES: WINTER 1992-'93;
TESTED ON JANUARY 1993 (FOR TWO WEEKS);
WINDOW WIDTH: 2 MONTHS AND A HALF; MSS = 0.001

5 NETWORKS WITH STRUCTURE INPUT-HIDDEN-OUTPUT
37-35-1 (1 HIDDEN LAYER).

INPUT DATA : HOUR, WIND (NORTH AND EAST COMPONENTS), VISIBILITY, SKY COVERING, HEIGHT OF THE LOWEST CLOUD GROUP, TEMPERATURE, DEW POINT, PRESSURE, VISIBILITY TIME DERIVATIVE.

OUTPUT: CIAMPINO VISIBILITY AT T+1, T+2, T+3, T+6, T+9.

4 STATIONS: ROMA CIAMPINO, ROMA FIUMICINO, LATINA, VITERBO.

Table 1.

CASE STUDY
MILANO LINATE (VISIBILITY)

DATA ARCHIVES: NOVEMBER '91 - FEBRUARY '92;
TESTED ON DECEMBER '91 - FEBRUARY '92;
WINDOW WIDTH: 1 MONTH; MSS = 0.01

5 NETWORKS WITH STRUCTURE INPUT-HIDDEN-OUTPUT
89-50-1 (1 HIDDEN LAYER AND 11 STATIONS), 5 NETWORKS
WITH STRUCTURE 9-8-1 (1 HIDDEN LAYER AND LINATE DATA ONLY).

INPUT DATA: HOUR, WIND VELOCITY, VISIBILITY, SKY COVERING, HEIGHT OF THE LOWEST CLOUDS GROUP, TEMPERATURE, DEW POINT, PRESSURE, VISIBILITY TIME DERIVATIVE.

OUTPUT: LINATE VISIBILITY AT T+1, T+2, T+3, T+6, T+9.

11 STATIONS: SEE FIG.5.

Table 2.

Two case studies concerning visibility forecast

We here present only a few results of short-range visibility forecasts obtained through application of our model; a more accurate presentation about our meteorological results will appear elsewhere (Pasini and Potestà, in preparation). The features of these case studies are summarised in tables 1 and 2.

As far as the Ciampino case study is concerned, in fig.3 a comparison between the neural forecast and persistence for all the period considered is shown, while in fig.4 the details of the very good performance at $T+1$ hour is presented. In the operational practice the results of subjective forecasts are very close to the persistence performance; therefore the results shown in fig.3 allow us to improve our operational forecasts.

As far as the Linate case study is concerned, we want to stress only the important role of the horizontal pattern of stations shown in fig.5. In fact, fig.6 presents the performance of the networks trained through data from 11 stations compared with the performance of the networks using only the Linate data.

The performance of the networks briefly presented here can be improved if data from vertical profiles and data on particulate matter concentration are available.

As it is well-known, low visibility values and the fog phenomenon are often related to episodes of intense pollution. Thus, studies just like the ones presented here can be relevant for atmospheric pollution. Furthermore, the important point is that these meteorological studies show the good performances of a neural network model in a field where the behaviour of the real system is dominated by non-linear relationships, feed-back processes and quick time-changing features, just as in the case of the chemical and physical environment in the boundary layer.

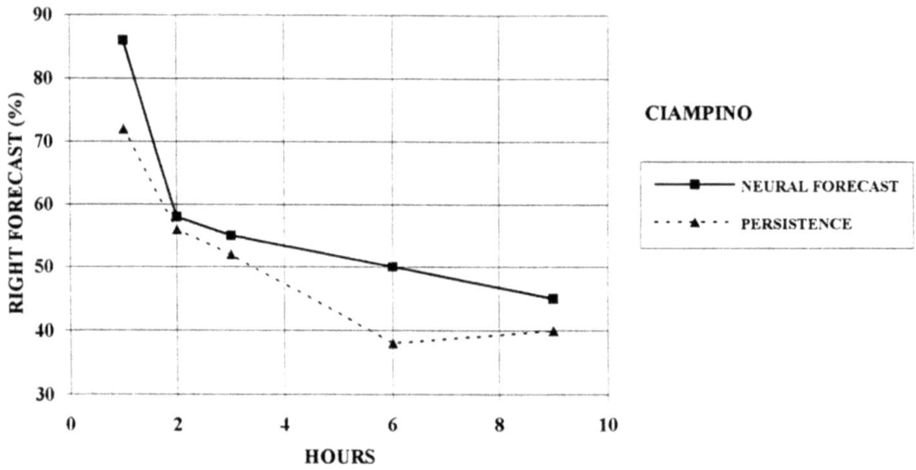

Fig.3. Comparison: neural forecast - persistence (the persistence is the forecast obtained considering the observed visibility at time 0 as a forecast for the following hours; in practice one predicts that visibility does not change).

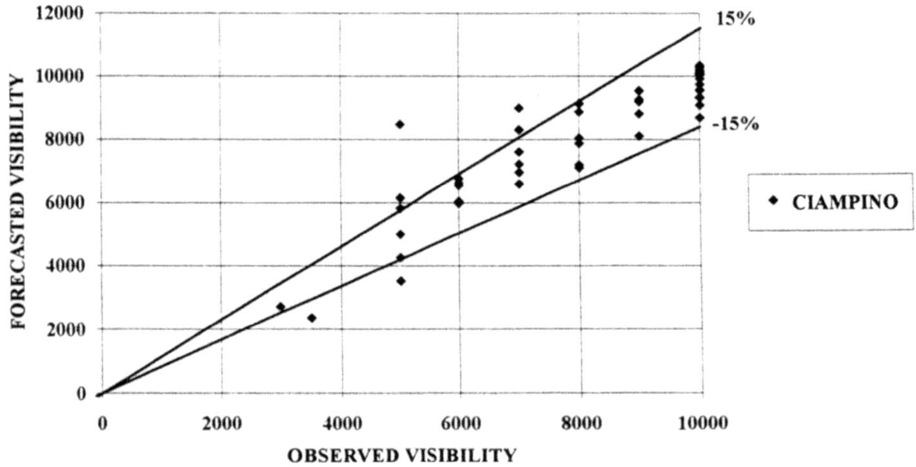

Fig.4. Neural - 1 hour forecast.

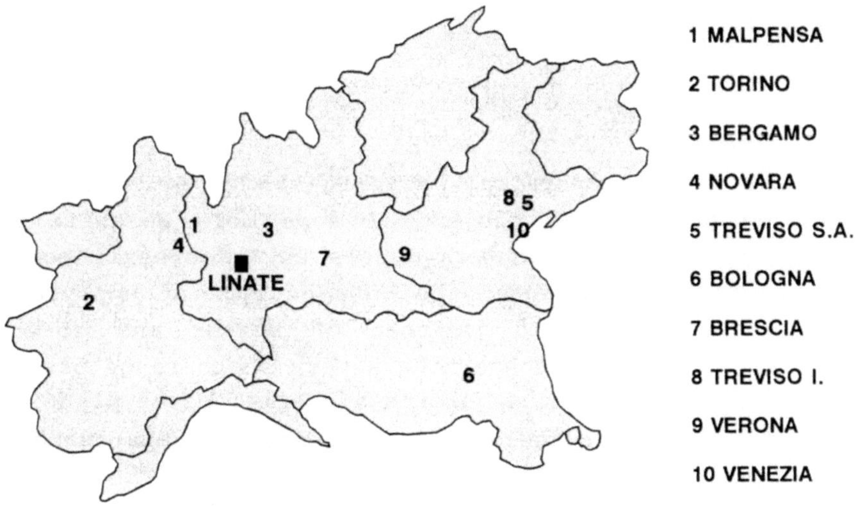

Fig.5. Horizontal pattern of stations for the Linate case study.

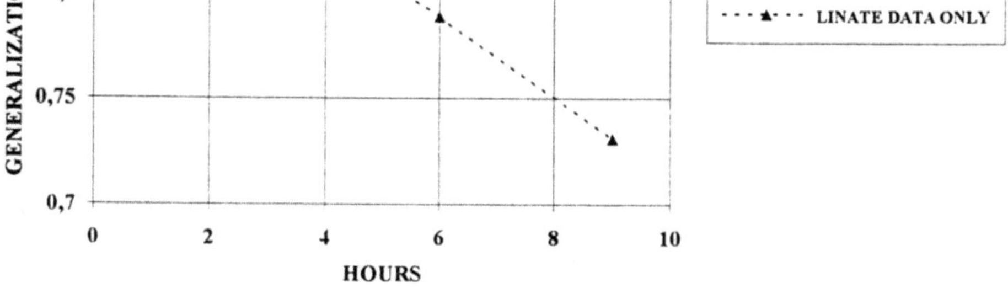

Fig.6. Sensitivity to the horizontal pattern.

Perspectives of applications of neural network models to atmospheric pollution problems

In the previous section we have presented an attempt at forecasting a local physical variable at the surface. In this chapter we present the project of some more specific future applications of neural network models to atmospheric pollution problems. To our knowledge, the only attempt at applying this technique in this field is that of Boznar and co-workers (Boznar *et al.*, 1994), concerning a very short-range prediction of SO_2 concentration values around a thermal power plant.

Here we limit ourselves to discussing three examples of application.

Example 1: monitoring a physical variable.

It seems to us that neural network modelling is a promising technique for the monitoring of the Mixed Layer Height (MLH) in distinct meteorological situations (advective, convective, stable). In the training procedure it is useful to consider the following inputs: vertical profile (wind, temperature, humidity), surface parameters (if necessary, also a horizontal pattern of temperature in the town), radioactive data (which can lead to a correct estimation of the MLH, as shown by Allegrini *et al.*, 1994). The target can be represented by the estimate of the MLH by means of known measurements or methods (SODAR and RASS measurements, potential temperature method, evaluation of the Richardson number). After the training phase, we could apply the networks structure with fixed weights to a practical use and obtain a correct MLH estimate without the use of expensive instrumentation or not uniquely determined methods.

Example 2: monitoring a chemical quantity.

In our opinion, our model could be useful for monitoring the concentration of a particular pollutant (a secondary pollutant, for instance). The following inputs can be used: concentrations of primary pollutants, physical variables of concern (vertical profile, MLH, surface parameters, solar radiation). The target can be represented by a secondary pollutant concentration measurement. We could obtain a correct estimate in real time of the secondary pollutant concentration without use of expensive instrumentation.

Example 3: forecasting a chemical quantity.
This case represents the most difficult task. Since we would like to obtain a forecast at $T+24$ hours, we cannot follow the example of Boznar and co-workers (Boznar *et al.*, 1994), whose prediction was at $T+30$ minutes, but we must associate the outputs of a dynamical meteorological model to our inputs. We can consider the attempt at forecasting the concentration of a particular pollutant (a primary pollutant, for instance) at time $T+\Delta T$. The following inputs can be used: primary pollutants concentrations at time T, meteorological forecast for elements of concern (wind, vertical profile, etc.) at time $T+\Delta T$. The target is represented by measurements of the pollutant concentration at time $T+\Delta T$. We could obtain a correct forecast for a primary pollutant concentration as post-processing of a dynamical model of the atmosphere and chemical data.

In short, a neural network model, tested on meteorological forecasts, has been presented. These results in forecasting a physical variable of the atmospheric system at the surface induce as to consider the applicationn of this model for monitoring and forecasting physical-chemical variables in the complex atmospheric pollution environment.

Acknowledgements

We wish to thank E. Setti (Polytechnic University, Milan) for helpful discussions, L. Graniero (Servizio Meteorologico Aeronautica) for his technical support, G. Punzo (IBM-SEMEA, ECSEC) and A. De Caro (Italian Air Force, DASRS) for some experimental studies concerning the Linate case study on their workstations

References

Allegrini I., Febo A., Pasini A, Schiarini S. (1994) Monitoring of the nocturnal mixed layer by means of particulate radon progeny measurement. J. Geophys. Res. 99 (D9), 18765-18777.

Boznar M., Lesjak M., Mlakar P. (1994) A neural network based method for short-term predictions of ambient SO2 concentrations in highly polluted industrial areas of complex terrain. Atmos. Environ. (in press)

Hertz J., Krough A., Palmer RG. (1991) Introduction to the theory of neural compotation. Addison Wesley.

Pasini A, Potestà S. Short-range visibility forecast by means of neural network modelling: a case study (in preparation)

Air Pollution in the Mediterranean: Atmospheric Dispersion Modelling

R. Salvador, E Mantilla, M. J. Salazar, and M. Millán,
CEAM, Palau de Pineda, Plaza del Carmen 4
E-46003, Valencia, Spain

Introduction

During the development of the EC's MECAPIP and RECAPMA projects, several aspects of the summer atmospheric circulations and associated air pollution dynamics over Spain and the Western Mediterranean were identified (Millàn, et al., 1992, Mantilla, et al., 1994). Threedimensional structures of the summer combined mesoscale circulatory systems were deduced. *During the day,* sea breezes develop in the coastal areas and can become reinforced due to the orographic uplift, depending on topographical orientation. In the peninsular areas, mesoscale circulation develops into a thermal low system. These circulations draw air inland from the lower atmospheric layers and they tend to further reinforce the subsidence over the colder sea. Associated return flows, influenced by topography, move towards the cold Mediterranean and Atlantic sea under a general level of subsidence aloof. *At night,* the Iberian Termal Low relaxes and subsidence predominates over the whole region. The coastal mountain ranges which surround the western Mediterranean area, including the Spanish Iberian system and the Atlas in North Africa, are higher than the main level of subsidence (between 700-1200 m) and part of the flow is confined to run along the coast parallel to the mountain ranges, towards Gibraltar. This flow is enhanced by the semi-permanent high in the Gulf of Lion, which forces a migration of the surface layer away from this area (Millàn, et al., 1992).
Under these meteorological conditions, the observed atmospheric pollutant concentration patterns show a non-homogeneous structure vertically and horizontally. It has been observed that anticyclonic subsidence does not represent a barrier for the pollutants to penetrate during the day, due to direct convective injection and/or to the topography aided injection.

NATO ASI Series, Partnership Sub-Series, 2. Environment – Vol. 8
Urban Air Pollution
Edited by Ivo Allegrini and Franco De Santis

The strong convective activity over the Iberian central plateau and/or orographic chimney effects provided by the coastal mountain ranges can easily penetrate and/or destroy the anticyclonic inversion during the day. During the afternoon, the convective activity over the central plateau can inject photooxidants from 3 to 5 km into the mid-troposphere and the observed mesoscale processes can transport ozone from 60 to 100 or more km inland from the coastal areas. Over the Mediterranean coastal strip, stratified reservoir layers form which are separed by weak inversions, stacked up to 2-3 km high. Depending on the height of the injection some of the ozone is able to escape and can be transported away from the area. At the same time, the lowest layers are brought inland by the sea-breeze during the day. In this way, recirculation of pollutants can persist for days adding new emissions and reaching episodic levels. Data analysis showed that the most relevant mesoscale processes are non-stationary. These include strong vertical and horizontal gradients of atmospheric parameters, coupling-decoupling cycles between upper and surface winds, and recirculatory flows. In this paper, some data collected during the 1989 summer campaign of the MECAPIP project were used to analyze the ability of the Lagrangian Adaptative Volume Plume Model (AVP), described in Ludwig et al. 1989, to simulate the dynamics of the power plant plume under the meso-meteorological conditions presented before.

Description of the study site.

The main airshed considered for the study, covering 50 by 46 km, is a flood plain that ends after approximately 30 km in a range of mountains which converge inland toward a narrow river valley (figure 1). The Castellòn power plant, within an industrial area, is located on the shoreline of the Spanish east coast. It is oil-fired and has two 540 MW units each feeding a 150 m stack. During the period presented here, only one unit was operating. The gathering of meteorological data relate to the plume dispersion behavior was begun by the power utility staff in 1982. The historical data set includes the standard meteorological information provided by the Spanish

National Weather Service (NWS) for the nearby station of Almazora (approx. 6 km inland). The analysis of the historical data (from 1983 to 1989) indicated that land-sea breezes occur during the whole of the year whenever the synoptic conditions are appropriate, although monthly windroses presented significant easterly winds associated with sea-land breeze from April to September. Seasonal windroses of the historical data show that two flow directions dominate: the first corresponds to the daytime sea-land breeze with east to south-southeast components, and the second to the nocturnal drainage flow induced by the topographical features of the site, coupled with land breezes, with west-north-west to north west components (figure 2).

Data analyses of summer campaigns show that the plume undergoes a diurnal cycle associated with the evolution of the prevailing atmospheric circulation (Millàn, et al. 1989). At about sunrise, the plume is transported approximately parallel to the coast in a SW direction, on a neutral to stable layer between 100-150 to 450-700 m decoupled from the surface layer. The onset of the sea breeze occurs early in the morning and the inland turning of the wind direction progresses, from ground upwards, and is completed by early noon. At about midday (11:00-13:00 UTC) when the Thermal Intemal Boundary Layer (TIBL) builds-up to plume height, the effluent suddenly fumigates downward toward the ground. During the late afternoon two plume behavior have been observed, they may be related to the formation and location of the Iberian Thermal Low. The plume either rotates counterclockwise toward a SW direction or rotates clockwise to end-up in a N-NE-N direction.

Application od the AVP model during a sea-breeze event.

The Lagrangian Adaptative Volume Plume Model (AVP) as described in Ludwig, et al., 1989 approximates a continuous plume by overlapping a series of discrete volumes that change their shape in response to vertical wind shear and changes of atmospheric stability with height. Each volume is described by five Lagrangian points at different altitudes. Horizontal and temporal changes in meteorological parameters are accounted for as the five points move and diffuse independently according to local meteorological conditions along their trajectories; thus each adaptative volume element consist of four segmented parts. Previous results.showed that the model reproduces the shapes, slopes, and general dimensions of plume vertical crosssections observed by airborne lidar at Kincaid, Illinois (Ludwig, et al., 1989). This model was improved by Bornstein et al., 1990 to establish a 3D sub-grid around the plume in which concentrations are computed, when the plumes are considered to be imbedded in 3D Eulerian dispersion models.

In this application, the AVP model was used to simulate sulfur dioxide concentration patterns in the surroundings of the power plant near the Spanish east coast, during sea breeze conditions. For this purpose, a typical sea breeze period was selected of the data collected during an experimental campaign of the summer of 1989. This intensive field campaign was conducted on a peninsular scale, in the frame of the MECAPIP project (as described in Millàn and Salvador, 1994). For this study, data collected 27 July 1989 was selected to simulate sulfur dioxide concentration patterns in the surroundings of the plant. Tethersondes were launched near the plant from early morning (04:00 TC) to late afternoon (18:00 UTC). An instrumented aircraft measured meteorological and air pollution parameters. The flight route was a horizontal fan pattern shown in Figure 1. A tracer experiment was also conducted and a mobile laboratory recorded SO_2 concentration every 10 minutes (CS-SUR in Figure 1). The synoptic condition for 27th July 1989 shows the Azores anticyclone centered over the Southwest of Ireland, with an associated ridge aligned NW-SE between the Gulf of Biscay and the east coast (figure 3). The Iberian Thermal Low (ITL) was located over the southern half of the peninsula. On the 500 hP weather map a divergence associated with the ITL was observed. The formation of the thermal low produces a significant pressure gradient over the northern coast to the center of the peninsula. Except for this gradient, the horizontal pressure gradient was weak over the peninsula, and sea breezes developed on the East coast.

Analysis of the input data.

The meteorological parameters that atmospheric dispersion models need are provided by an objective analysis of observational data or by Boundary Layer models. Although these models were developed originally to be use with such a model, in these applications hourly wind and temperature profiles recorded by tethersonde were objectively analyzed to provide model meteorological input data, assuming horizontally uniform wind field. The terthersonde was located at approx. 4 km from the shoreline (CS-SUR, in figure 1).

Vertical air temperature gradient was used to calculate atmospheric stability following Ludwig et al. (1977). First, the meteorological variables were calculated in 20 levels in the vertical using a weigh interpolation scheme, and then they were interpolated linearly in time, giving values every hour. Each hour the meteorological values were updated, assuming that during this period the values are uniform. They were modelled every 14 hours, from 04:00 to 17:00 UTC, this is the typical period of the sea breeze.

Source emission parameters, such as emission rate, exit temperature and velocity of the gas SO_2 were given by the utility staff. In this study, it was considered a buoyant plume. Some modifications in the plume rise parameterization of the original version of the model were made to account for the atmospheric stratification observed on this day. Effective plume height was between 300-400 metre, except for the last two simulated hours,16:00-17:00, when the effective height reached up to 600 m due to tile light wind at stack height.

Analyzed Wind Field Pattern.

The analyzed resulting wind field were represented in figure 4. In this simulation, it was assumed a vertical wind equals to zero. The analyzed wind field showed that from 7:00 to 13:00 UTC a progressive increase in wind speed was observed, associated with seabreeze/upslope winds (Figure 4). During these hours, the growth growth of the TIBL, indicated by a higher wind speed, increased very rapidly during the morning and decreased slowly during late afternoon. During the sea-breeze period, the wind shifted in a clockwise direction, attributed primarily to the Coriolis force action on the sea-breeze. After 16:00 UTC, light return flows were observed above 400 m. For successive hours, the return flow layers became deeper as a consequence of persistence of large scale subsidence over the cold sea.

Analyzed Temperature Profiles.

The analyzed temperature and wind profiles showed that early in the morning atmospheric stratification presented several elevated inversions, which could be attributed to the different time arrival and heights of the sea-breeze/up-slope return flow layers of the previous days (Figure 5). At 12:00 LTTC, an unstable surface layer and strong wind shear were observed along with return flow around 800 m to the top. A weak subsidence inversion was observed throughout all the soundings which sunk progressively towards the ground along the evening and the subsidence became more intense after 15:00 and by 17:00 its base was detected to around 200 m.

Results and discussion

Under the prevailing meteorological conditions, the following simulated plume evolution was as follows: early in the morning the plume was transported decoupled from the ground, with northeasterly wind, parallel to the coast. For successive hours, the plume rotated clockwise, became channeled along the river valley and produced a rotating fumigation from ~7 km west of the plant to 32 km, until 16:00 UTC (Figure 6). Anticyclonic tuming of the plume continued during the whole simulated period. It ended up almost decoupled from the ground in a northerly direction at 17:00 UTC (see Figure 7).
Significant concentration levels, at 10 meters, were not predicted until 10:00 UTC, when the onset of fumigation occurred (Figure 6). By 13:00 LTTC, the concentration distribution showed a multiple peak pattern. Maximum concentration values of 72 ppb were predicted around 10 km west of the plant and a second peak of 62 ppb around 17 km west from

the plant. By 16:00 UTC some surface concentration was predicted, but by 17:00 most of the material was aloft (Figure 7). Modeled concentration patterns only reached the ground-base monitoring station, CS-SUR, at 13:00 UTC (predicting 62 ppb) while the observed concentration peak was around 12:00 UTC (measuring 234 ppb). Observed concentration decreased slowly until 17:00 UTC. Differences between observed and modelled values at this station could be attributed to overprediction of plume rise, to inaccuracy of the wind field or to the measurement of concentrations from other sources downwind.

Isopleths of the concentration cross-section were plotted in perpendicular planes to the plume centerline (Figure 7). At 07:00 UTC the plume was mostly confined to 200-450 meter layer, between the two inversion detected during morning (Figure 4). Some evidence of turbulent mixing was predicted in lower levels. By 10:00 LTTC wind shear was reflected in the concentration pattern cross-section, showing a slanted shape and two maxima at different levels, due to higher wind speed at the lower levels while plume rotates northward. By 12:00 UTC the plume was fumigating down to the ground. At 17:00 UTC plume was decoupled from the ground and had a ">" shape showing the effect of wind shear. Material was mostly over the 200 m inversion, at 19 km away from the plant.

Sulphur dioxide concentrations were measured by the instrumented aircraft at 500 points at different heights from the m.a.s.l. at nearly constant height following the terrain (Figure 8). Maximum concentrations were observed close to the source. The general plume shape was well reproduced by the model and multiple maximum zones were well predicted, although modeling results failed to simulate the maxima near the source. This may have been caused by inaccurate calculation of the terrain height over the flight route. In the future those heights would be digitalized precisely. Concentration values calculated by the model were, in general, overpredicted.

Conclusion

These preliminary simulations show that the model can reproduce the temporal plume evolution. The model predicted rotating fumigation of pollutants around midday according to experimental observations. In the future, the model will be applied with data collected during the RECAPMA project, during which experimental measurements of plume rise and dispersion parameters of the plume were conducted (Alonso et al. 1993). Comparison with this data will permit adjustment of plume rise and diffusion parameterization. It will also be using outputs generated by a PBL model to provide more accurate meteorological parameters.

References

Alonso, L. A., G. Gangoiti, M. M. Navazo, J. A. Garcia, C. M. Gutierrez Canas, J. L. Ilardia. 1993. Experimental Determination of σ_y, σ_y and Plume Rise from a Tall Stackin the Western Mediterranean Coast (Castellòn) Spain. Paper presented at the 6th European Symposium on Physics-Chemical Behaviour of Atmospheric Pollutants, Varese, Italy, October 18-22.

Borstein, R, Z. Boybeyi, and F. Ludwig. 1990. Further Developments of Adaptative Volume Plume Model. In Air Pollution Modelling and Its Applications VIII (ed. by H. van Dop and D.G. Steyn). Plenum Press, New York and London.

Mantilla,E.,J.F.Bea, R. Salvador, and M. Millàn. 1994. Meteorological Processes on the Western Mediterranean. Experimental Analysis. Air Pollution II. Vol. 2. Editors. J.M. Baldasano et al., Computational Mechanics Publications. Southampton. pp 441-448.

Millàn, M. and R. Salvador. 1993.Field Experiment Design for the EC project MesoMeteorological Cycles of Air Pollution in the Iberian Peninsula (MECAPIP). Paper presented at the Conference on regional Photochemical Air Quality Measurement and Modeling Studies. 7-12 November 1993. San Diego. Califomia.

Millàn, M., R. Salvador, B. Artinano and I. Palomino. 1989. Power Plant Plume Dispersion Study in a Coastal Site with Complex Terrain. In Air Pollution Model. and its Appl. VII (ed. H. van Dop). Plenum Press, New York and London. pp 361-371.

Millàn, M, B. Artinano, A. Alonso, M. Castro, R. Fernandez-Patier, and J. Goberna. 1992. Mesometeorological Cycles of Air Pollution in the Iberian Peninsula. Air Pollution Research (ed. M. M. Millàn) Report 44, CEC DG XII/E- 11, B-1040. Belgium.

Ludwig, F. L., L. S. Gasiorek, and R E. Ruff. 1977. Simplification of a Gaussian Plume Model for Real-Time Minicomputer Use. Atmos. Environ. 11: 431-436.

Ludwig, F. L., R. Salvador, and R. Bornstein. 1989. An Adaptative Volume Plume Model. Atmos. Envir. 23: 127-138.

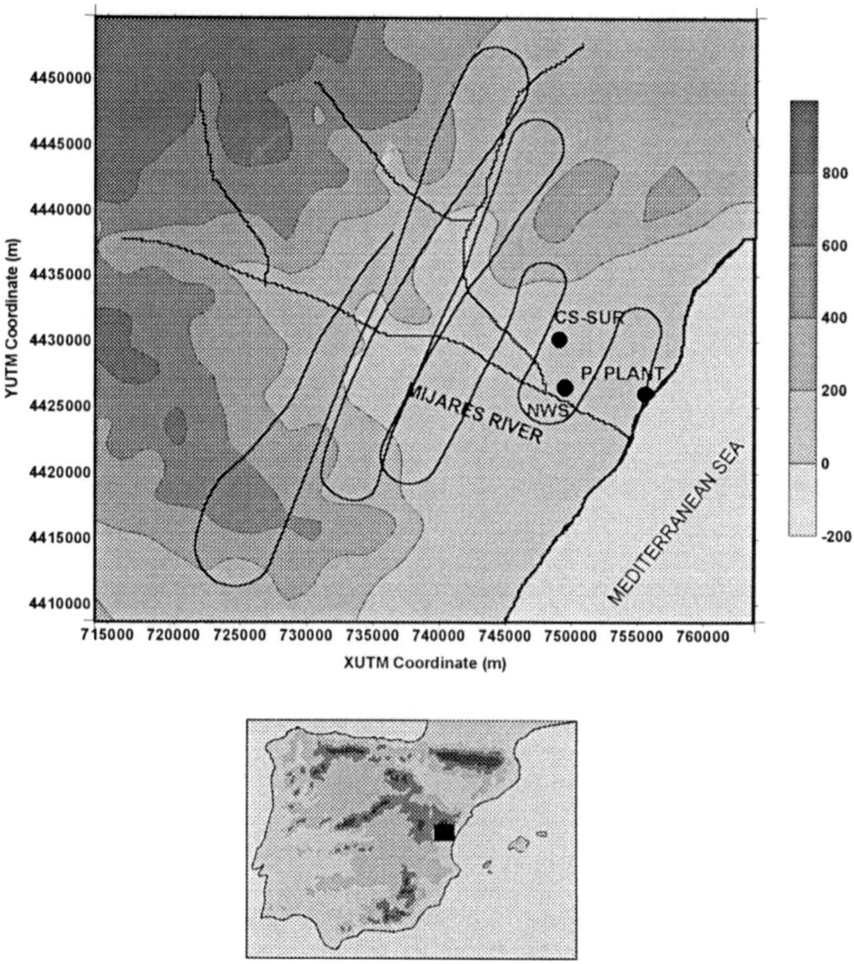

Figure 1. Map of the site showing main topographical features, ground-base monitoring station (CS-SUR), Almazora NWS, location of the power plant and tethered-sonde (P. Plant). Dotted line is aircraft flight route.

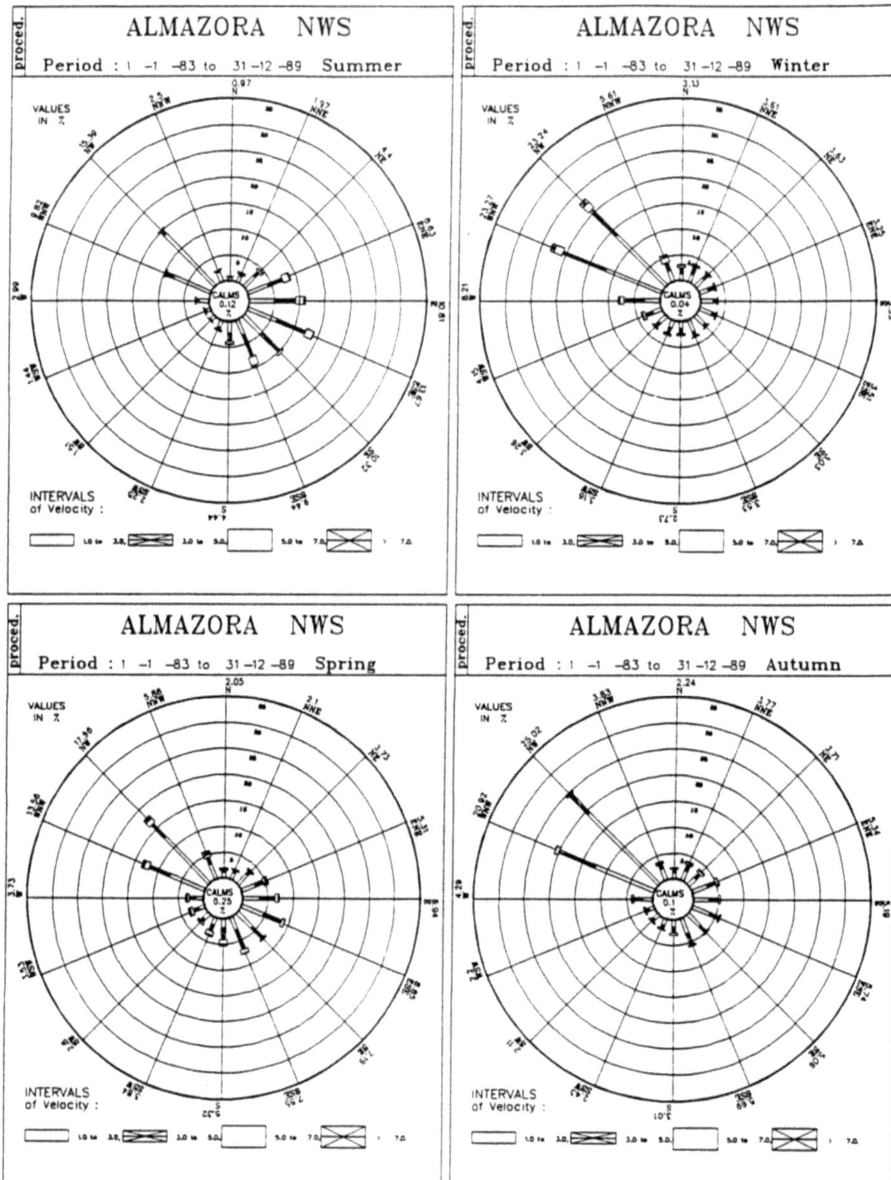

Figure 2. Seasonal windroses at Almazora National Weather Service.

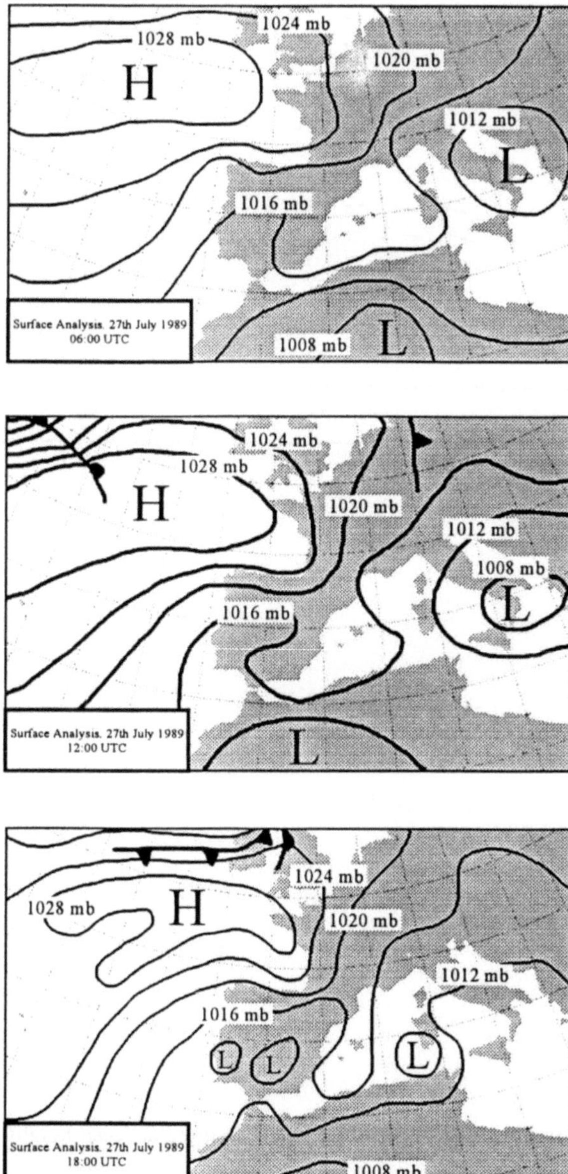

Figure 3. Synoptic surface pressure map.

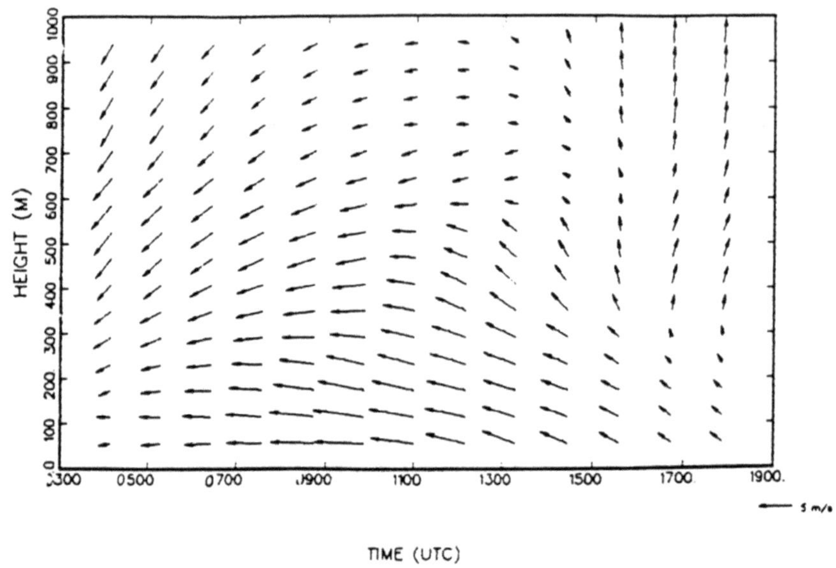

Figure 4. Analyzed wind profiles for 27 July 1989 obtained at fixed heights, from soundings near the power plant.

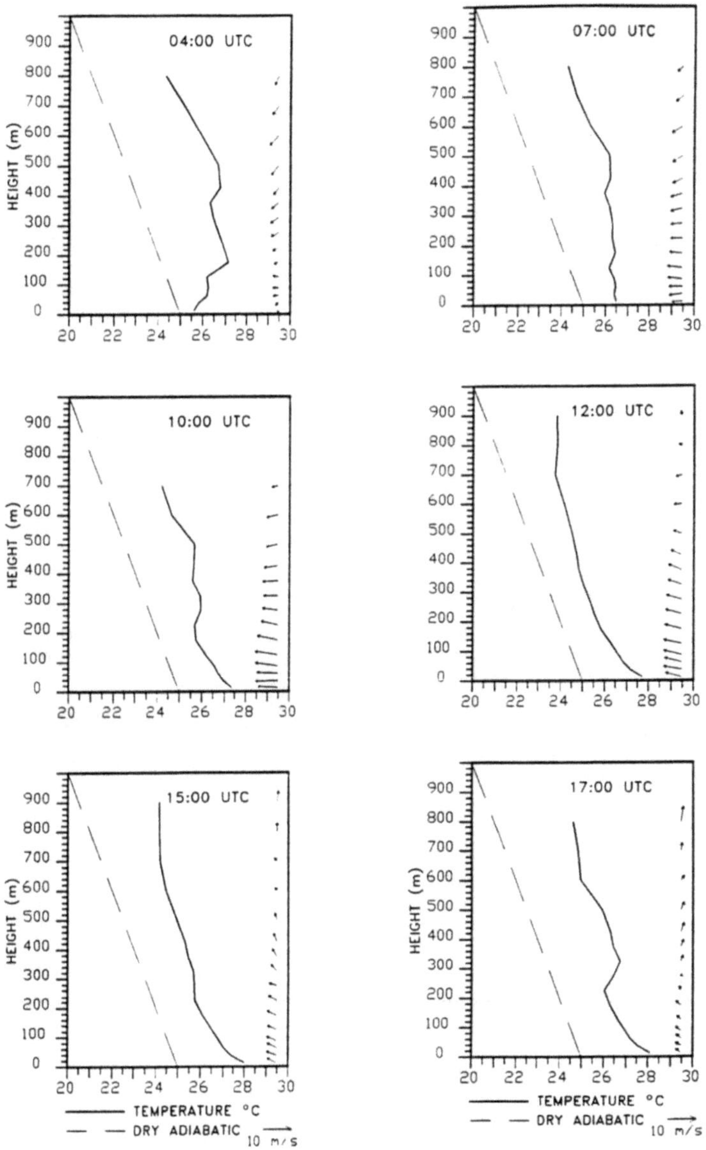

Figure 5. Analyzed temperature and associated wind profiles for 27 July 1989 obtained at fixed height from the soundings. Dry adiabatic has been plot for reference.

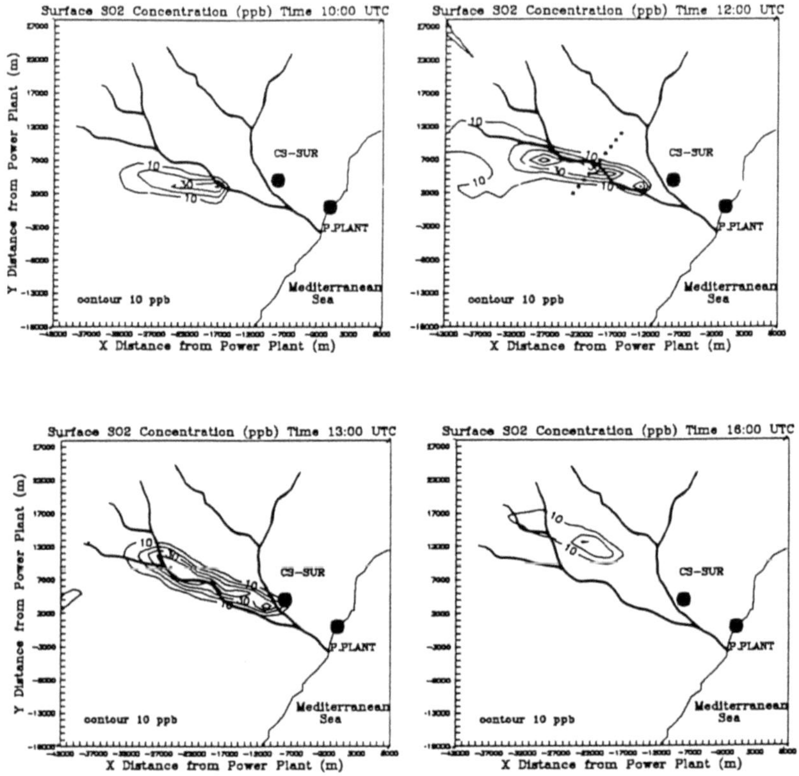

Figure 6. Calculated sulfur dioxide concentrations pattern for 27 July 1989 at 10 meter above the suface, for some hours with significant level of sulfur dioxide concentration. Dotted line refers to cross-section of figure 7.

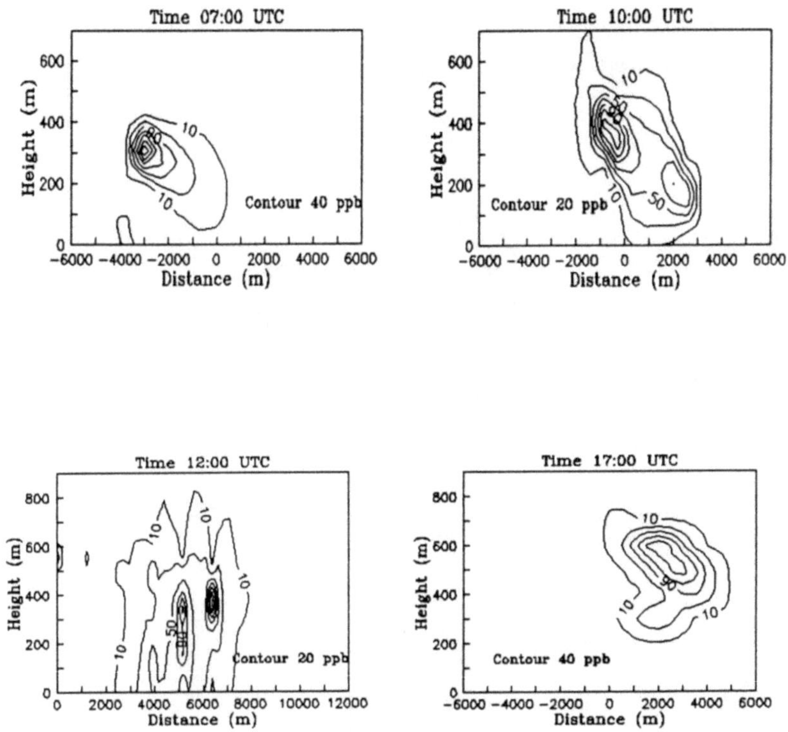

Figure 7. Calculated sulfur dioxide concentration pattern for 27 July 1989 in a N-S plane about 16 Km W of the source at 07:00 and 10:00 UTC, in a NE-SW plane about 17 Km NW of the source for 12:00 UTC, and in a W-E plane 19 km N of the source for 17:00 UTC.

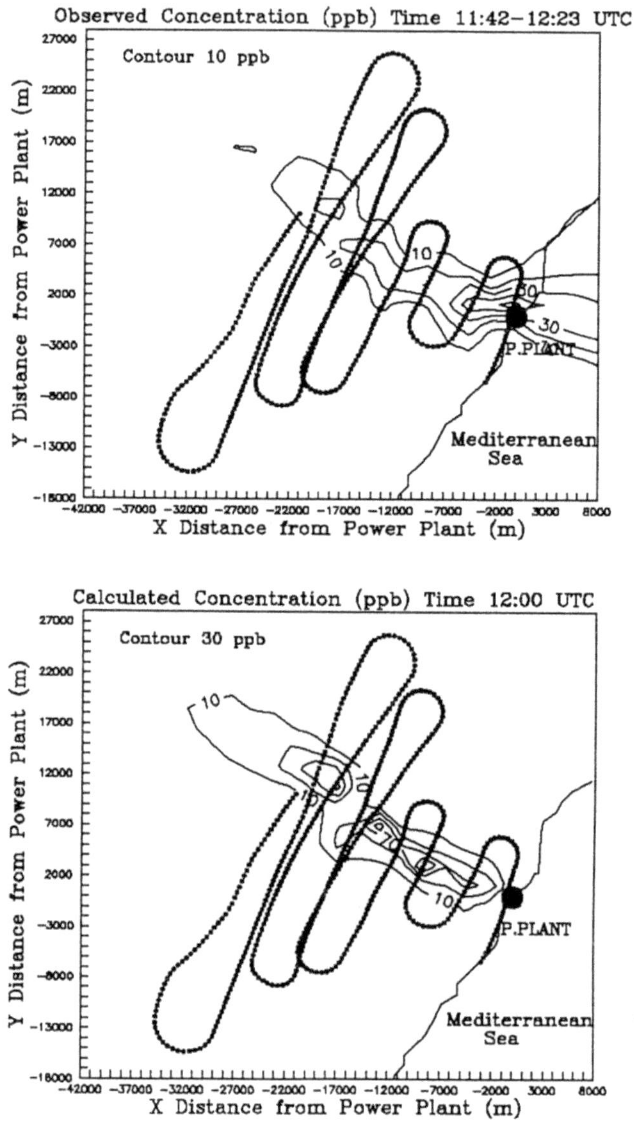

Figure 8. Comparison between aircraft observation and calculated sulfur dioxide concentration pattern in a approximated constant height following terrain.

System Approach to Air Quality Monitoring - Identification of Main Problems

Jacek Walczewski
Institute of Meteorology and Water Management
ul.P.Borowego 14
30-215 Cracow, Poland

Air Quality Monitoring in the system of relations and feedbacks

Air-quality monitoring is a type of system, which is also part of a higher-order system of relationships and dependences. It is necessary to be aware of this fact in order to be able to rationally plan and design the monitoring systems. Fig.1 presents a general scheme of relationships which influence the monitoring activity. This scheme reveals a system with 2 feedback loops: the "outer" one and the "inner" one. Let us follow the "outer" feedback loop. As the result of human activity, emissions of different substances are released into the atmosphere. Here they are dispersed and their chemical composition is often changed. The final effect is the distribution of concentrations of different pollutants at ground level, in other words - air pollution.
The monitoring of air pollution records the types of pollutants and their concentrations in their spatial and time variability. This information is eventually used for the adoption of some control strategies. If the air quality is found to be acceptable or quite acceptable, development of emissions is controlled accordingly to local land-use plans (resorts, parks, intensification of urbanization and industrializzation); if air quality is poor, emission abatement activities are necessary, and they can be of long-range or short-range character.
At this point the "outer loop" of the feedback closes - the information on emission effects is used to control the emissions.
The "outer loop" may have a side-branch. Sometimes the response to poor air quality cannot be active (emission abatement), but only passive: a form of adaptation to the existing situation.

NATO ASI Series, Partnership Sub-Series, 2. Environment – Vol. 8
Urban Air Pollution
Edited by Ivo Allegrini and Franco De Santis

In this case, too, the information from a monitoring system remains important and useful.

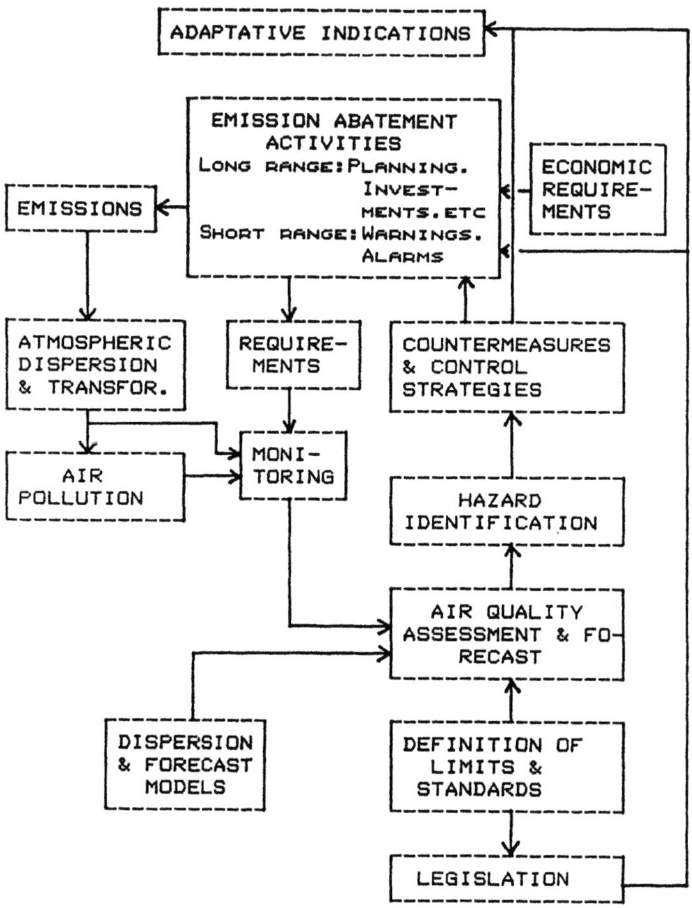

Fig.1. Scheme of the system of relations and feedback influencing Air Quality Monitoring.

The "inner loop" of feedback may be followed from the point in which air-pollution control strategies have been formulated. Accepted emission-control schemes, or other forms of response to the information on air quality, may superimpose new requirements on the monitoring system. These requirements may refer to the spatial and temporal scheme of measurements, to the measurement accuracy (or general data quality) and others. In most cases only the average levels of concentration are inadequate for data users. In order to describe and apply the cause and effect relationships which create air pollution it is necessary to have enough information on the processes of the atmosphere.

In other words, the "inner loop" reflects the simple statement that the type of the monitoring system depends on the goals it should reach. Although simple and obvious, this point of view is sometimes overlooked, and existing monitoring systems often appear as isolated, or independant bodies. The closed-loop system shown in Fig.1 receives the input from the outer environment at 3 points (assessment of air quality, set of economic requirements and third use of models) The first input is the assessment of air quality which needs a definition of limits and standards to be compared with measured data of air pollution. Here there is a connection between the broad field of investigations and analyses of what is acceptable, what is harmful and what is hazardous for human health, vegetation and cultural heritage. This is an open field of discussion which is far from reaching a final conclusion; the differences of air quality standards in different countries testifies the complexity of the problem.

The setting of standards also has its legal aspect. In general, environmental legislation can have far-reaching consequences on emission abatement activities, and indirectly - on monitoring system operations. The second input to the system is the set of economic requirements, superimposed both on pollution control strategies and on the functioning of monitoring systems. Here we meet a more general problem: how much are we ready to pay for our environmental safety? The methods of cost-benefit analyses need quantification of the cost of elements, like human life and health, living natural resources and cultural heritage. Another practical problem in many countries is the following: the financial support for air quality protection is strictly limited and inadequate to relative needs; it is thus necessary to make the difficult choice between support for monitoring or spending money on abatement technologies. Here it

should be mentioned that investments in abatement technologies may miss the target if they are made without sufficient recognition of the situation, and recognition of situations which need monitoring systems.
The third input to the system are models which are used for interpolation, extrapolation and forecast of air pollution. They also play an important role in the designing of monitoring networks, in planning long-term emission abatement activities (models for simulation studies), as well as in the operation of warning systems (pollution forecast models).
When designing a monitoring system, first of all goals must be formulated. They can be divided into 3 groups:
- assessment of air quality;
- determination of cause-effect dependences;
- warnings.

It is very difficult to design a monitoring network without any previous data on air quality. If these data are indeed lacking, it is necessary to carry out a preliminary recognition by means of studies of emissions, inflow, meteorological conditions and phenomena, which can contribute to the air pollution.

Internal structure of the Air Quality Monitoring System

Looking inside the "Monitoring" box in Fig.1 we find an internal structure, whose general aim is to transform the environmental input into output information (Fig.2). The form of output information is defined by "Requirements", which prescribes the types of pollutants monitored, time and space resolution of observations and the format of output communications to be used outside of the system.
The environmental input is essentially a system of physical signals, which should be received and recorded by sensors of the Measurement Block. The Measurement Block consists of monitoring stations and laboratories, meteorological stations (or links to a standard meteorological network), and - sometimes - separate instrumentation for the measurement of pollution inflow to the area of interest (mobile stations, "dedicated" stations, remote-sensing systems, etc.).
The operation of the Measuring Block needs continuous technical

support: maintenance, repairs, materials, spare parts, calibration routines. Moreover, it should be subjected to Data Quality Assessment.

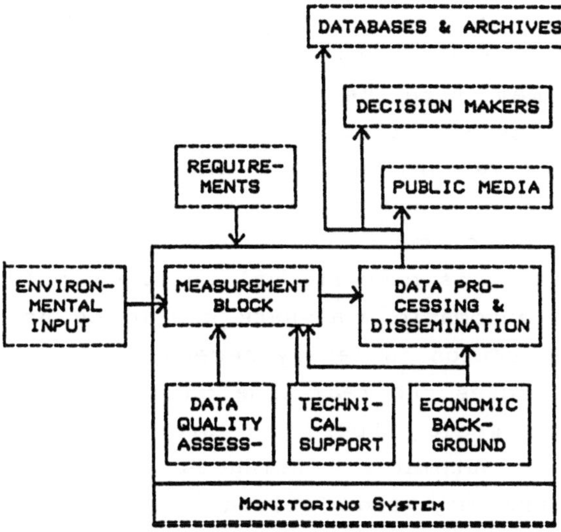

Fig.2. The general structure of Air Quality Monitoring System.

The role of these two elements: Technical Support and Data Quality Assessment cannot be overlooked, because running the measuring networks without adequate technical support and without a Data Quality Assessment System, may lead to the collapse of the whole monitoring system.

The data from the Measurement Block are transferred to the block of Data Processing and Dissemination. Data transfer and processing techniques may have different forms depending on communication and the computer systems used. Careful selection of these systems at the very beginning plays an important role in achieving the goals of monitoring. The wrong choice may jeopardize the effects of very good measurements.

The Processing and Dissemination Block should produce the final information which will be used: messages for decision-makers and public media, bulletins and reports, data files for databases and archives.

The last element of the monitoring system is the "Economic Background", in other words - the source of financial support for the functioning of the whole machine. There is a diversity of practical solutions here: national funds, local funds, incomes from data users ("user-pays-principle"), sponsorship by industries or by consortia of emitting plants and citizen organizations. The most common source of problems is the situation which may here be called "the trap of initial sponsor". It happens, that a rich sponsor (e.g. an international organization) supports the set-up of the system and then the system is abandoned by the sponsor and then it has to find its own means of survival.
The cost of operation is proportional to the complexity of the system. Thus, the automatic analysers should be used only when their continous operation is really necessary, e.g. for warning purposes. In such cases the real-time data output to decision-makers and public media should be provided, otherwise the effort and cost of automatic continuous measurements will be lost. These remarks may seem trivial, but the fashion for automatic analysers and the relatively ready access to "initial sponsors" quite often lead to a waste of time and money due to the installation of systems which are not able to survive longer than 1-2 years.

A practical example an air monitoring system in Cracow,Poland

It seems opportune to illustrate the theoretical considerations given above, by mean of an example from experience, showing how the system of relationships and dependences works in real life. The case of Cracow may be instructive, because it joins together the structural background of Eastern Europe with elements of Western know-how and technology. The regular monitoring of air quality began here in the late sixties and a network of 24 stations, measuring mean-24 h concentrations of SO_2 and fine dust was established. This network was complemented by several tens of dust collectors. The network - slightly reduced and modified - continues to operate at present, too. It is managed by the Sanitary Inspectorate, belonging to the Ministry of Health. Since the early seventies, efforts have been made to prepare 24-hour forecasts of SO_2 pollution in wintertime; this has only been partially successful, because the method of manual analyses of samples in laboratories makes it practically impossible to supply the data on concentrations in time for the preparation of forecasts. It should be mentioned, that statistical methods, used for predictions, make use of- among other things - of concentration data from the previous day. Nevertheless, the annual and seasonal pollution data from the network have been extremely useful for many practical purposes; they have also revealed the phenomenon of winter pollution episodes, quite severe in the mid-seventies (up to 1000 $\mu g/m^3$ diurnal mean values of SO_2 concentration).

These observations gave impulse for the preparation of a new monitoring system, oriented to give warning signals, at least in the period of winter episodes.

The cost of putting this idea into effect was very high, in relation to the funds available for air quality protection from national sources. Setting up the new automatic measuring network had to wait until some aid came from abroad. One automatic station was installed in 1988 thanks to a donation from Sweden, and a network of 8 stations became operational in 1992 as the result of support from the USA. At the same time, a World Bank programm helped to create the remaining infrastructure, complementing the support from national sources. As a result, a new automatic system emerged as a joint effort of national and foreign know-how, with a substantial contribution from US Environment Protection Agency and counsellors of the World Bank.

A simplified scheme of the system, connected to the automatic monitoring network, is shown in Fig.3. The real-time concentration data are telemetered to the Pollution Data Center, managed by the Environment Protection Inspectorate (belonging to the Ministry of Environment). They are transmitted to the Local Forecast Office of Meteorological Service, where they meet meteorological data from standard and dedicated networks, including vertical profiles (inversion layers, upper winds) from a specially organized station.

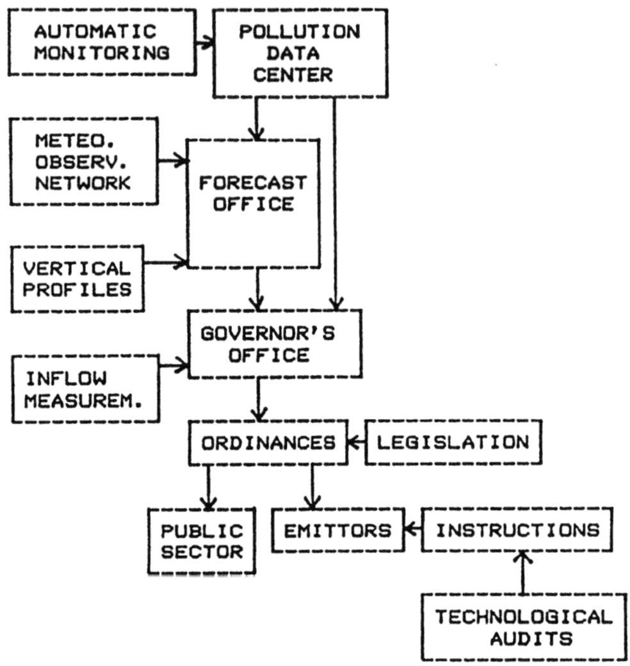

Fig.3. Structure of the Air Quality Monitoring and Warning System in Cracow, Poland (a simplified scheme).

The Forecast Office is a round-the-clock working unit, with the capability of issuing pollution forecasts or warnings based on current measured data.

The body, legally authorized for the issue of ordinances, obligatory for the public and industrial sector, is the Governor of the Province and his staff. The Governor receives summarised messages from the Pollution Data Center and current information from the Forecast Office (the Forecast Office has been authorized to transmit selected information directly to press, radio and local TV). In the case of winter episodes, emission abatement activities should be enforced. A group of industrial and energy-producing plants is prepared for emergency reductions of emission of SO_2 and dust. Technological audits have been performed in these plants by groups of specialists, and individual instructions for each plant have been provided. It has also been taken into consideration the plan to limit traffic in case of smog episodes. Putting this emergency system in operation needs adequate legal background. The local legislation of the Province (Governors's ordinances) must be adapted to the legislation at national level (Environment Protection Act and/or Act for Extraordinary Environmental Hazards).

Here is one of the biggest problems in making the whole system operational as an Alarm System, enforcing compulsory actions in industrial and public sectors. Till now, the system has produced and disseminated communications, which are for information only: current pollution reports and forecasts, regular monthly bulletins, annual reports. They are complemented by publication of indications referring to behaviour and health protection in cases of smog episodes ("adaptative measures").

Moreover, in the meantime the occurence frequency and severity of winter episodes has become much lower (rare events with 24-h means of SO_2 concentration up to 350 $\mu g/m_3$) and the new problem of summer smog is becoming quite serious.

Summarizing, the main problems of the system are:

(a) Financing - At present the system is supported by the Local Environment Protection Fund,whose resources are variable from year to year;

(b) Communication systems - their dependability and cost;

(e) Integration of computer system of participating units;

(d) Supplies of spare parts for instrumentation repairs;

(e) Matching the legislation to the system potential and objectives.

Development of smog warning systems in large industrial metropolises based on the example of Upper Silesian Industrial Region.

Leszek Osródka
Environmental Control and Monitoring Department
Institute of Meteorology and Water Management
Katowice Branch
Ul. Mariacka 6
40-014 Katowice
POLAND

Air quality monitoring is usually carried out taking into consideration the boundaries of particular cities. However, urban development processes aim at the creation of large urban-industrial complexes, occupying large areas and many cities wich cause various emission profiles. An example of such a metropolis is the Upper Silesian Industrial Region located in southern Poland, which can be a perfect area to carry out research work and to identify problems.

Most of these problems are still far from being solved scientifically and practically. This study is an attempt to show these problems and suggest the direction of research work aimed at explaining such problems.

The Upper Silesian Industrial Region is the largest urban-industrial complex in Central Europe. Furthermore, the geographical conditions and the nature of housing development are different than those in other known urban and industrial complexes.

It occupies approximately 1% of the total area of Poland, while it is inhabited by more than 10% of the population of Poland. As regards the total emissions, the region has reached the first position in Poland emitting approximately 20% of all the pollution in the country.

The Upper Silesian Industrial Region (GOP) is located in the south of Poland, within the area of the Silesian upland. The average altitude above sea level is between 200 and 400 meters, the drops

NATO ASI Series, Partnership Sub-Series, 2. Environment – Vol. 8
Urban Air Pollution
Edited by Ivo Allegrini and Franco De Santis

within the Silesian upland reach 80 meters. There is a watershed crossing the area of GOP between two major rivers of Poland (Wisla and Odra). The central part of Gop has been heavily transformed from an anthropogenic point of view through the creation of urban-industrial complexes mixed with wasteland, forests, landifills etc. Most of GOP'S natural environment is heavily contaminated including air, water and soil pollution (fig. 1).

Fig. 1. Location of monitoring stations in the Upper Silesian Industrial Area. 1.- neighbouring industrial areas, 2 - ambient air monitoring stations, 3 - meteorological stations, 4- central meteorological station (sodar), 5 - central data acquisition and processing station, 6 - borders of the Upper Silesian Industrial Area.

The climate of the Upper Silesian Industrial Region belongs to a group of temperate climates influenced by the continental and oceanic impact.

Over the years, high industrialization and compact housing development in the central part of Katowice region called here the Upper Silesian Industrial Region, have caused those particular climatic features, peculiar to large urban and industrial centers. As a result of air pollution, the access of direct solar radiance has been reduced by approx. 25% over the whole spectrum. It has also been observed that the center of GOP is clearly overheated in comparision with the surrounding area creating an "urban heat island" with an average intensity of 1°C a year (in extreme cases the differences reach 5°C). The structure of an urban heat island has a multicellular nature, and the highest intensity has been recorded in the most industrialized and urbanized areas (Bytom, Zabrze, Ruda Slaska) [Osródka, 1991].

It is also necessary to take into consideration the influence of GOP on the change of wind structure in the region. While the general anemologic conditions of the region are mainly determined by general circulation factors (the prevailing wind direction is NW in hot season and SW in cold season) and forced air flow through Moravia Gateway occurring 30% of the time, the wind structure over GOP is modified by local factors. In GOP an average slackening of wind speed reaches the level of 40% of the original speed.

An important geographical element complicating the aerosanitary relations in GOP is the presence of other large industrial centers in its neighbourhood.

About 20 Kilometers West of the western edge of the region there is a large coke oven and chemical industry complex. In the South-West, in northern Moravia, the Ostrava coal basin is located. A certain topographic system causes the air flow to be forced through Moravia Gateway and the transboundry pollution is transported to GOP even in some circulation circumstances that are not connected with the wind inflow from the South-West [Walczewski,1976, Hlawiczka S. et al., 1994].

In the East, GOP is in the neighbourhood of the Cracow urban-industrial complex which, even though to a smaller degree, also has an impact on GOP's aerosanitary conditions. According to a study of Walczewski and Tomaszewska [1992] approx. 3.5 to 7% of annual pollution emitted by Sedzimir Steel Mill in Krakow reaches GOP.

Therefore, the following factors have an influence on the total pollution in GOP:

- high pollution background,
- industrial point source and fugitive emissions,
- tramsportation emissions,
- local emissions.

There is a particularly important problem of low level fugitive emissions from households that is predominant in the centers of old GOP towns.

The most important emission problems in this region include emissions of SO_2, NO, CO and aromatic hydrocarbons.

The first air quality monitoring network in Katowice region was established in 1950 by SANEPID. However, the passive methods of air quality monitoring were established to a greater extent in 1970. The information obtained from these measurements allows us to establish a considerably complete picture of air pollution with more than 20 chemical substances, on an annual basis as well as for hot and cold seasons.

An addition to this monitoring system is the automatic air monitoring system that was established in 1993.

Due to the complicated nature of GOP, the Regional Air Pollution Monitoring System in Katowice region has to take the following factors into consideration at the same time: air pollution background, high pollution contribution of industrial emissions, pollution from transport and fugitive low level emissions.

At present the Regional Air Monitoring System consists of four elements:

- emission monitoring unit (equipped with mobile emission monitoring stations),
- ground level pollution monitoring unit (equipped with 10 automatic air quality monitoring stations located in representative points (as regards ground level pollution concentration).of the Katowice region
- metereological unit (consisting of the Central Observatory Station in Katowice-Muchowiec and 5 automatic meteorological stations which evaluate the meteorological background of the system)
- forecasting - optimizing unit, which is designed to work out the forecast conditions for ground level pollutant concentrations in different time-limits, including forecasting of emergency conditions.

The unit that links the operation of the system is the Central Station of Data Collection, which receives the information from ground level monitoring stations and from metereological stations via the radio.

The permanent data customers are:

- government and local governments
- environmental control organizations
- general public.

These organizations have computer terminals of the system which are the stations where the data are made available (fig.2).

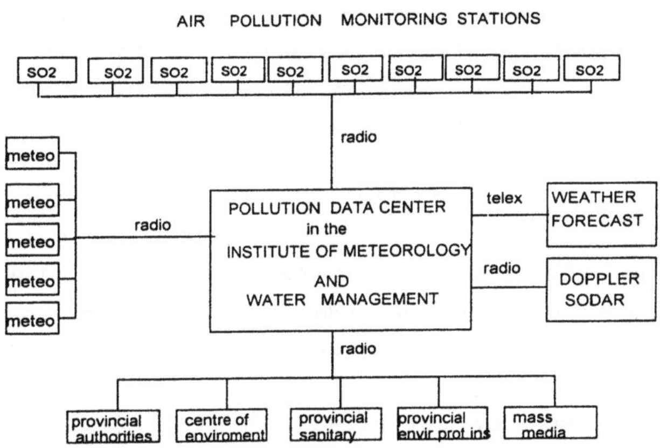

Fig. 2. The automatic air pollution monitoring in the GOP.

The operation of the Regional Air Monitoring System is carried out on a continuos basis (reports on momentary - 30 minute pollutant concentration and metereological data are transmitted to the central unit of the system through the radio).
Fig.3 shows air quality parameters measured automatically by system stations and the name of the analytical method used.

Pollutant	analytical method	number of sites
PM 10	β - absorption	10
NO and NO_2	Chemiluminescence	10
SO_2	Pulsed fluorescence	10
CO	Gas filter correlation	8
O_3	UV absorption	3
Aliphatic hydrocarbons	Photometry	3

Fig. 3. Automated network in the GOP

Moreover, in order to evaluate the impact of local pollution sources, the stations monitor some selected meteorological parameters:

- air temperature and humidity,
- wind speed and direction,
-total radiation.

The stations operating in the meteorological unit carry out the same meteorological measurements plus precipitation. Additionally, the Central Observatory Station carries out the measurements of physical properties of the limitary strata including the inversion and boundary layer by means of Doppler sodar.

An addition to the monitoring system is the use of correlation spectrometer COSPEC V to monitor SO_2 and NO_2 emitted by point sources in GOP.

As a result of the studies and experiments on the issue of atmospheric protection in GOP [Walczewski, 1989] a three-part concept of forecasting harmful aerosanitary situations in the region has been worked out:

- preliminary forecast of potentially harmful situation based on meteorological information,
- statistic model of air pollution forecasting in a monitoring point based on the pollution deposition data in that point and meteorological situation development,
- air pollution (mainly SO_2) spatial dispersion forecast based on a model using the diffusion-advection formula.

At the moment work is being done on the implementation of a statistic model.

One of the most important objective of the system is its warning function and therefore the determination of the emergency thresholds for smog alarm.

In this regard, work is being done in two directions: with the use of statistic criteria to raise the alarm (i.e. the frequency threshold of occurrence of specific pollution is determined by the level of smog alarm) and with the use of medical criteria (i.e. considering that the concentration of a specific substance and the time of its occurrence can be a potential hazard for the health of citizens).

The annual monitoring results from Upper Silesian urban complex show that, a very high complexity of emissions in this area such as high pollution background, influence of industrial emissions, influence of transportation and that municipal emissions have the greatest impact on ground level concentration results. There is evidence of much higher concentrations of almost all of the monitored pollutants (except ozone) during the winter time in comparison with summer time. This experience is not too optimistic from the point of view of establishing a smog warning system, and moreover, from the point of view of emission control in smog situations. It is known that controlling the emissions can basically affect only industrial point source emissions and possibly (to a smaller extent) transportation. There is no effective way of influencing municipal emissions, and these are the ones which are the potential sources of smog hazard in Upper Silesian urban complex.
An additional monitoring problem is the fact that the network of Regional Air Monitoring System in Katowice region was planned to be the general control network. Its warning function was established much later. That is why the location of monitoring points was devoted to serve some other purposes and the monitoring network has to perform several functions at the same time.

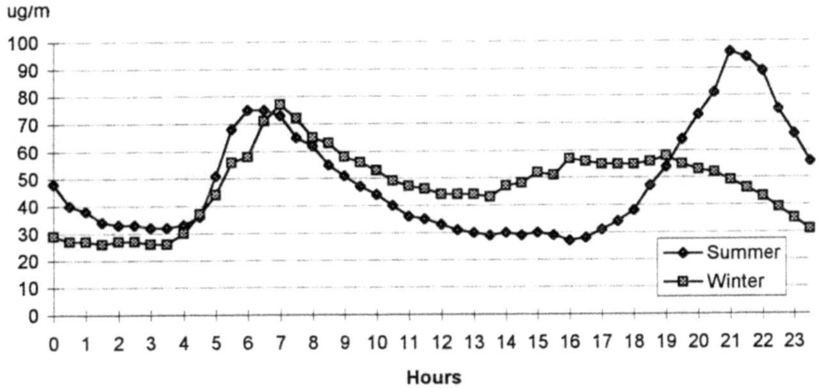

Fig. 4. Daily trend of NO2 in Sosnowiec

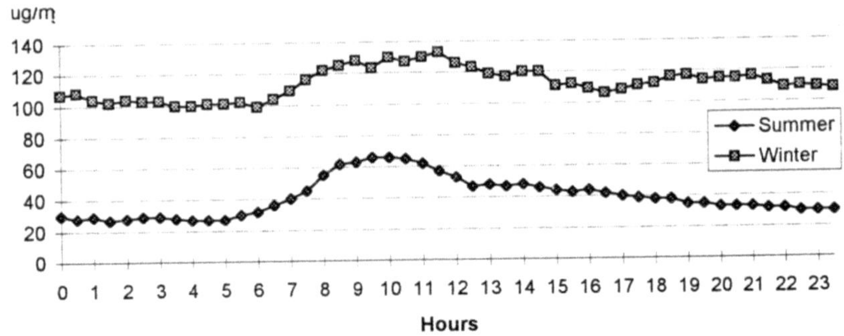

Fig. 5. Daily trend of SO2 in Bytom

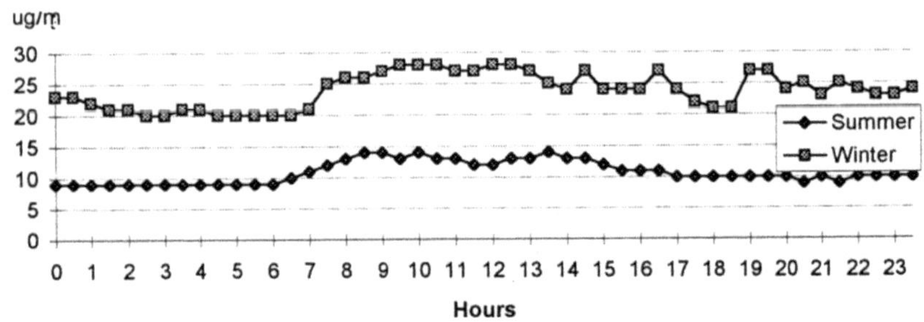

Fig. 6. Daily trend of SO2 in Kuznia Nieborowicka

The results achieved confirm the influence of the impact of particulate emission types on the ground level concentration resulting in particular monitoring points. One could distinguish monitoring points being under the predominant influence of:

transportation emissions (fig.4),

municipal emissions (fig. 5),

pollution background (fig. 6).

This situation, makes it necessary to look for some other ways of establishing conditions to annunce smog alarms rather than those commonly used. It may, therefore, happen that particular alarming thresholds will be considerably exceeded in only one station representing the impact of one pollution factor. Therefore, it is necessary to aim at defining the representativity of monitoring points not necessarily through the interpolation of results between the stations. However, there is a question that arises. How can this be done? Can we assume that the pollution occurring in a given place e.g. from transport pollution (nitrogen oxides) monitored at a station being under the impact of this type of pollution, can be interpolated to the environment of all roads of similar nature? In other words, if it is noticed in the monitoring stations that the pollution threshold of communal emissions was exceeded in one part of the metropolis, does this entitle us to announce the smog alarm in the whole monitored area?

Therefore, it seems that the monitoring network in such metropolises should only perform the function of verifying the results obtained by means of appropriate modelling. However, there is a problem of selecting the model and estimating its predictors. Obviously, in the case of a complex structure of pollutant emissions it is difficult to choose an appropriate tool. Although research of the natural environment of Upper Silesian metropolis considerably exceeds the standard measurements of this type, it will take a long time before we will be able to get answers to most of the presented problems.

Bibliography

Osrodka L.,(1991), Zroznicowanie klimatu Gornoslaskiego Okregu Przemyslowego na przykladzie warunkow termicznych powietrza, Ph.D. thesis, Krakow, Jagiellonian University,

Tomaszewska M., Walczewski J., (1992), Zasieg smug dymowych widocznych na zdjeciach satelitarnych a stan rownowagi atmosfery, Zesz.Nauk. AGH, Sozologia i sozotechnika, 35, Krakow, 35-59,

Walczewski J., (1976), Some characteristics of industrial smokes propagation in the Silesian Region, Poland, Artificial satellites, vol. 11, No 3, 71-74,

Walczewski J., (1989), Systemy prognozy zanieczyszczenia atmosfery nad Gornoslaskim Okregiem Przemyslowym - faza eksprymentalna, Wiad.IMGW, Tom XII, z. 1-2, Warszawa.

The state of the energy system and pollution control policy on the background of Poland

Józef Gega
School for Environmental Engineering and Protection
University of Mining and Metallurgy
Al. A. Mickiewicza 30
30-059 Cracow
Poland

Introduction

The problem of enviromental protection is and will be one of the greatest challenges to humanity. The intensive growth of the population all over the world determines the level of the environmental hazard. This is due to the fact that each inhabitant of our planet occupies and uses an increasing part of the environment with its basic elements: air, water, soil, landscape. Only the way in which a man uses his space determines the possibility to satisfy his needs. At the moment we can observe the threatening geographical changes of the earth, formation of concentrated urban-industrial agglomerations with the whole series of more developed anthropogenic structures that reduce our natural space - the basis of our existence. These changes produce a growing need for different forms of energy. Not only there is a demand for energy such as coal, gas or electricity, but there is a necessity to produce cheap and rational energy services. This, however, is closely related to the recuperation of the energy carriers and fuels, which is the greatest danger to the environment. It is very difficult to predict how the world's energetic requirement is going to change in place and time, as it depends on the rate and structure of population growth, life-style (related to the environment), economic level, political structures, and finally, the level of technological development.

In order to be able to make a prediction about the Polish development of the power industry in the coming 21st century, the vast context of the surrounding European Countries should be taken into account. All the parameters of the achieved level of

NATO ASI Series, Partnership Sub-Series, 2. Environment – Vol. 8
Urban Air Pollution
Edited by Ivo Allegrini and Franco De Santis

economic development of the United Europe are a starting point for planning the coming transformations of Central and East European power industries. It can be assumed that the perspective of a positive transformation in Polish power and heating industry in the above mentioned context, with the availability of the effective solutions will depend on the rate of economic development of a country to a great extent, whereas less on the country's economic or political specificity.

Future of European power industry

In a special edition of the European Directorate General for Energy - DG XVII, of September 1992, chapter 1 "Towards the New Century" and chapter 2 "The 21st Century Energy System" the basic parameters of the EC energy system transformation up to the year 2005 are presented as follows:
- base thc development of the international EC market on the pan-European framework plan,
- support economic growth of the EC in the central and eastern countries of the former Soviet Union,
- provide energy security, accompanied by the management and maintenance of reserves as a continuation of the customer-producer dialogue,
- gradually reduce the CO_2 emission to the level of 1990, gradually by the year 2000,
- set up and bring into effect the European political and monetary union.

In compliance with the assumptions for the future, the foreseen average annual growth (by the year 2005) of energy is expected to be 1.3% - 1.6%, and for energy and oil - 2% and 1%, respectively; the need for coal is expected to drop by ca. 1.3% annually. A considerable gas consumption increase (3.3% yearly), depending on the rate of gas introduction to power stations, is expected too. Currently, the price of the sold coal is higher than the market price, and it will remain on the same level up to the beginning of the 21st century. This will most probably be followed by a drop in the coal production in the EC countries (to reduce losses resulting from the surcharge of the difference between the costs of production and the world prices). Out of the fuels used, coal will gradually decrease from the present level of 22% to 17.8% by 2005. The average gas consumption in the same period will increase by about 60%. Gas as a fuel is advantageous because:

- it is highly competitive compared to coal and nuclear energy,
- it provides ca. 50% lower costs of investment as compared to the coal-operated power stations.

There are plans for the development of energy recuperation from renewable sources, biofuels included. The percentage of energy from such sources is expected to increase considerably from about 2% now to about 50% in 2005. It is now possible to prepare a basis to foster these sources in the future.

At present, electric power energy production rate increases by about 2.4% annually, but in the coming years it will be about 2.0% annually by the year 2005. The percentage of electric power in the calculation of needs is about 18%. The production of energy from hydropower stations and nuclear stations will remain on the same level.

The principle of a varying increase, defined in the Treaty of Maastricht, and the convention on climate changes assumed in Rio also embrace a sustainable development of power industry. The principle of a sustainable development aims at finding possibilities of meeting social and economic aspirations of a higher standard of living with the concurrent protection of the environment.

Considerations over the predicted level of energy production up to the year 2050 are based on the following elements:

- growth of population from 5.3 mld now to about 10 mld,
- level of hazard resulting from climate changes and the necessary quantitative and structural limitations in the planned development,
- series of developmental decisions taken in the 1990s, strongly determining the strategy of development of the power industry in the 21st century.

Yet it is necessary for the image of the 21st century power industry to pay more attention to the issue of:

- possibility of economic growth of societies,
- common accessibility of energy services,
- environmental protection,

- strategy of generation-to-generation continuity of producing the necessary quantities of energy.

In the future many possible solutions for power industry should be analysed and various possible scenarios considered. In the face of a possible shortage of sufficient amount of fuels, the share of coal, now estimated negatively, may increase. In such a case coal would have to be cleaned and used predominantly in the local power and heating systems, with all the negative impact on the environment. Hydrogen and electric energy will probably play a more significant role in the future.

Electrochemical processes, heat pumps, catalytic methods of energy conversion may turn out to be common processes in the nearest future. Bearing in mind the present state of the European power industry, current activities and trends, one can say this is not a good way to achieve a sustainable development of the energy industry in the 21st century. It is possible that due to energy consumption, CO_2 concentration in the air may increase from 350 ppm to 500 ppm by the year 2005. This could result in disastrous effects in the form of climate changes.

To eliminate the highest ecological risk coming from the energy industry it is necessary to:

- achieve the highest level of energy efficiency as far as engineering is concerned,
- enter the realms of intensive energy growth recovered from reproducible sources to the level of 40% to 50% in the total consumption,
- act in favour of development and high share of fuels (except coal) in energy industry,
- lower energy consumption in each economic process,
- advance science and scientific research fostering new energy technologies.

Polish way to power industry of the 21st century

The development of the Polish power industry will probably aim at obtaining similar standards as those of well developed West European countries. The state of our power industry, being a reflection of the bad economic state of the country, fails to meet standards now achieved in the modern energy systems of

Western Europe. Energy consumption per average Pole (for all his needs) is about 4.7 t of conventional coal. Owing to high losses during transformation processes in technological power systems, caused by leakage of fuels and energy carriers, as well as heat losses during energy transport and in the distribution networks, produce of low process-efficiency. The losses are higher because of the lack of instruments with automatic control of power systems, or their inertia in adjusting to the changing energy demand, or lack in the thermal insulation of the heating systems. Low efficiency and heat losses are greatly caused by our fuel structure, based on the combustion of coal at low efficiency (about 80%). Yet, the biggest potential source of energy profits lies in the limitation of the excessive energy consumption of the Polish economy expressed as a price of the used energy per unit of national income. This value, expressed as a per cent share of the costs of energy in price of a unit of the National Product (12 - 13% in Poland), is 3 - 4 times higher as compared to the West European countries. The amount of electric energy can be significantly reduced in a household by the application of energy-saving equipment, e.g. refrigerators, washing machines, TV sets, lighting. The value of energy consumption is lowest in the places where energy services are carried out directly by the users. One of the greatest users of energy is industry, the consumption level of which depends on the production structure (e.g. metallurgical industry - high energy consumption, electronic industry - energy-saving) as well as the applied techniques and technologies. Planned restructuring of industry and modernization of technological processes would increase the profitability and competitiveness of energy-saving products. Energy efficiency can be increased in many branches of economy. One of the conditions of improvement of the state of energy industry is to definitely change the structure of the applied fuels, especially for heating houses, eliminating coal and coke in favour of highly calorific, highly efficient and clean natural gas. It appears from recent data that the average heat consumption used for heating in Poland is 2 - 2.5 times higher than in Scandinavian countries, having more severe climate conditions.

Moreover, our energy industry is the biggest threat to the Polish environment with its dust and gas emission, and to a certain extent to the European environment as well. This is due to the dominanting role of coal as a fuel in energy systems (no suitable technical-technological protections against such emissions). The perspective of changing this state in the Polish energy industry in the coming 21st century is connected with the following possible transformations:

- significant economic advances,
- changed structure of the fuels used (decrease in the percentage of coal in fuel calculation) in favour of gas, methane and sulphur-free liquid fuels,
- restructure Polish industry to low energy consumption,
- restructure and build new, highly efficient energy installations,
- build and equip the Polish energy industry with reliable systems of ecological protection,
- promote and stimulate the improvement of fuel-technological systems directly at the energy receivers .

These positive transformations will result in the correction of the existing parameters promoting the Polish energy industry systems to the European level. Favourable changes in energy industry could be measured by an intensive economic development of Poland at the beginning of the 21st century, with the present level of energy consumption maintained.

Ecological barriers and transformation of the Polish energy industry of the 21st century

The years 1995 - 2005 is the predicted time of necessary and intensive economic development of Poland. This is also the time to restructure our economy and industry and to carry out pro-ecological transformations mainly in the energy industry. The Polish energy industry is the main emitter of atmospheric pollution in Poland, constituting an ecological hazard on a European scale.

Participating in the European economic system, Poland will have to comply with the limitations and conventions, e.g. on SO_2 and NO_x, and in the nearest future also on CO_2 emissions, respected by a majority of countries. This will result in taking certain measures, above all, a change in the structure of fuels currently used in the Polish energy industry from hard and brown coal, dominating energy carriers, to natural gas and energy from renewable sources.

Simultaneously, to lower the energy consumption of our economy economic initiatives must be undertaken so that energy consumption of National Product could drop to the level comparable with that obtained in the countries with the highest economic efficiency. Sources of electric and thermal energy should be gas-oriented, or in the case of coal-oriented ones, there must be full protection against dust and gas emissions. At the same time, due to the necessity to lower hazards caused by greenhouse gas emissions, CO_2 in particular, and relative climatic changes, the Polish energy industry will be forced to reduce CO_2 annual emission from 15 t per inhabitant to the required level. Those activities favouring our development, connected with a zero- or low-emission energy consumption, should result in a reduction of the ecological risk to a safe minimum.

The 21st century will be a challenge to restitute ecosystems and degraded resources to the natural enviroment. This will be evidenced by the survival of forests, oceanic and sea phytoplancton and natural resources, in the general sense.

Some current data in the Polish energy system

Around eighty per cent of the power engineering in Poland is based on hard and brown coal burning.The total fossil fuel usage in Poland equals around 180 million conventional coal per year. The energy usage structure in Poland, using 1992 as an example, can be presented as following:

1) coal fuels 77.6%
2) oil and oil products 12.9%
3) gaseous fuels 7.6%
4) solid waste fuels 1.3%
5) unconventional energy sources (water, wind, sun)0.3%
6) imported electric energy 0.3%.

In 1992 the total production power was 134 TWh; the structure of energy production was: 55% - from hard coal, 42% from lignite, 3% - hydro. Gross domestic electricity demand was 132 TWh, and international exchange was 9.3 TWh of export and 6.7 TWh of import.

Enviromental emission data of main air pollutants

SO_2	total emission	2.82	mln tons/year
NO_2	total emission	1.13	mln tons/year
CO_2	total emission	363.00	mln tons/year
particulate matter	total emission	1.58	mln tons/year

of this:

emission from:		SO_2	%	NO_2	%	particles	%
professional energy	mln tons/year	1.31	46.4	0.37	32.7	0.42	26.6
industrial energy	mln tons/year	0.42	14.9	0.12	10.2	0.64	40.5
other sources as: local power stations, workshops, agriculture	mln tons/year	0.75	26.6	0.10	8.90	0.52	32.9
mobile sources	mln tons/year	0.09	3.20	0.40	35.4	-	-

Presently, there is an urgent need to work out the diagnosis of the actual state of Polish energy and to establish a strategic course of Polish activity in this area.

Scenarios of development of energy sector

In compliance with the elaboration of the department of Energy Problems of the Institute of Basic Technical Problems, Polish Academy of Sciences, two macroeconomic scenarios were distinguished by the year 2010:
- low,
- high.

They differ in:
- assumed rate of increase of the National Income,
- changes in the economic structure,
- degree of rationalization of energy consumption.

The most significant lowering of energy consumption of our economy could result in:
- substitution of old technologies with new ones,
- change of the production structure (restructuring).

To show the impact of these factors better on the size of the demand for energy, energy-consumption and emission of pollutions, three variants were presented for each scenario:
- reference variant, i.e. "continuation",
- variant of technological rationalization,
- variant of complex rationalization.

1. Variant of "continuation" assumes that the economic development up to the year 2010 will be based on the present structure of production. The increase of the demand for energy, and the pollution connected with the assumed dynamics of the National Income.

2. Variant of "technological rationalization", with the assumed dynamics of the National Income growth and the economic structure unchanged, encompasses in its programme enterprises lowering the energy consumption as well as a change of the structure of energy carriers (technological-energy rationalization; substitution of fuels).

3. Variant of "complex rationalization" assuming that there will be a change in the Polish production economic structure with the aim of making it work in the same way as in the developed countries, as well as the technological rationalization as in the previous variant.

Rationalization of use of fuels and energy refers to:
- industry (metallurgic, electromechanical, chemical, mineral etc. industries),
- municipal-living sector.

In order to eliminate energy waste in industry, the existing technological processes should be optimized. The organization of work, services and use of reclaimed materials should be improved, and above all, technological and energy rationalization, with modernization of production methods of energy-consuming products should be introduced.

Technological modernization leads to the formation of a new structure of energy consumption by the reduction of the percentage of solid fuels in favour of gaseous fuels and electric energy and by lowering the demand for energy in the course of rationalization.

The municipal-living sector, being one of the biggest sectors responsible for a considerable amount of emissions (e.g. from the combustion of coal in households), also enhances such emissions in other sectors through the demand for energy (electric energy, heating etc.).

In this sector energy is consumed for:

- heating households, shopping-servicing centers, public utility buildings,
- preparation of warm water in households,
- cooking,
- lighting and household appliances,
- use of such municipal services, as e.g. local transport.

The municipal-living sector can be rationalized:

- by reduction of the fuel consumption;
- changing the structure of energy carriers;
- improving the quality of fuel ,
- purifying the waste gases,
- recuperating energy from renewable sources.

Modernization lies in reducing the fuel and energy consumption through:

- buildings with a lower heating demand,
- insulation,
- double glazing,
- automation of heaters (thermal gauges),
- automation of central heating and boiler houses (weather regulators, heating programmers),
- application of a quantitative-qualitative regulation in heat distribution networks,
- production of more energy-saving household appliences and lighting.

Foreseen demand for energy in the years 1990-2010 in [PJ/year]

	1990	1995		2000		2005		2010	
Variants		N	W	N	W	N	W	N	W
Continuation		2929	3607	3478	4714	4103	5873	4859	4058
Technological rationalization	2728	2955	3404	3223	4012	3568	4623	3448	5069
Complex rationalization		3075	3277	3162	3548	3225	3848	3397	4058

In the ecological variant of Polish power industry, the following reduction of emissions foreseeable by the year 2010, as compared to year 1990, is the following:

- dusts by 75%,
- sulfur oxides by 45%,
- nitrogen oxides by 10%

Technical infrastructure management in urban areas

Rapid geographical and demographic changes, and the relevant spatial-economic transformations, require the maintenance and development of technical infrastructures in the highly urban areas, along with a rational system of their management.So far the management of the infrastructure of: water and sewage systems, waste, city transport and communication, industry, power industry, environmental monitoring, has a fully distributive character. Because of the lack of a collective system and field identification, with the overlapping elements of infrastructures, together with a complex, many stage management, the procedures for analyses and decision-taking, are extremely delayed and this can often followed by inter-structures clashes.
Therefore, to make the management both rational and efficient, it is necessary to elaborate and implement a system of an integrated set of data to be used for the identification of field components of urban infrastructures, (strategic and operational sets). Thanks to the uniform numerical cartographic notation, based on a geographical information system (GIS) with field data, it would be possible to immediately identify all infrastructure elements, at any time and place of the city

analysed, with the possibility of taking objective, rational and quick decisions. Of special importance is the city heating system from centralized and local sources, from thermal-electric power stations as well as housing estates or local boiler-houses.

The power industry and the system of ecological safety

The evaluation of the atmospheric air pollution shows that the traditional processes of production and use of energy in the form of electric energy and heat are still the main sources of contamination. Demographic and geographical changes make the scale of the hazard even bigger. This is connected with the growing population and density of building development, with the accompanying necessity to safeguard energy sources. The inevitability of migration processes and the increasingly broadening urban area lead to the increment of concentrated dust-gaseous emissions with a transboundary propagation.
Therefore, the strategic policy of ecological safety cannot be limited purely to the identification of parameters characterizing the state of changes without any reference to the work of the main sources of hazard, that is the power industry.
High emission from power industry often corresponds with the geography of a high percentage of solid fuels, that is hard and brown coals, in the fuel structure. Poland belongs to the areas with an unfavourable structure of prevalence of solid fuels. In order to reach a higher level of ecological safety, activities must be undertaken to eliminate or limit emissions, especially from these sources.
In Polish thermal-electric power stations, local boiler houses and household stoves (hard coal-, brown coal- and coke-fired in most cases), the protection of the people from the extreme impact of high dust-gaseous emissions from these sources is the basic strategic problem now.
Industrial and communication emissions occupy a considerable position in the balance of emissions to the air. Hence the necessity to set up a monitoring system, that apart from the current record of emissions, would be able to transmit signals of short-term meteorological forecasts about the possibility,

time of occurrence and level of hazardous smog states in the city as a whole or parts of the city. The predicted levels of pollutant concetration should be the basis for undertaking suitable measures with the objective to lower the level of the dust-gaseous emissions from the most intensive sources. Hence the necessity to work out scenarios of activities at each level of the predicted hazards.

In the procedure of reducing the level of emissions into the atmospheric air, the possibility of limiting production or the intensity of traffic should be taken into account.

The states of smog hazard are most common in winter, when the need for heating is greatest. Bearing in mind the high dust-gaseous emission intensity from the sources of thermal and electric power production, it would be advisable to use fuels with lower sulphur and ash content, especially prepared for emergency periods.

Therefore, the centralized city heating system would profit from the construction of peak heat-generating plants fired with natural gas or sulphur-free heating oil to cover the increased demand for heat in the periods of minimum temperatures, or when increased levels of ecological risk are observed. Furthermore, if ther were peak gas- or oil-fired boilers, then it would be possible to eliminate the inertia of reaction whenever there was an increased demand for thermal energy (result of a long distance transport and the time used for starting up the boilers) in the area of high density of heat load.

References

Energy in Europe. A view to the future (1992) Special Edition of Directorate General for Energy (DG XVII)

Gêga J (1993) Koncepcja racjonalnego systemu ogrzewania miasta Krakowa (A Concept of Rational Heating System for City of Cracow). Gospodarka Paliwami i Energia, 6: 3 - 7

Gêga J (1992) Ways to an Efficient Heating System for the City of Cracow. Air Pollution and Climate Series, Published by The Swedish NGO Secretariat on Acid Rain, 2

Ochrona srodowiska 1993 (Environmental Protection in Statistics 1993) (1993) Central Bureau of Statistics (GUS), Warsaw

Ochrona srodowiska w panstwowej polityce energetycznej (Environmental Protection in the Polish Energy Policy) (1994) Energy Department of Polish Academy of Sciences, Warsaw

Urban Air Pollution Monitoring in Lithuania

A. Girgždys, S.Trakumas, V. Ulevičius, and A. Juozaitis
Aerosol Research Laboratory
Institute of Physics
Goštauto 12
2600 Vilnius, Lithuania

Air pollution has been regularly monitored in ten Lithuanian cities at 22 air quality monitoring stations (Vilnius - 6 stations, Kaunas - 3, Panevėžys - 3, Klaipėda - 3, Šiauliai - 2, Jonava - 1, Kėdainiai - 1, Mažeikiai - 1, N. Akmenė - 1, Venta 1). The measurements of 20 atmospheric pollutants have been carried out at each air monitoring station. The list of monitored pollutants includes SO_2, NO_2, NO, CO, NH_3, H_2S, HF, Cu, Zn, V, Pb, Fe, Cr, Mn, Ni, dust, benz(a)pyrene, formaldehyde, and sulfates (Environmental Protection Department, 1994). Furthermore, episodical field measurements have been performed to solve certain problems.

Air monitoring data showed that up till 1991 concentrations of different air pollutants were continuously increasing. Recently, they have dropped down from 20 to 60% due to the economical crises which is common to all post communist countries. The comparison of the atmospheric pollution indexes calculated in 1988 and 1992, taking into account concentrations of various atmospheric pollutants, is shown in Table 1. Everyone can see that the atmospheric pollution indexes significantly decreased in all cities except Vilnius where it remained stable.

Despite the fact that environmental state in Lithuania has improved, there still remain certain problems. For instance in many cities concentrations of NO_2 frequently exceed accepted ambient air standards in Lithuania. The mean daily value of this pollutant was found to be in the range from 0.02 to

NATO ASI Series, Partnership Sub-Series, 2. Environment – Vol. 8
Urban Air Pollution
Edited by Ivo Allegrini and Franco De Santis

0.05 mg/m^3 in different cities, see Figure 1. However, maximal values of NO_2 concentrations in Vilnius were found to be up to 8 times higher than accepted ambient air standards (AAS) in Lithuania. The statistical analysis showed that in Vilnius in about 5% of studied cases NO_2 concentration exceeded standard value, whereas in about 1% of cases it was two or more times higher.

Table 1.

City	1988	1992
Vilnius	3.60	3.62
Kaunas	9.65	5.30
Klaipėda	6.10	4.00
Šiauliai	10.15	5.05
Kėdainiai	2.55	1.70
Jonava	4.85	2.30

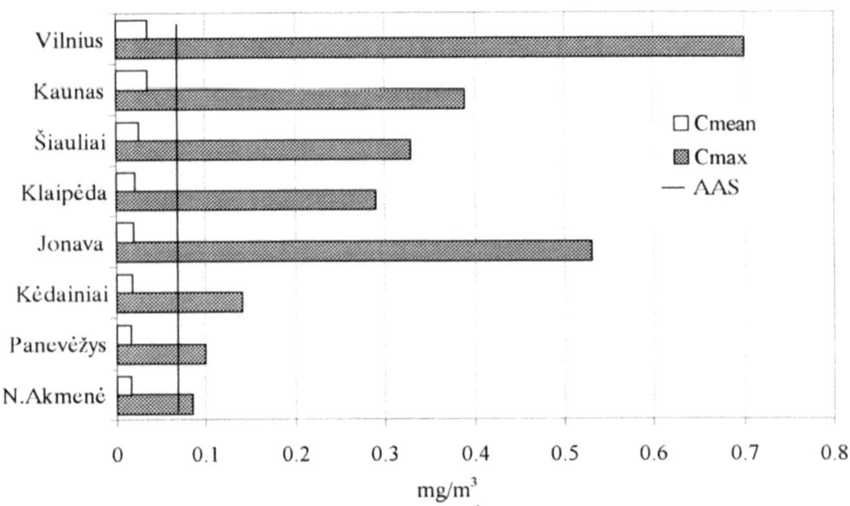

Fig.1. Mean and mmaximal NO_2 concentration.

Vehicular emissions

The inventory of emission sources in Lithuania shows that up to 58% of air pollutants in Lithuania are emitted into the atmosphere by transport (Environmental Protection Department, 1994). Figure 2 shows that industry and energy production accounts for 25% and 12% of air pollutants, respectively. Vehicular emissions contribute up to 80% of CO, 60% of hydrocarbons, and up

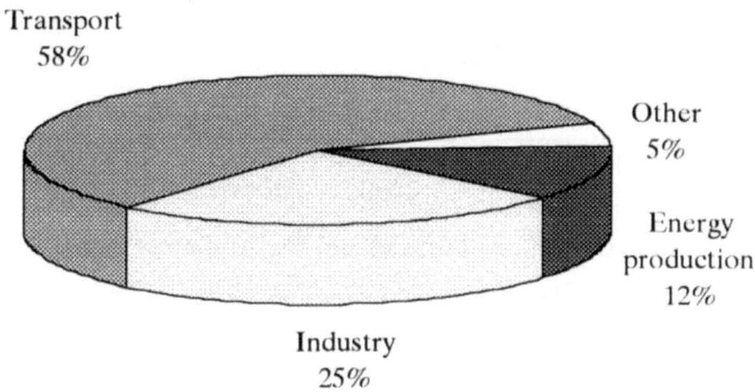

Fig.2. Inventory of emission sources.

to 48% of NO_x released into the atmosphere. Therefore, a special attempt has been focused on the estimation of the strength of vehicle emissions in main city street intersections and the assessment of their dispersion in the nearest vicinity.

The CO and NO_2 concentration values during the traffic peak hours (7.30 - 9.30 a.m.) have been measured in seventeen Vilnius' city street intersections. These measurements started in August and finished in September, 1991. The short characteristics of all intersections investigated are summarized in Table 2.

Sampling sites were set up two meters from the driveway and one and half meter above the ground level. CO samples were collected into glass containers of 500 mL capacity and later analyzed by gas chromatography. NO_2 samples were collected into an absorption tube at 0.25 L/min sampling flowrate and 20 min sampling time. Inner walls of the sampling tubes were covered by a thin film of sodium iodine with an addition of potassium arsenic. To increase adsorption efficiency of the NO_2 gases sampling tubes were filled with a glass beads covered with the same absorbent. The concentration of the NO_2 in samples was analyzed using photometrical method. Four repeated samples of CO and NO_2 were taken each time and twenty measurements were performed in each street intersection.

Table 2.

Sampling site	Short characteristic	Intensity of traffic (hr^{-1})
1	Y-intersection near bus station. No heavy trucks	1878
2	Midpoint of a tunnel with a length of about 200m	1956
3	T-intersection	5436
4	T-intersection with a few heavy trucks	4854
5	Four way intersection	5172
6	T-intersection in old city. No heavy trucks	1386
7	Circular intersection with a few heavy trucks	3476
8	Circular intersection	2956
9	Circular intersection in old city	2472
10	Y-intersection in suburb	2532
11	Four-way intersection in suburb	2124
12	Four-way intersection in suburb	3414
13	Four-way intersection	2802
14	Y-intersection in suburb	1998
15	Y-intersection	3024
16	T-intersection	2814
17	Four-way intersection	2214

The values of NO_2 concentrations obtained in seventeen Vilnius' city street intersections are displayed in Figure 3. The range of the NO_2 concentrations varied from 0.08 to 0.15 mg/m^3. The accepted NO_2 standard value of 0.085 mg/m^3 was exceeded in fifteen street intersections. At the sampling sites 2 and 12 measured NO_2 concentrations were near to two times higher than AAS value, and only at the sampling site 6 it was slightly lower than that value.

Figure 4 demonstrates that measured CO values ranged from 6.0 to 18.5 mg/m^3. The accepted CO standard value of 3.0 mg/m^3 was exceeded in all seventeen city street intersections investigated. At the sampling site 5 CO concentration was more than six times higher than AAS value.

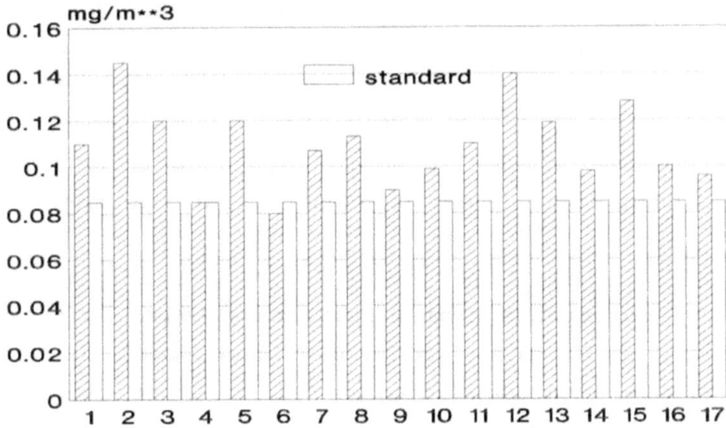

Figure 3. Comparison of the measured NO_2 concentration values with the accepted ambient air standard in Lithuania.

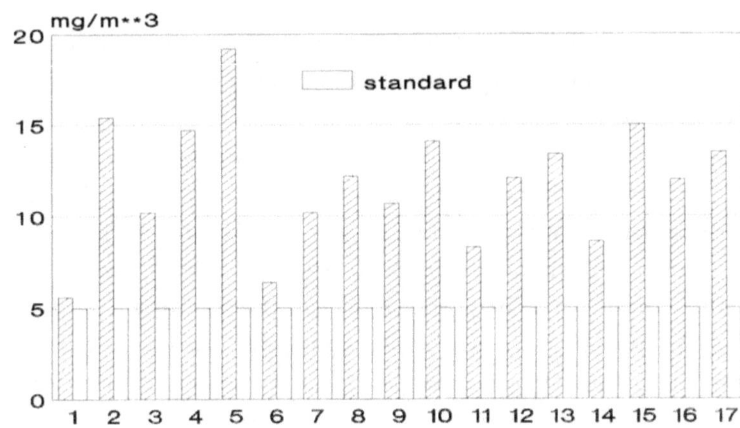

Figure 4. Comparison of the measured CO concentration values with the accepted ambient air standard in Lithuania.

The analysis of the traffic flow indicated that in the street intersections, where vehicles with the diesel type engine dominated, NO_2 concentrations were relatively higher. While in the measurement sites, where dominated vehicles with the gasoline type engine, relatively higher CO concentrations were recorded. The measured absolute values of CO and NO_2 concentrations did not necessarily follow the intensity of the traffic flow. Surface roughness together with meteorological conditions were essential parameters which determined the dispersion of pollutants and subsequently their concentrations in the intersections.

Dispersion of pollutants from a stationary emission sources

An operational puff dispersion model described by Verver and Leeuw (1992) has been adopted to predict dispersion of air pollutants from a stationary emission sources. This model takes into account the following processes: relative diffusion by wind shear and turbulence, advection, the diurnal cycle of boundary layer height and stability, dry and wet deposition and chemical transformations. The emission characteristics for each source can be entered manually or retrieved from a built-in emission database. The computer's communication module automatically collects meteorological data: air temperature and wind speed and direction.

Figure 5 shows the mapping of Vilnius city used in calculations. Vilnius is situated in a valley on the confluence of Neris and Vilnia rivers. This is a jagged and hilly area. The city has three thermoelectric power plants, seven regional boilers, and about 300 big and small enterprises.

The main Vilnius city pollution sources, emitting SO_2 and NO_X gases have been analyzed. According to the pollution impact, these sources were divided into three groups: 1) thermoelectric power stations; 2) regional boilers; 3) the most polluting enterprises. The main SO_2 and NO_X emission source is the third thermoelectric power station which emits more than 80% of SO_2 and 60% of NO_X.

The concentration fields of SO_2 and NO_X pollutants presented in this paper were investigated by using mean annual meteorological data of 1990 of the Vilnius city. It has been supposed that the main pollution sources operate in a maximum regime.

Figure 6 shows the calculated SO_2 concentration values for thirty Vilnius city squares under the influence of winds from various directions. This figure is useful when we need to get information about favorable or unfavorable wind directions for

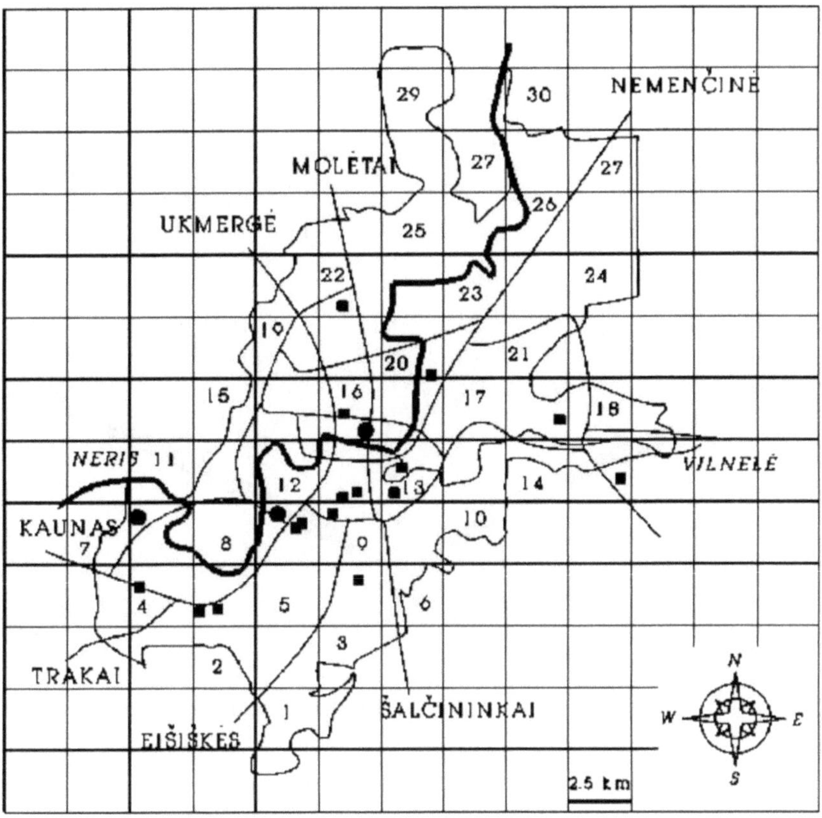

Figure 5. Mapping of Vilnius city used in calculations.

certain city districts. For example, in the inhabited area of Pašilaičiai which is located in square 19 unfavorable wind directions are from South-East, South, and South-West. The maximal SO_2 concentration of 0.27 mg/m^3 occurs when the wind blows from the South-East direction.

Figure 7 shows the calculated NO_x concentration values in Vilnius under the influence of winds from various directions. Let us look at the same Pašilaičiai area. Maximal NO_x concentrations there are observed again when wind directions are from the South-East and South-West. All the other wind directions are favorable for this area: NO_x concentration values are near to zero. The similarity between SO_2 and NO_x concentration distribution fields is not surprising because these two species are emitted into the atmosphere mostly from the same sources.

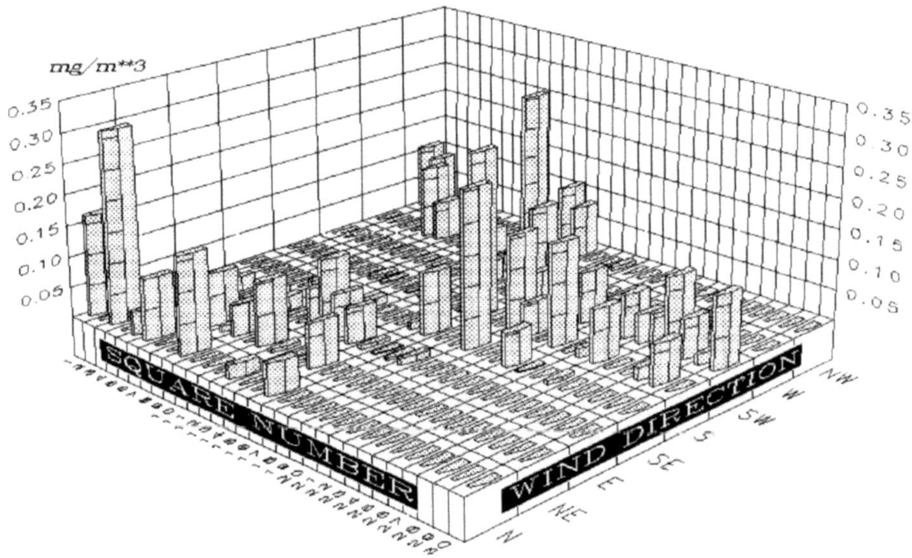

Figure 6. SO_2 concentration distribution in Vilnius city.

The above discussed model is able to do real time calculations. Dispersion pictures can be presented graphically on the monitor or printed out.

Concluding this discussion it should be stressed that when performing urban air pollution monitoring it is necessary:

- to carry out inventory of pollution sources;
- to gather information about variability of meteorological conditions;
- to perform modeling of the dispersion of pollutants;
- to perform field measurements of the concentrations of these pollutants;
- to compare experimental and calculated data;
- if there is disagreement between measured and calculated data, to find the reason for this disagreement and to make corrections.

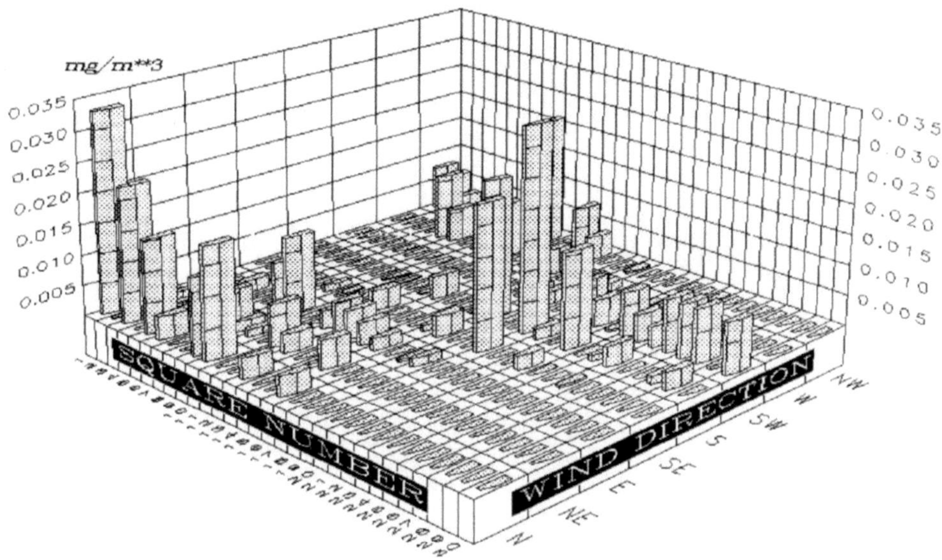

Figure 7. NO_x concentration distribution in Vilnius city.

REFERENCES

Environmental Protection Department (1994) The State of the Environment and Implementation of Environmental Policies in Lithuania. Vilnius, Lithuania.

Verver G.H.L. and De Leeuw F.A.A.M. (1992) An Operational Puff Dispersion Model. Atmos. Environ. 26A:3179-3193.

The partecipation of the Hungarian Metereological Service in the meteorological establishment of the smog alarm of Budapest.

Valéria Sandor
Institute for Atmospheric Physics
of the Hungarian Meteorological Service
H-1675 Budapest, P.O. Box 39.
Hungary

Introduction

In the program of reducing air pollution, the most important part of the metereological activity is to determine the frequency of high pollution concentrations at the surface level, and to estimate the expected concentration values. Research studies in connection with the smog alarm of Budapest started in 1980 at the Hungarian Meteorological Service. At our Service, a forecast model has been developed, which is used to establish of the smog alarm in Budapest. In this model a forecast of the air pollution level is made, which gives an estimation of the concentration levels on the basis of the concentration data of the previous day and the metereological situation. This metereological service makes it possible for the environmental protection experts to take appropriate decisions, to garantee less pollution.
Taking the metereological situation into consideration this forecast is made for the regions of the city. The forecast is then forwarded to the Health Institute and, subsequently, the experts of the Mayor's office decide whether to issue a smog alarm having a complete picture of the immission data and the meteorological forecast.

Air pollution measurements

The pollution of ambient air is measured in Hungary by different measurement networks. One of these networks operates in Budapest. The monitoring system has been operated by the Health Institute

NATO ASI Series, Partnership Sub-Series, 2. Environment – Vol. 8
Urban Air Pollution
Edited by Ivo Allegrini and Franco De Santis

since 1978. The system was renewed in 1991, the sites of the stations were changed and new instruments were installed.
The monitoring system has two important tasks. The first is to continually check the air quality downtown and in the outskirts as well. The second task, which has a greater importance, is to indicate dangerous situations.
The system has only 8 stations because of financial reason (Fig.1).

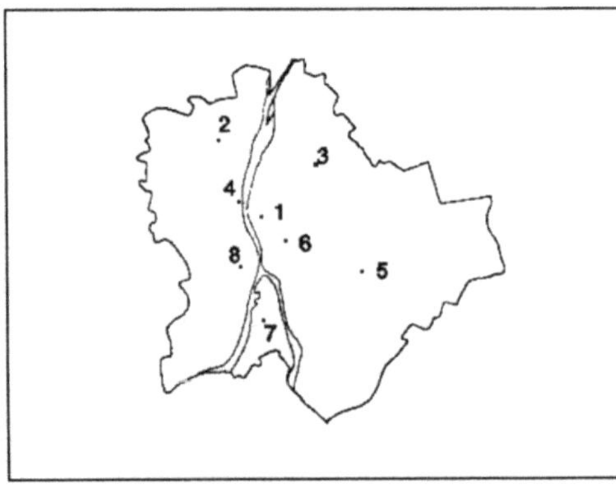

Figure 1:The stations of the monitoring system in Budapest

Four of them are in the centre of the city (stations 1,4,6 and 8), and the other 4 are outside the city (2, 3, 5 and 7) to characterize several types of the environment (blocks of flats in less polluted areas, block of flats in industrial areas districts with family houses, industrial zones.

Assuring the representativity of the measurements was a very important aspect. The stations lie in the dominant wind directions: NE-SW and NW-SE. The monitoring system is suitable for both winter and summer smog situations.
The measured air pollutants are:
all stations: NO, NO_2, CO and dust,
two stations (4 and 6): hydrocarbons (TNMHC only),
two stations (2 and 6): ozone,
all stations: wind direction, wind speed, temperature and relative humidity,
Stations No. 6: global radiation, air pressure.

The data are trasmitted from the centre to the Meteorological Service and to the Lord Mayor's office.

The connection between the meteorological elements, parameters and the air pollution.

The formation, the deposition and decay of air pollution, are strongly influenced by meteorological elements, parameters and meteorological situations. Meteorological parameters have a different importance on air pollution, and have a different importance in certain periods of the year.
In the following, we deal only with the influence of the wind and orography on the air pollution in Budapest. The position of Budapest is characterised by the Danube, which divides the city into two parts. The eastern part is flat while the western part is hilly; we can find here the Buda-mountains of a few hundred meters high. In the city - depending on weather conditions - the air quality is changeable. The closest connection seems to be between the concentration and wind speed. This is an inverse relationship. However, this connection is not so obvious in all wather conditions or over the whole city. To illustrate this phenomenon we present the connection between the sulphur-dioxide concentration and the wind direction and speed on Széna square, stations No. 4, at the foot of the Buda-mountains (Fig. 2).It is apparent that with increasing wind speed, the concentrations generally decrease, except for the case of the SE winds, when the concentration generally increases. Namely, in the case of this wind direction a polluted centre is formed, due to the flow on the lee and upwind side of the mountains. The several 100 m high Buda-mountain hinders the air flow, and high concentrations can be measured at the foot of the mountain.The concentration values are very different in the case of different wind directions. The most polluted situations in the city can be detected in the case of SE, S and E Winds, while the least polluted is in the case of NW wind is the prevaling wind direction in Budapest, but SE wind has the second priority. So the wind direction forecast is a very important part of our forecast model.

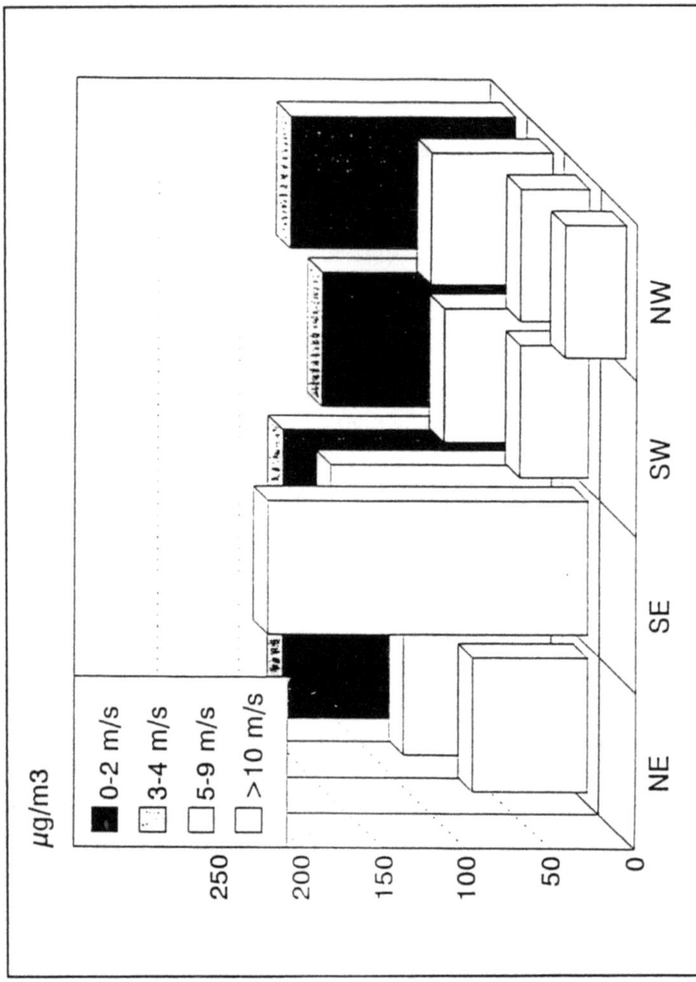

Figure 2.: The SO_2 concentration depending on wind speed and wind direction (Budapest, Széna tér, 1978-1982).

The relationship between air pollution and meteorological situations

During the examination of the macrosynoptic weather situations we observed that the weather situations developing in the permanent anticyclones over the Carpathian-basin favour the accumulation of air pollution in the winter and summer.

Winter smog

Winter smog is formed in anticyclones when the conditions for forming cold "air-pads" are favourable. These situations are formed over the Carpathian-basin in winter after cold fronts have passed through. In the lower regions, several hundred metres, the weather is almost stable, misty, dry, cold and in consequence of the stable layered air mass, the vertical exchange is weak. This air mass is suitable for the accumulation of air pollution. However, in such weather conditions high air pollution must be expected not only in urban areas, but throughout the Carpathian-basin. Therefore, it is necessary to know the level of background pollution too, namely in the cities any emission reduction which could result in lower concentrations than those of the background level.

The cold "air-pad" situations are the most frequent in the Carpathian-basin in January and February; they exist for 2-3 weeks or even one month. These weather conditions come to an end when a strong cold front passes through and the air mass over the city, is replaced by relatively fresh air.

As an example, such a period was formed in January and February, 1993. The weather in the Carpathian-basin was detemined by a large anticyclone over Western- and Central - Europe, with 1040 hPa pressure. In this period the concentration level was very high. On Fig. 3 we can see the daily average

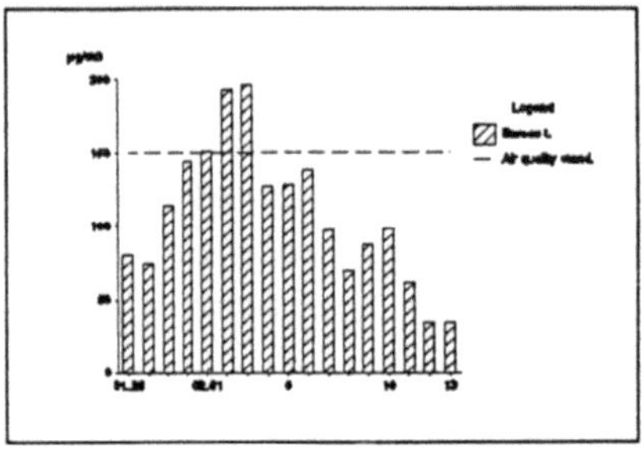

Figure 3.a: SO_2 daily average concentration

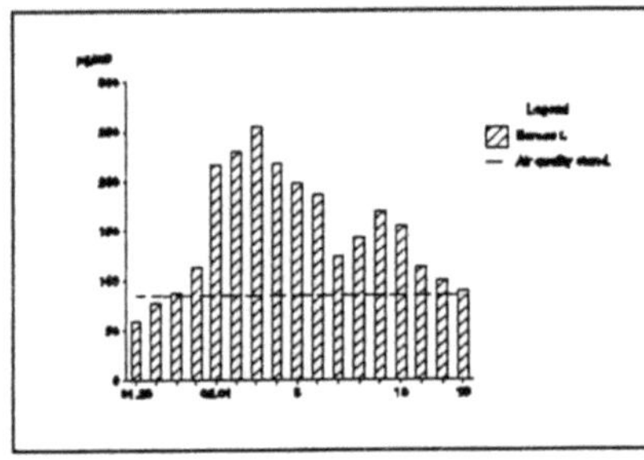

Figure 3.b: NO_2 daily average concentration

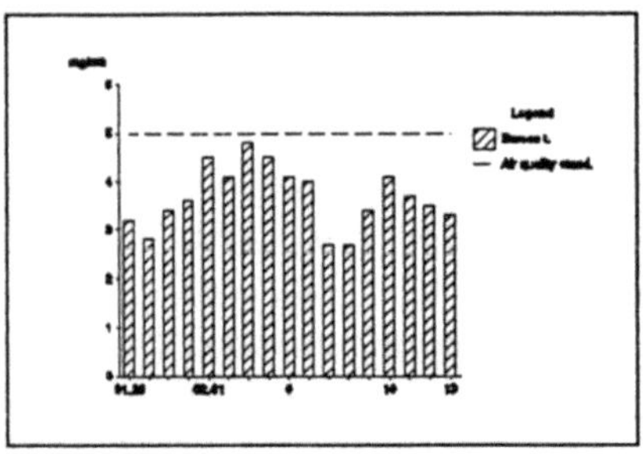

Figure 3.c: CO daily average concentration

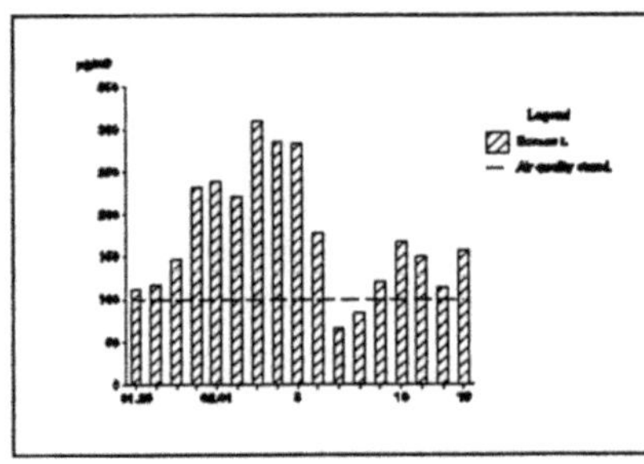

Figure 3.d: SPM daily average concentration

concentration of sulphur-dioxide and nitrogen-dioxide in the period from 28th January till 13th February. In the period determined by the antyciclone from 28th January till 6th February the air quality standard of nitrogen-dioxide was exceeded. The nitrogen-dioxide concentration on the 3th February was almost the triple of the air quality standard. The air quality standard of sulphur-dioxide was exceeded on two days,in spite of the fact, that exceedance is not very common, because the emission of sulphur- dioxide has been reduced during the last years. The important components of the winter smog are nitrogen-oxides and carbon-monoxide and SPM. On the 7th February a weak cold front passed over the Carpathian-basin resulting in the temporany decrease of the concentration level.

The photochemical smog

In the permanent anticyclones in summer we can take into consideration the formation of photochemical smog. In these anticyclonic situations, which are connected with upper level warm ridges, the formation of photochemical smog is favoured by the strong solar radiation and the high daily temperature.

In Fig. 4 the diurnal variation of the formation of the photochemical smog can be seen, calculated from the data of monitoring network in Budapest. In the morning rush hours the emission of nitric oxide, carbon monoxide and hydrocarbons increases.

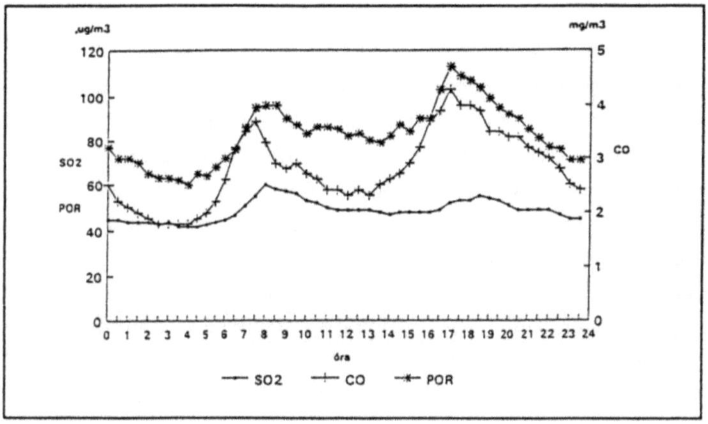

Figure 4.a.: The average daily change of air pollutant in Budapest

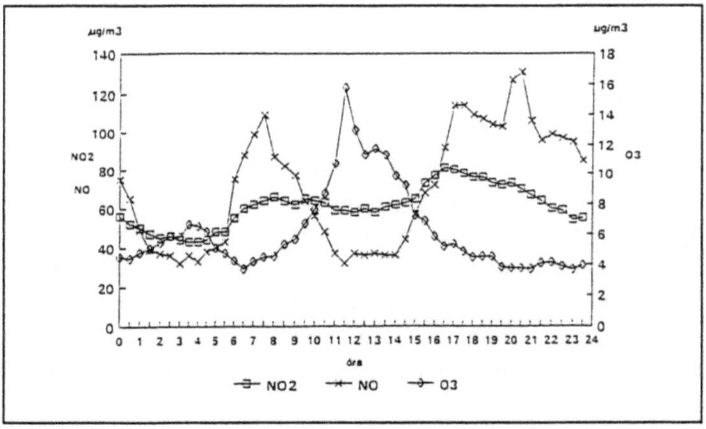

Figure 4.b.: The average daily change of air pollutant in Budapest

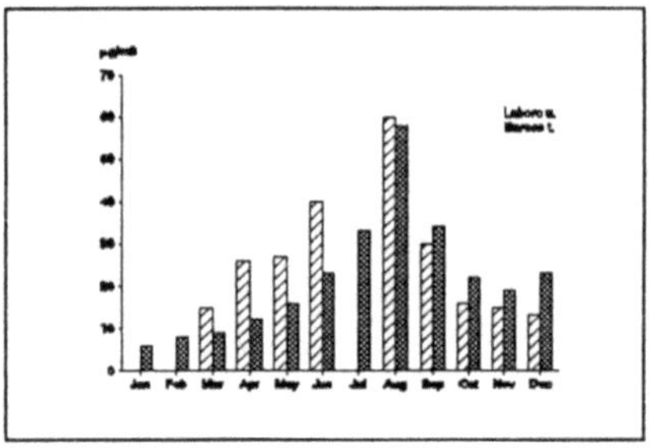

Figure 5: Monthly average ozone concentration (1992)

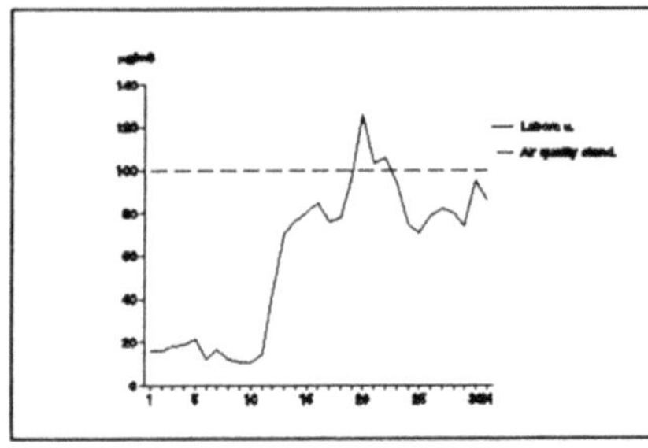

Figure 6: Daily average ozone concentration (Aug. 1992)

The peak of the nitrogen dioxide concentration is reached in the afternoon, because the formation from nitric oxide to nitrogen-dioxide needs some time. In the morning hours the ozone-concentration is low. The reason for this is, on the one hand that oxidation of nitric oxide uses much ozone, and on the other hand in the morning, the speed of ozone formation is low because of the low solar position. The maximum ozone-concentration is detected in the early afternoon. The nitrogen-dioxide concentration decreases in the afternoon, because of chemical processes, and the ozone-concentration decreases, too. The speed of the photochemical processes slows down.

The high ozone-concentrations are not usually formed in the cities due to the abundance of reductive species reacting with ozone. High ozone-concentrations can be expected 50-100 km downwind of Budapest.

In the follwing case another situation is presented, when high ozone-concentration levels developed in the city. The August of 1992 was very hot, dry with plenty of sunshine in Hungary. The daily average temperatures were between 23 and 27 degree centigrade, the climatological averages were exceeded by 4-6 degrees. Such high average temperatures had not been observed in August in Hungary since the beginning of the measurements. The duration of the sunshine was near to record values, too. The monthly precipitation amount over most of the country was only 70% of the usual.

In consequence of the hot, sunny weather the ozone-concentration in Aug. was high (Fig. 5).

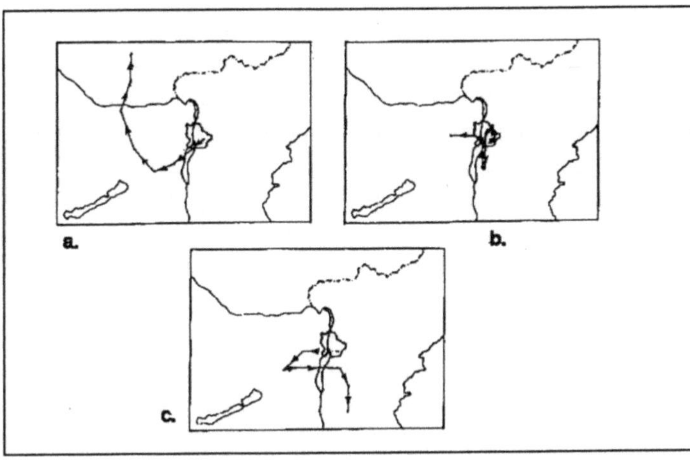

Figure 7: Forward trajectories a. 19. Aug 1992, b. 20. Aug 1992, c. 21. Aug 1992

The highest concentration was measured on the 20th of August. The weather in the Carpathian-basin on this day was determined by a weak anticyclone. The weather was sunny and warm with day-cumulus clouds. The temperature maximum was 32-38 °C, the sunshine duration was 11-12 hours. All these circumstances were suitable for an increase in photochemical processes (Fig. 6).
The weather conditions on the previous and the following days were similar to those of the 20th, but the wind direction and wind speed were different. The daily average wind speed was 2.6 m/s on 19th, 1.7 m/s on 20th and 2.0 m/s on 21th. the wind direction on the 20th was variable.In Fig. 7 we show the air mass trajectories calculated from the surface wind data.
On the 19th and 21 st the air mass left the city, but on the 20th the air flow transported the poluted air mass outside the city, and some hours later brought it back, so this new source of air pollution added to the primary high pollution. The high concentration is the result of this process.

The forecast method of air pollution of Budapest

The Hungarian Meteorological Service worked out a forecast method for supporting the smog alarm plan of Budapest. In the course of the forecast we have to answer the question:

- can weather conditions be expected with high air pollution ?
- can further increase of air pollution be expected?
- can a rapid concentration decrease be expected in a short time?

1. The air pollution input data are:

The data from the monitor network of the Health Institute: the daily average immission data from the previous day, and the concentration values from the previous week (SO_2, NO_2, CO and dust).

2. The meteorological input data are:

- wind data from Budapest - previous day,
- stability value and wind data from 00 UTC radiosonde measurements,
- surface wind forecast on 12 UTC,
- wind forecast for the lower 3000 m layer,
- precipitation forecast for 12 UTC.

The model uses regression techniques and decision method.

The air pollution data are in some cases incomplete and have some mistakes or errors, so the first part of the program is devoted to data substitution and data correction; in the course of this, the program substitutes data on the basis of statistical calculations, so the input data are processed after filtration, correction and substitution.

The seasonal, weakly and daily changes of the concentrations are taken into consideration, too.

In the forecast model among the meteorological elements and parameters we took into consideration those which have the highest correlation with the air pollution.

The output of the model is the expected level of air pollution, the comparison, the immission level and the air quality standards and the spatial distribution of air pollution in Budapest. The experts of the Mayor's office decide whether to issue a smog alarm having a complete picture of the concentration data and this meteorological forecast.

Space-Time Statistical Structure of the Pollution Field over Sofia, Bulgaria

D. Syracov* - M. Syracova°
* - Ntl. Institute of Meteorology & Hydrology, Sofia, Bulgaria
° - Department of Meteorology, Sofia, Bulgaria

Introduction

The space-time statistical structure of the pollution field over the industrial and residential regions, as well as the climatic regime of the pollution, provide valuable information for monitoring and control strategy development. In this study, data from two measurement campaigns in Sofia have been used to extract such information concerning two main urban pollutants - SO_2 and NO_2. Each campaign consists of two experimental periods, a winter and a summer run, respectively. The first campaign took place in the period June 10 - 25, 1976 (summer run) and November 25 - December 9, 1976 (winter run). Air samples were taken in 10 (respectively 12) points every 3 hours. The second experiment was more complicated. The winter run took place in the period November 20 - December 20, 1981, and 16 sampling points were used. The summer run took place in the period June 15 - July 15, 1982 and 15 stations were used. The time resolution of the measurements was 1.5 hours. The wet chemistry (colorimetric) method was used to determine the pollutant concentrations in all runs. Both experiments covered almost the same area including the centre of the city. As one can see during the second campaign, a more dense net of points was used. The time lag was twice shorter and the campaign duration was twice longer compared to the first one.

Mathematical tool

The statistical structure of the air pollution field over Sofia is characterised by the space-time correlation and structure functions, determined as

NATO ASI Series, Partnership Sub-Series, 2. Environment – Vol. 8
Urban Air Pollution
Edited by Ivo Allegrini and Franco De Santis

(1) $$B(\rho,\tau) = B(\mathbf{r}_i,\mathbf{r}_j,\tau) = \overline{f'(\mathbf{r}_i,t)f'(\mathbf{r}_j,t+\tau)} \ ,$$

(2) $$D(\rho,\tau) = D(\mathbf{r}_i,\mathbf{r}_j,\tau) = \overline{[f'(\mathbf{r}_i,t) - f'(\mathbf{r}_j,t+\tau)]^2} \ .$$

Here $f(\mathbf{r},t)$ denotes the variable under consideration, in our case the concentration of a particular air pollutant, at the point $\mathbf{r}$ at a given moment t, f' is the fluctuation, i.e. the deviation of f from its mean value $\overline{f}$, ρ is the distance between points $\mathbf{r}_i$ and $\mathbf{r}_j$, τ is the time lag, the bar denotes statistical averaging.

It is well known that at $\rho = 0$ and $\tau = 0$ the correlation function is equal to the variance

(3) $$\sigma^2(\mathbf{r}) = B(\mathbf{r},\mathbf{r},0) = B(0,0) = \overline{f'^2(\mathbf{r},t)} \ .$$

The mathematical tools of correlation and structure functions are well developed for stationary, homogeneous and isotropic fields (see Panchev, 1980). It is obvious that the pollutant fields do not strongly fulfil these conditions since the mean field $\overline{f(\mathbf{r})}$ and the field of the variance $\sigma^2(\mathbf{r})$ are not homogeneous. Nevertheless, in the averaging process the stationarity as well as spatial homogeneity and isotropy of the pollutant fields are supposed *a priori*. In fact, it can be proved that the assumption above is true for the fields of normalised fluctuations $\varphi(\mathbf{r},t) = (f - \overline{f})/\sigma$ with $\overline{\varphi} = 0$ and $\sigma_\varphi^2 = 1$.

The calculation method for (1) and (2) is described in details in Syrakova, 1975. According to the theory both functions (1) and (2) satisfy the equation

(4) $$D(\mathbf{r}_i,\mathbf{r}_j,\tau) + 2B(\mathbf{r}_i,\mathbf{r}_j,\tau) = \sigma^2(\mathbf{r}_i) + \sigma^2(\mathbf{r}_j) \ ,$$

which is valid at any values of ρ and t, so special measures are taken to ensure the fulfilment of this condition in the process of calculation. In spite of the assumed homogeneity and isotropy of pollutant fields, the functions $B(\rho,\tau)$ and $D(\rho,\tau)$

can have anisotropic shape when $\rho \neq 0$ and $\tau > 0$ due to the prevailing wind direction. For Sofia it is W-E (or WNW-ESE) and this direction was chosen as positive when ordering the sampling points.

Due to the inhomogeneity of the mean field $\overline{f(\mathbf{r})}$ and of the variance field $\sigma^2(\mathbf{r})$ the normalised correlation and structure functions are calculated as

(5) $$b(\mathbf{r}_i,\mathbf{r}_j,\tau)=\frac{B(\mathbf{r}_i,\mathbf{r}_j,\tau)}{\sigma(\mathbf{r}_i)\sigma(\mathbf{r}_j)} ,$$

(6) $$d(\mathbf{r}_i,\mathbf{r}_j,\tau)=\frac{D(\mathbf{r}_i,\mathbf{r}_j,\tau)}{\sigma^2(\mathbf{r}_i)+\sigma^2(\mathbf{r}_j)} .$$

It can be easily seen that (5) and (6) are the correlation and structure functions of the normalised fluctuations, which also satisfy the condition (4). For this reason, the structure functions have not been analysed.

Mean concentration and variance fields

The space distributions of the mean values of SO2 and NO2 concentrations for all runs are presented in Fig.1 and Fig.2. It can be seen that the mean concentrations for both pollutants are greater in winter, the differences being manifested more markedly for SO2. The winter fields show their maximum in the central part of the city. The summer concentrations are not so clearly distributed. Some tendency to deformation in SW-NE direction can be noticed which is probably related with the orientation of the main highways. The comparison of the mean concentrations, obtained from the two experiments, shows some peculiarities. The concentration fields are not identical. Only the winter SO2 concentration fields for the two periods are approximately identical. The summer SO2 concentrations for 1982 turn out to be greater than the 1975 summer ones. The NO2 concentrations for both seasons show a decrease from 1976 to 1981-2, which is

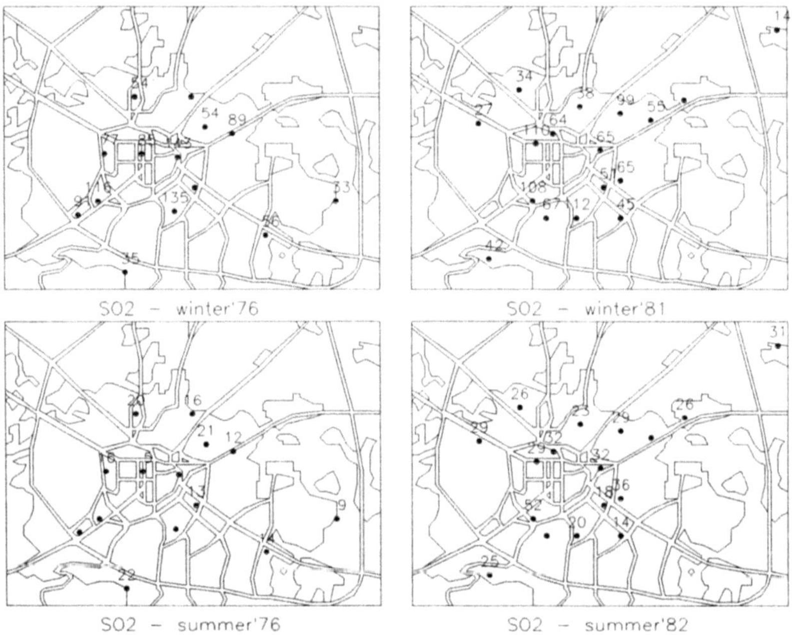

Fig.1. Mean concentrations of SO_2 in µg/cub.m

difficult to explain bearing in mind the increase of traffic during that period. The variance fields, which are not presented here, show features similar to those of the mean concentrations. The only exception is the NO2 winter '76 field, which has a minimum in the central part of the city.

Correlation functions

Further on, only the normalised correlation functions of SO2 and NO2 will be discussed. As these functions represent two-dimensional surfaces, it is more convenient to regard one of the variables (ρ or τ) as an argument and the other as a parameter and viceversa. Thus, we get two sets of cross-sections denoted as $b^{\rho}(\tau)$ and $b^{\tau}(\rho)$, respectively.

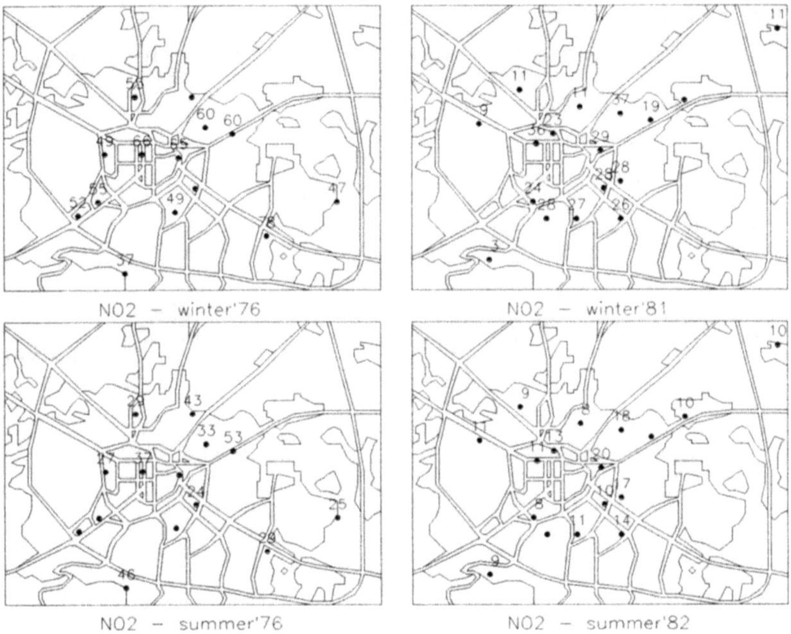

Fig.2. Mean concentrations of NO_2 in μg/cub.m

In Fig.3 the time-dependent correlation functions of SO2 are presented. As one can see, the empirical points do not lie on smooth curves. This can be caused by various reasons which are mainly:

- the small volume of the samples;
- the use of not more than one time series;
- the great variability of the pollution field;
- observation errors (errors during the sampling and chemical analysis);

For this reason the results can only be treated qualitatively. Nevertheless, some basic features and tendencies can be observed.

The most essential feature of the family of curves $b^{\rho}(\tau)$ for SO2 is the rapid decrease of the correlation function with time. The time correlation functions $b^{0}(\tau)$ (i.e. $b^{\rho}(\tau)$ at $\rho=0$) reach the value of 0.5 within time intervals of less than 3 hours, i.e. the time radius of correlation is less than 3 hours. Only in the winter run of the second campaign this quantity reaches a value of about 4 hours. At $\rho \neq 0$ the time correlation functions decrease more rapidly in summer than in winter.

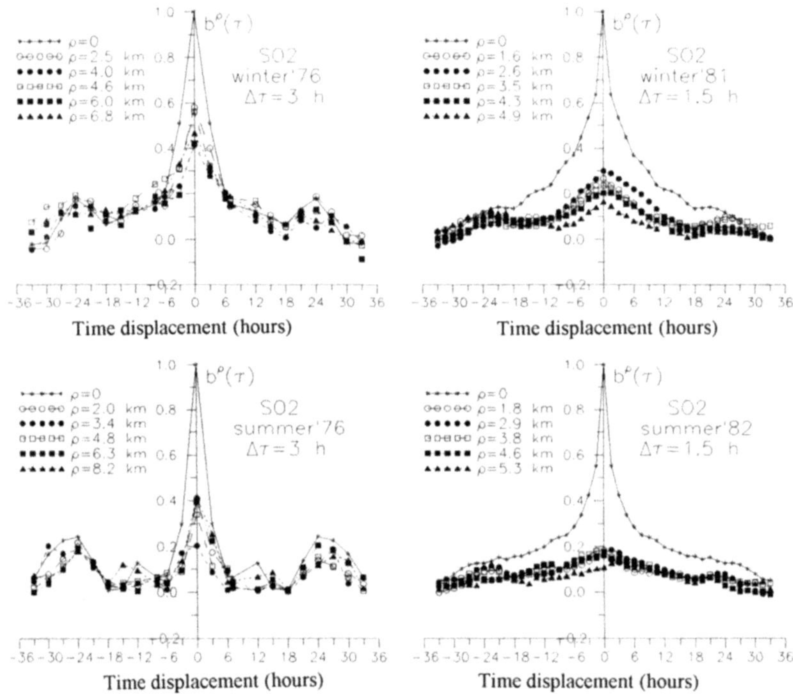

Fig.3. Normalised space-time concentration functions for SO_2 as functions of τ at fixed ρ

The time functions in the first campaign possess a well-manifested secondary maximum at $\tau = 24$ hours. It is better pronounced in summer than in winter and is probably related to the well-manifested diurnal variation of the SO_2 concentrations. Despite the better time resolution, the results from the second experiment, do not show such a diurnal component of the time correlation function; a secondary maximum at a time interval of 24 hours is difficult to be noticed.

As it was already mentioned, the method of calculating correlation functions takes into consideration the anisotropy in W-E direction due to the prevailing winds. This would cause a displacement of the correlation curve's maxima to the East (right-hand side in the figures) at $\tau > 0$. This effect is well pronounced for the macro-scale meteorological fields and is explained in detail in Panchev and Syrakova, 1975. However, the family of curves in Fig. 3 does not show such a maxima displacement. Yet, the anisotropy leads to some asymmetry of the curves at $\rho > 0$ - the values of the functions at $\tau > 0$ are greater than those at $\tau < 0$. This effect is better pronounced in the second period.

The displacement of the maximum of the correlation functions to the right for $\rho > 0$ takes place when certain inhomogeneity (perturbation) in the pollution field located in a certain point at moment t turns out to be located in a downwind point at moment $t+\delta t$ as a result of its transportation. The measurement of the wind velocity during the experiments shows prevailing westerly and easterly winds, the westerly being more frequent. This yields mean W-E transport with mean speed of 1-2 m/s for all runs. If we assume this to be the characteristic mean velocity of the zonal wind, this will imply that for a tame lag of 1.5 or 3 hours the mentioned inhomogeneity together with the entire field of pollution will be transported to a distance of about 10-20 km, i.e. it will be moved out of the investigated area. In other words, after one time lag a new pollution field will be formed by the multiple sources. That is why the maximum values for all the curves $b^{\rho}(\tau)$ turn out to be at $\tau = 0$. For this reason, the existence of local homogeneity and isotropy of the pollution field over the town can be assumed, i.e. the influence of the macro-scale anisotropy decreases with the scale decrease.

Fig.4 presents the normalised space-time correlation functions for SO2 as functions of distance ρ for fixed time lags τ, i.e. $b^{\tau}(\rho)$. As one can see, the space correlation functions (at $\tau = 0$) also decrease very rapidly. The space radius of correlation ρ_c ($b^{\tau}(\rho_c) = 0.5$) is about 2 km for the first experiment and about 1 km for the second one. The winter runs show a slightly greater radius of correlation than the corresponding summer ones. Here the anisotropy in W-E direction also leads to a slight asymmetry of the curves, values at $\rho > 0$ being a little bit higher than those at $\rho < 0$.

This implies rather irregular distribution of the pollutant, irregularity being greater in summer.

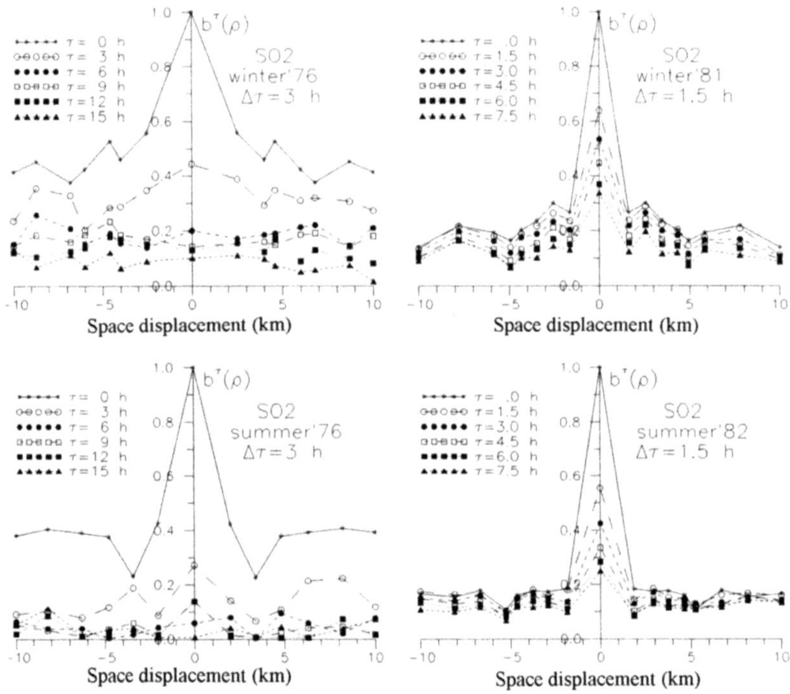

Fig.4. Normalised space-time concentration functions for SO_2 as functions of ρ at fixed τ

The location of the extremes of the space correlation function gives a certain idea about the structure of the pollution field. If the distance corresponding to the secondary maxima of $b^{\tau}(\rho)$ is assumed as a characteristic distance between the perturbations and the minima between the main and the secondary maxima is assumed as a characteristic scale of perturbations, one can estimate the characteristic distance to be about 7-9 km and the characteristic scale about 4-5 km, being a little bit greater in winter than in summer.

Fig.5 present the correlation functions $b^{\rho}(\tau)$ for NO2. The family of curves shows a different behaviour for both experiments. In the first experiment, the winter radius of time correlation τ_c ~ 2 h is about twice less than the summer one. In the second experiment the situation is reversed. The winter radius of time correlation (~5 h)

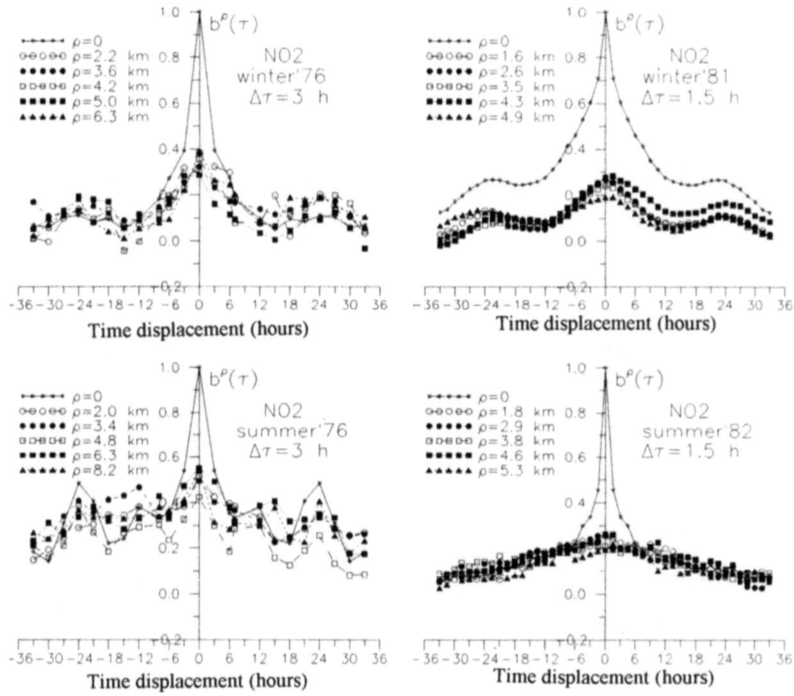

Fig.5. Normalised space-time concentration functions for NO_2 as functions of τ at fixed ρ

is greater than the summer one (~1 h). The curves in this figure are similar to those for SO_2 (Fig.3). The secondary maxima, related with the diurnal variation of the NO2 field, are well pronounced for all runs, except for the summer'81. A slight asymmetry of the curves at $\rho > 0$, due to the prevailing wind, can be noticed as well.

The family of functions $b^{\tau}(\rho)$ for NO_2 are presented in Fig.6. These curves also show differences for the two campaigns. The radius of space correlation in 1976 is about 2 km in both seasons while the second experiment radius is 1 km. Again, no summer-winter differences have been found. The curves do not give the possibility of estimating characteristic scales of the pollution field. During the second field campaign the NO and phenol concentrations were also measured. The calculated correlation functions show much less space and time correlation than SO2 and NO2 ones. The time radius of correlation is less than one lag (1.5 h) and the space function reaches zero immediately at the first space gradation

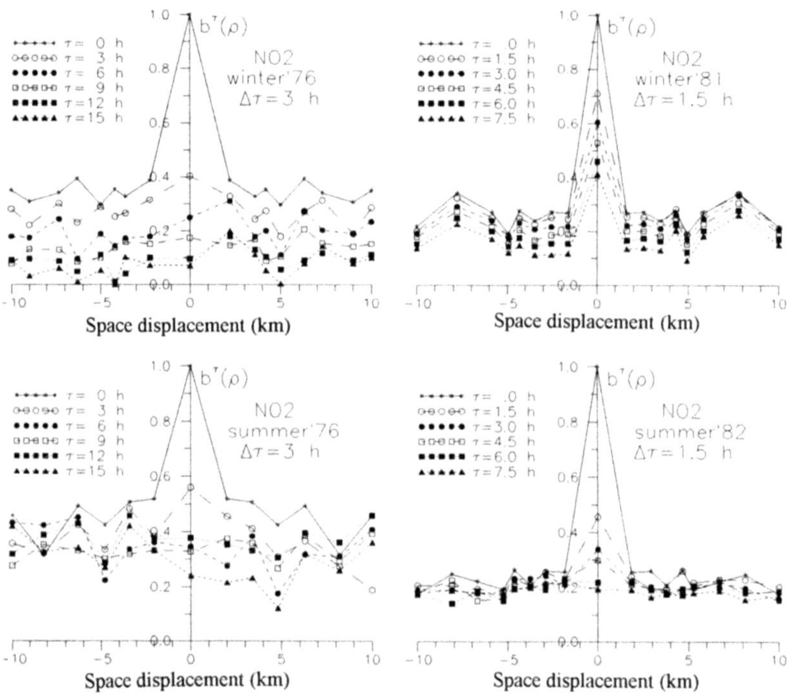

Fig.6. Normalised space-time concentration functions for NO_2 as functions of ρ at fixed τ

A probable reason for this bad correlation is the great relative error of observation, due to the small concentrations of these pollutants.

Conclusions

As a result of the considerations above, the following conclusions can be drawn:

1. The mean concentrations of both SO_2 and NO_2 are greater in winter than in summer.
2. The space distribution patterns of mean concentration and variance for both pollutants have better pronounced features in winter than in summer.
3. The correlation of pollution concentrations decreases very rapidly both in space and time.
4. The curves $b^{\rho}(\tau)$ show well-pronounced diurnal variations, better seen for SO_2. In some cases the curves $b^{\tau}(\rho)$ can give certain information about the characteristic scales of pollution field inhomogeneities.
5. When the area is not very large, as it is in our case, the field of pollution can be considered locally homogeneous and isotropic.

Compared to the first experiment the space-time correlation functions obtained using the data from the second experiment show better time but worse space correlation. But even this big experiment (with a greater number of stations, longer period of measurements and better time resolution) is not representative enough from a statistical point of view. This means that many experiments of this kind with much longer duration are necessary to obtain reliable information for the space-time structure of the air pollution field in urban areas.
From the point of view of monitoring and control of urban air pollution the following conclusions can be drawn on the basis of the presented results:

1) The great observation errors of the wet chemistry method decrease the value of such kind of data assessment. The results would be more valuable and reliable if data from automatic analysers were used, which besides greater accuracy allows for better time resolution as well.

2) The high price of this equipment, however, restricts the possibility of achieving a good space resolution. Even the most developed countries can afford a limited number of such apparatus, which usually are placed in critical (from the point of view of air pollution) points. So, such a network is used more for control purposes than for monitoring ones. A possible solution to this problem is to invest more effort and funds

for investigation and real time monitoring of pollution sources. This information together with the necessary meteorological information can serve as an input to adequate numerical dispersion models, providing a real time picture of the space distribution of the main air pollutants. The numerical results can be continuously compared to and calibrated by the measurements of the few automatic stations for air pollution control.

References

Gandin, L.S., and V.P.Boltenkov, 1964: On the methodology for investigation of three-dimensional macrostructure of meteorological fields. Proceedings of GGO, iss.165, Gidrometeoizdat, Leningrad (in Russian).

Panchev, S., 1971: Random Functions and Turbulence. Pergamon Press.

Panchev S., and M.Syrakova, 1975: Space-time statistical macrostructure of the geopotential field at the 500mb level. Hydrology and Meteorology, v.XXIV, No 4 (in Russian); The GARP Programme on Numerical Experimentation, Report No 8.

Syrakov, D., M.Syrakova, and A.Dragoeva, 1980: Space-time structure of the field of pollution over Sofia based on data of the experiment "Sofia-76". Hydrology and Meteorology, v.XXIX, No 1. (in Bulgarian).

Syrakova, M., 1975: Determination of the space-time statistical structure of meteorological elements using observational data. Hydrology and Meteorology, v.XXIV, No 2. (in Russian).

Syrakova, M., 1976: Space-time statistical macrostructure of the temperature field at the 500mb level. Hydrology and Meteorology, v.XXV, No 3 (in Russian); The GARP Programme on Numerical Experimentation, Report No 9, 1976

Towards the Assessment of Air Quality and its Possible Health Impacts in European Cities

Rob Sluyter, Wim den Tonkelaar[1]
National Institute of Public Health and Environmental Protection (RIVM)
Laboratory for Air Research
P.O. Box 1
3720 BA Bilthoven
The Netherlands

Introduction

Europe is a highly urbanised continent. In 1990, more than 70% of its total population lived in cities. The concentration of human activities on a relatively small area causes enormous pressure on the natural system and has led to numerous environmental problems. In the last decade air quality problems and associated topics like exposure of citizens have been addressed through case studies. One of the first attempts to describe the air quality in a number of cities and investigate (dis)similarities between them was made within the GEMS-AIR project. In 1992, WHO/UNEP published a report "Urban Air Pollution in Megacities of the World", summarising air quality data for the 20 most populated cities of the world (WHO/UNEP, 1992). WHO recently published their Concern for Europe's Tomorrow study (WHO, 1994) in which an assessment of urban air quality and potential population exposure is given for European cities with more than 50.000 inhabitants; however this was not done through comparative city analysis but through statistical analysis of European air pollution concentration and population density fields. In the first part of this paper, attention will be focused on the research RIVM conducts to come to an

[1] TNO Institute of Environmental Science. P.O. Box 6011, 2600 JA Delft, The Netherlands

NATO ASI Series, Partnership Sub-Series, 2. Environment – Vol. 8
Urban Air Pollution
Edited by Ivo Allegrini and Franco De Santis

assessment of urban air quality and its possible health impacts for large sets of European cities. This research focuses on *city background concentration fields*, i.e. the concentration of pollutants at places within cities not directly influenced by sources such as industry and traffic. City inhabitants are exposed to at least these levels whenever they are outdoors.
Here, some results will be presented of a study RIVM, NILU and MGO have conducted in the framework of the *Europe's Environment programme* on the air quality situation in all 105 European cities with more than 500 000 inhabitants[2] (Sluyter, 1994); in the second part of the paper, a method will be presented for the on-line extrapolation of hourly CO, NO_x and NO_2 measurements from 13 Dutch street stations to *busy streets* of major Dutch cities.

Methodological framework

Air quality in cities is the result of a complex interaction between natural and man-made environmental conditions. To be able to identify (dis)similarities in air quality conditions between cities and assess the impact of policy options, information on, and tools to evaluate environmental conditions determining urban air quality are essential. When making comparative air quality assessments for a large selection of cities, it is impracticable to make a very detailed assessment of the situation per city. Therefore, a methodological framework has been developed in which the urban air quality situation is assessed using *Pressure*, *State* and *Impact* modules (PSI-chain). The modules consist of, and are linked through simple indices which can be calculated from the most common and basic data (Figure 1). The *pressure* module describes the natural and man-made environmental conditions. In the *state* module air pollutant concentrations are evaluated against WHO-Air Quality Guidelines (WHO-AQGs). In the *impact* module, exposure and (possible) health effects on citizens are

[2] Approximately 148 million people live in these cities (21% of the total European population).

estimated. Health impacts will only briefly be covered in this paper. Two complementary tools describing the PSI-chain are currently being developed at RIVM for the assessment of city background concentration levels:

- A set of urban environmental indices which will be empirically linked to air pollutant concentrations;
- A simple urban air quality box model which is able to calculate (annual) average city background concentrations and the number of exceedances of air quality guidelines for a broad selection of cities on the basis of relatively simple input data (city area, regional background concentrations and meteorological observations).

Both tools will be discussed using data from the *Europe's Environment* study to illustrate the methodology. First, European urban environmental data availability, necessary for urban PSI-assessments will be discussed.

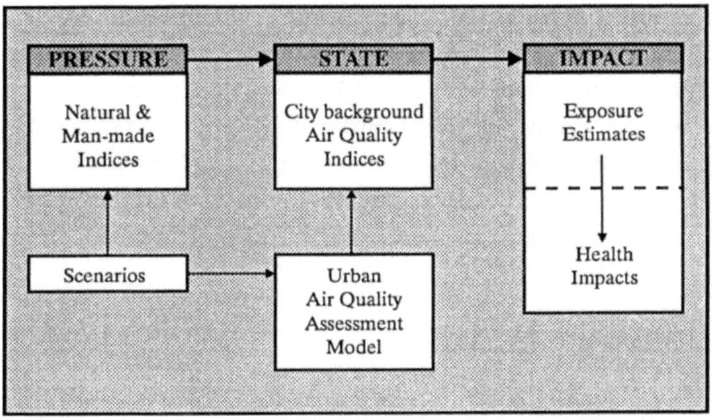

Figure 1: The Pressure-State-Impact framework used for comparative assessment of urban air quality.

European Urban Environmental Data Availability

To date, no complete integrated urban environmental database exists such as would be necessary for integrated urban air quality assessment studies. However, for cities within the EU a database is available containing air quality data (APIS; EC, 1992) and a database containing information on monitoring stations (GIRAFE; EU, 1994). Six-hourly meteorological observations for all cities are available from an ECMWF database (described in Potma, 1993). Population data is available through UN-IEDS. In the framework of the Europe's Environment programme a questionnaire was sent to municipal authorities to collect (additional) data on size, population and siting of the city, urban emissions and air quality data. Information was requested on the basis of the "physical" or "morphological" city. The definition of a city or conurbation on the basis of physical reality rather than on administrative boundaries called for an unconventional breakdown of existing statistics. It proved to be difficult to generate figures from existing material. As far as population, size, meteorology and siting of the city are concerned, a more or less complete data set is now available for the 105 selected cities. The availability of emission data varies, dependent on the component, from 29% (Pb) to 46% (NO_x) of the selected cities. Concentration data is available for 33% (O_3) to 79% (SO_2) of the cities. Besides these data gaps, differences in inventory methods and monitoring techniques often make comparing information between countries (or even between cities in one country) difficult.

Urban Environmental Indices

Figure 2 presents the indices which were defined to reflect both the natural and the man-made environmental conditions "determining" urban air quality[3] (the *pressure* module). It should be noted that far more indices could be thought of. The indices were calculated for the 105 cities which were selected

[3] In fact the current scheme focuses on winter and summer type smog. Winter smog episodes are generally characterised by high concentrations of SO_2 and PM, summer smog by O_3.

in the framework of the Europe's Environment programme. Indices were ranked in 5 classes.

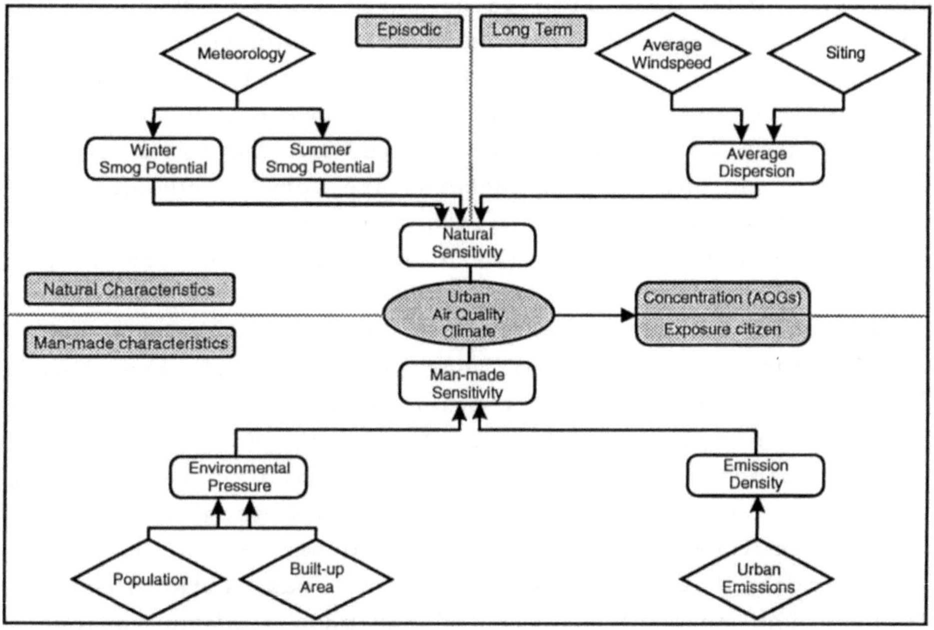

Figure 2: Together with the topographical siting of the city, the average wind speed is used as indicator for the city's *average dispersion* conditions. An index called *Meteorological Smog Potential* (MSP) has been defined for both the winter (MSP-W) and summer (MSP-S) half year to describe the probability of enhanced concentrations on the basis of the synoptic situation independent of emissions. The meteorological smog potentials are at the moment based only on the local 6-hourly synoptic situation. The MSP-W is based on atmospheric stability, temperature, precipitation and wind velocity. MSP-S on temperature, cloudiness, residence time and wind velocity Total urban emissions are correlated with population size and population density. City size also influences advection and vertical exchange of pollution and thus residence time of pollutants in the urban atmosphere. *Environmental pressure* is defined as the combination of population size and density. Since emission data are fragmented, *emission indices* have been defined only for winter smog pollutants (SO_2 and/or particulate matter) and summer smog precursors (VOCs and NO_X), both as density per square kilometer.

No pattern in the distribution of siting characteristics is apparent in Europe. Because cities located near Europe's north and west coasts and in Ukraine experience highest average wind speeds, the average dispersion index tends to be more

favourable in these regions. Lowest average wind speeds and subsequently less favourable dispersion conditions are generally found away from large water bodies; Northern Italy and the Balkans are similar. MSPs are linked to climatological regions. Cities located in the marine west coast temperate climatic region have relatively low MSP-W and MSP-S. Meteorological conditions are tempered by the vicinity of the Atlantic ocean and weather patterns are very changeable. Cities found in the continental climatic region with cold winters obviously experience high MSP-W values. Cities found in the Mediterranean region have a low MSP-W in general but, because of prolonged and hot summers a high MSP-S. MSP values can vary considerably from year to year according to the observed weather patterns. Neither population nor built-up area show regional differences in Europe. The winter type smog emission index is highest in the coal-burning cities of Central Europe and in some British cities. The summer type smog emission index shows no regional differences, but information is limited.
The validity of the pressure-module was first tested by plotting the defined natural and man-made indices against the average number of exceedances for SO_2 and PM (indicative of winter smog, average city background concentrations) (Figure 3). Both sensitivity indicators correlate with measured exceedances. This implies that cities with the same emissions and size but located in different parts of Europe (different MSP-W values) can experience completely different exceedance figures. In collaboration with WHO-ECEH, the environmental indices will be calculated for European cities with more then 50.000 inhabitants (approx. 3650 cities) . Indices will be empirically linked to measured air pollutant concentrations through multiple regression analysis. The empirical relations between the natural and man-made characteristics and air pollutant concentrations will enable simple scenario analyses.

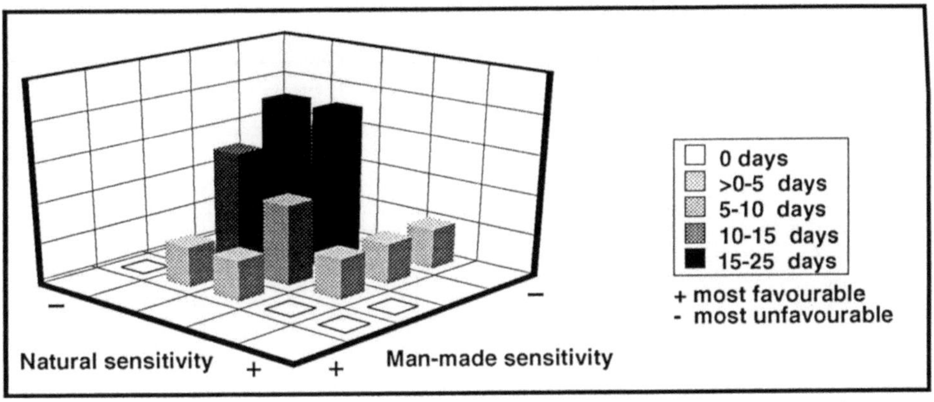

Figure 3: Average number of exceedances of short term WHO-AQGs for winter smog pollutants in recent years. SO_2: 24-h > 125 $\mu g/m^3$; TSP: 24-h > 120 $\mu g/m^3$; black smoke: 24-h > 125 $\mu g/m^3$. Based on 78 cities.

Concentrations and exceedances of WHO-AQGs in major European cities

Table 1 presents a set of WHO-Air Quality Guidelines (WHO-AQGs) for common components which are used as indicators for typical urban processes. These internationally agreed AQGs are taken as reference values against which the urban air quality is evaluated and from which population exposure will be estimated (*state* module).

Table 1: WHO-AQGs for relevant pollutants and expected effects (RIVM, 1992;WHO, 1987)

Pollution type	indicator	WHO-AQG risk level $\mu g/m^3$	Effect level	Effects
Short term effects				
Summer smog	O_3	150 (1 hour)	200 $\mu g/m^3$ classification mild	Lung function decrements, respiratory symptoms, inflammation
Winter smog	SO_2+ TSP	250 (1 day)	400 $\mu g/m^3$ classification moderate	Decraese lung function; increased medicine use for susceptible children
Urban/traffic	NO_2	150 (1 day)		
Long term effects				
Traffic	Benzene	2.5 (1 year)	10^{-7} yearly risk on carc.	Leukaemia; neurologic symptoms
Traffic/industry	Lead	0.5 (1 year)		Effects on blood formation, kidney damage; neurologic cognitive effects
Combustion	SO_2+TSP	100 (1 year)		Respiratory symptoms, chronic respiratory illness
Combustion/industry	B(a)P	0.0001 (1 year)	10^{-7} yearly risk on carc.	Respiratory tract and lung cancer

TSP: Total Suspended Particulates

In Table 2, exceedances of these AQGs for the city background are tabulated for the 105 cities from the Europe's environment study for which data were available. Collected air quality data suggests that in almost all cities one or more WHO-AQGs will be exceeded in a typical year. Average SO_2 and particulate matter concentrations have dropped considerably in a large number of cities. Despite this downward trend, exceedances of the short-term WHO-AQGs for SO_2 and particulate matter, indices for winter smog, are still widespread in urban areas. Ozone, indicative of photochemical smog, is at present monitored only in a limited number of (West) European cities. Available data suggests that high ozone concentrations may occur in many cities, especially in residential districts. Reported annual average NO_2 concentrations do not show a clear trend; nor are there regional differences. The short term WHO-AQG was recently exceeded in 6 of the 40 cities with data. Data on Pb and CO were collected for *hot spots* only. City background levels will be substantially lower. The lower limit of the WHO-AQG for Pb was recently exceeded in only 5 of the 49 cities for which data are available. The short term WHO-AQG for CO is exceeded in almost all cities for which data are available, sometimes even by a factor 4. No clear trend in concentration levels is apparent.

Table 2: Exceedances of WHO-AQGs on city background locations in European cities with more than 500.000 inhabitants

Indicator	WHO-AQG $\mu g/m^3$	Percentage of cities with exceedances observed	Number of cities with data[1]
SO_2	50 (year)	13	78
SO_2	125 (day)	43	76
Black smoke	50 (year)	13	15
Black smoke	125 (day)	-	-
TSP	70 (year)	63	55
TSP	120 (day)	86	77
SO_2 + PM	250 (day)	70	74
NO_2	150 (day)	15	40
O_3	150 (hour)	81	27
Pb[2]	0.5 (year)	14	49

1 104 cities selected
2 Highest observed on "hot spots"

Population exposure

In the *impact* module, exposure of, and (possible) health effects on citizens are estimated. The actual exposure of the urban population to air pollution is difficult to estimate. Besides determining the spatial and time variation of ambient and indoor pollutant concentrations, it is necessary to know the location and physical activity level of inhabitants. Since this information, especially for a large set of cities, is not available, rough first order exposure estimates were made by calculating the number of citizens exposed to exceedances of WHO-AQGs using a simple model. The model presumes a circular shaped city with evenly distributed population and emission densities. The concentration gradient in the city is derived from the maximum observed concentration and mean maximum concentration. The gradient is compared to WHO-AQG in relation to the number of stations on which the gradient was based. Then the area of the city where the AQG is exceeded is calculated. Map 1 presents the geographical distribution of the percentage of citizens exposed to exceedances of WHO-AQGs for winter smog pollutants. Extrapolated to all 105 selected cities, approximately 50% (75 million) people are exposed to these exceedances. It is thought that the (potential) population exposure estimates can be refined by use of statistical models of population behavior. A first estimate of effects on population health can be obtained by extrapolating results from epidemiological research to the exposure estimates.

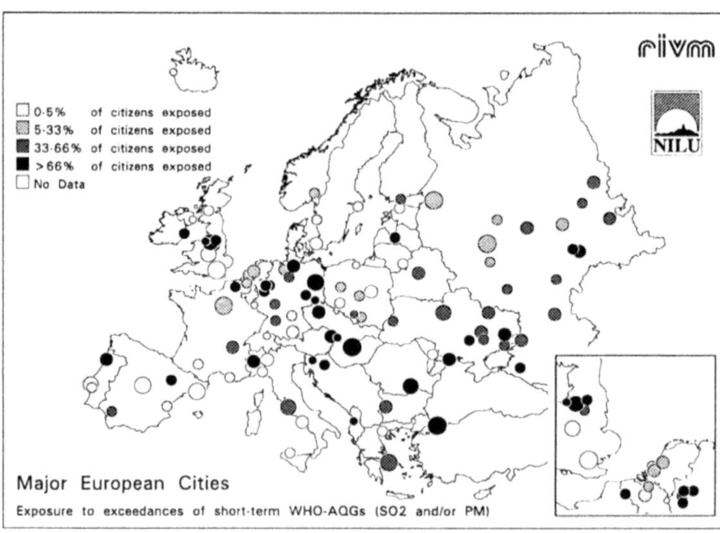

Map1: Rough population exposure estimates to exceedances of short term WHO-AQGs for winter smog pollutants (SO_2 and or particulate matter (PM)). Bullet sizes represent the population numbers of the cities.

Urban Air Quality Assessment model

RIVM is developing an Urban Air Quality Assessment Model (UAQAM), a simple box model which must be able to calculate (annual) average city background concentrations and the number of exceedances of AQGs for a broad selection of cities. This model will be used primarily to estimate urban air quality for smaller European cities lacking a monitoring network and to conduct scenario analyses. Since urban environmental data, especially for smaller cities, is often incomplete, inputs required for the model have been intentionally limited to routinely available data only (meteorological observations from a WMO station (Potma, 1993), city area, regional background concentrations and emissions).
Emissions from the source categories traffic, industry and power plants, and space/domestic heating are considered to be mixed instantaneously throughout the Urban Boundary Layer (UBL). A statistical daily emission pattern is used for the three source categories. Moreover, space/domestic heating emissions in the model are linked to ambient temperature. The dimensions of the box are given by the city area (built-up area) and the UBL-height. This UBL height varies with time, dependent on 6-hourly meteorological observations from a WMO station which are linearly interpolated to 1-hourly values. A software library by Holtslag and Beljaars (1990) is used to calculate surface fluxes. The translation of wind speed from 10 to 50 meters is performed using so-called flux profile relationships. The UBL height for neutral and stable conditions is calculated according to Nieuwstadt (1981), for unstable conditions according to Tennekes (1973). The urban (background) concentration is obtained by summing the regional background concentration and the city contribution.
The model is currently being tested for SO_2 + NO_x in 17 cities for which data are available through the *Europes Environment* project. Test runs show the current version of the model is able to predict annual average city background concentrations within a factor of 2.

Carsmog.

In busy streets, exceedances of AQGs for traffic related components are to be expected. However, continuous monitoring of air quality in busy streets would require a very extensive network of measuring stations. Because of the impracticability and high costs of an extensive measuring network in all cities, a calculation system has been developed as an alternative. This system (CARSMOG) makes use of the hourly measurement results (CO, NO_x and NO_2) from 13 street stations from RIVMs national air quality network (LML) and determines on-line air pollution levels in busy streets in the larger Dutch cities where measurements are not available (den Tonkelaar; Wildschut, 1993).

In the model the hourly concentrations of a component at a particular place in a city street are assumed to depend on the following factors: the annual average emission in the street; the relative emission for that hour of the day and that day of the week; the emission pattern over the year; the wind direction and velocity; the hourly background concentration and an annual average dilution factor representing the degree of dilution of traffic emissions to a concentration at the measuring point.In order to calculate concentrations in streets it is necessary to have available hourly traffic data (number of cars, proportion of goods traffic, average speed). Since such data are continuously registered nowhere, a different approach is taken here. Starting point for the system is the assumption that the traffic patterns (and thus also emissions from traffic) in all Dutch streets fluctuate in more or less the same way from hour to hour. This traffic pattern can be deduced from a detailed analysis of the measured concentrations at the thirteen street stations of the LML. A semi-empirical relation, developed in the TNO-traffic model (van den Hout and Baars, 1998) and further developed for the CAR-model (Eerens et al, 1993), is used to calculate NO_2 concentrations from the calculated local contributions to the NO_x contribution and background ozone concentrations.The model is used to deduce the fluctuating traffic pattern by normalising hourly concentrations per LML-station to a relative emission strength by correcting for background concentrations, for the influence of the wind direction on the concentrations at that measuring station, the wind velocity in the street, the emission pattern over hour and day of the week and for month of the year. What is left is a factor that represents the fluctuation of the traffic emissions for that station and hour in the street concerned. By averaging these relative factors over all street stations, one obtains a national picture of the complex of traffic emissions, dispersion characteristics and NO_2-formation in that specific hour.The model is calibrated by choosing the model variables in such a way that the calculated average traffic contribution is within 1 percent of the measured traffic contribution. For the calculation of air quality in Dutch streets, the statistical traffic contribution is determined on the basis of the calibrated model variables (the contribution that may be expected for that hour, day and month and the given wind direction and velocity) for each hour per measurement station. Subsequently, the relative deviation of the calculated traffic contribution with regard to the measured traffic contribution is determined for all stations. By averaging these factors a measure of the general combined (traffic, dispersion, NO_2-formation) trend in the Netherlands is obtained for that particular hour. With the national traffic picture as input it is possible - taking into consideration the specific emission data and dispersion parameters for a street, and the valid hourly data on background concentration and wind velocity for the city - to calculate the hourly air pollution levels in large cities.In order to determine the air quality in the Dutch cities, existing streets are not taken as the starting point; an approach with theoretical standard streets has been chosen. Standard streets are regarded as being representative of the total of busy streets in the center of the city concerned. Characteristics are obtained from a weighted averaging of the traffic-specific and geometric features of the streets concerned. CARSMOG has been tested by calculating the street concentration at a monitoring station using the national traffic picture determined from the other stations as input and comparing these calculations with the concentrations measured at the station. Table 3 presents the results of this test for 1991. The table shows that the performance is sufficient under "average" conditions. The performance of CARSMOG during smog episodes is not worse (den Tonkelaar; Wildschut, 1993).Calculated "standard street" concentrations are available on-line within one hour to municipal authorities of the biggest 8 cities in the Netherlands through a computer based system called VIDITEL.

Table 3: Performance of CARSMOG in 1991. Concentrations on a station were calculated using the national traffic picture determined from the other stations as input for the model.

Conclusions and recommendations

The Pressure-State-Impact methodological framework promises to be a valuable tool for comparative assessments of the air quality situation in a large set of cities, since it relies only on the most basic routinely available data, which are translated into clear and simple indices. First results obtained using this method show distinct regional differences in environmental *pressure* and *state* factors. Furtermore, the results confirm the ability of this simple approach to correlate *pressure* indices well with the actual air quality situation where known.

Municipal authorities should be encouraged to collect urban environmental data, necessary for urban air quality assessments, in a standard way. Tools (e.g. standard questionnaires, dispersion models) to collect urban environmental data and to assess urban air quality on the basis of this information should be disseminated, results preferably gathered by an international organisation. Harmonisation of monitoring activities between cities, not only as far as technology is concerned, but also with regard to the siting of monitoring stations, should be encouraged.

Collected urban air quality data suggests that in almost all cities at least one WHO-AQG will be exceeded in a typical year. Similar conditions may occur in smaller cities. These exceedances indicate possible health risks to which citizens are exposed, especially in the heavily polluted cities of Central and Eastern Europe.

RIVM is developing a simple box model for city background urban air quality assessments, using only simple and routinely available data. First test results are promising. A description of the model, including a justification of model choice and first test results will be published shortly.

CARSMOG shows that it possible to calculate concentrations in "standard" streets using a normalised annual average street contribution, the distribution of traffic emissions over hours of the day, days of the week and months of the year and the fluctuating traffic pattern obtained from a comparison between measured and modeled street concentrations for operational street stations. The model can be a valuable tool for municpal authorities for the control and even short-term prediction of their street air quality. The model could easily be applied to cities in other countries with a monitoring network. As far as the Netherlands is concerned, a system analogous to CARSMOG will be developed to calculate city background concentrations in cities where no monitoring network is operational.

References

EC (1992) APIS: Air Pollution Information System, reference. CEC DG XI/B3 (PC database).

EU (1994) GIRAFE: Air quality monitoring networks in the European Community, CEC DG XI/B3 (PC database).

den Tonkelaar, W.A.M., H. Wildschut (1993) Development of CARSMOG: A system for the on-line calculation of air pollutant concentrations in busy streets through extrapolation of measurements in other streets (in Dutch). IMW-TNO, report 93/084, Delft, The Netherlands.

Eerens, H.C., C.J. Sliggers, K.D. van Hout (1993) The CAR model: The Dutch method to determine city street air quality. Atmospheric Environment, Vol. 27B, no.4: pp. 389-399.

Holtslag, A.A.M., A.C.M. Beljaars (1989) Surface flux parametrization schemes; Developments and experience at KNMI. Proceedings workshop on parametrization of fluxes over land surfaces, ECMWF, Reading 24-26 October 1988.

Nieuwstadt, F.T.M. (1981) The steady-state height and resistance laws of the nocturnal boundary layer; Theory compared with Cabouw observations. Boundary layer meteorology, Vol. 20: pp. 3-17

Potma, C.J.M. (1993) Description of the ECMWF/WMO Observational Data Set, and associated data extraction and interpolation

procedures. RIVM, report no. 722401001, Bilthoven, The Netherlands.
RIVM (1992) The environment in Europe: a global perspective. RIVM, report no. 481505001, Bilthoven, The Netherlands.
Sluyter, R.J.C.F. (ed) Urban and local air quality in major European cities, scientific background document to Europe's Environment. RIVM/NILU, report no. 722401004, Bilthoven, The Netherlands (in press).
Tennekes, H. (1973) A model for the dynamics of the inversion above a convective boundary layer. Journal of the atmospheric sciences, Vol 30: pp.558-567.
van den Hout, K.D., H.P. Baars (1988) Development of two models for the dispersion of air pollution by traffic: the TNO-traffic model and the CAR-model (in Dutch). MT-TNO, report 88/192, Delft, The Netherlands.
WHO (1987) Air quality guidelines for Europe. World Health Organization, European series no. 23, Copenhagen, Denmark

Acknowledgements

Thanks are due to James Burn (RIVM) for editorial assistance. Hans Eerens, Frank de Leeuw, Karel van Velze, Addo van Pul and Esther van Zantvoort (all RIVM), and Steinar Larssen, Knut Grönskei (both NILU) are acknowledged for their comments on early drafts of this paper and their participation in the Europe's Environment programme.

Urban Air Quality Measurement: Organizational principles and Procedure Aspects

G.M. Tishchikov, B.B. Kozeruk
Main Administration of Hydrometeorology of Belarus
16 Komsomolskaya Street,
220600 Minsk
Belarus

The present paper is concerned with the air monitoring system, operated in the industrial centres of the Republic of Belarus. It should be noted that the system has been designed as a part of the former USSR National system of observation control and is currently revised in new social-economic conditions of the independant state.
Belarus belongs to industrial-agrarian countries and represents a region, with a dominant power-intensive and material-consuming machine building, developed chemical and petrochemical, light wood-working and food industries. In this connection the main part of population (68%) lives in towns, subjected, as a rule, to a considerable anthropogenic impact, which proves the urgency of the problem of interest.
Monitoring of radioactive pollution, resulted by the Chernobyl NPS disaster is a special problem for Belarus. Radiological impact is one of the main factors influencing the environment of eastern and south-eastern of Belarus.
Atmospheric air quality is monitored in Belarus industrial centres by two agencies: the Main Admistration of Hydrometeorology (Glavhydromet), which makes up to 95% of all observations, and the Ministry of Public Health (Figure 1).

The leading role of Glavhydromet is dictated by the fact, that atmospheric air monitoring is patterned after meteorological network principles: regularity, unity of the programmes and observation methods, representativeness of observing points. Moreover, monitoring results may be interpreted, notable on-line, only with a use of meteorological data. It is noteworthy, that the

NATO ASI Series, Partnership Sub-Series, 2. Environment – Vol. 8
Urban Air Pollution
Edited by Ivo Allegrini and Franco De Santis

responsabilities of the two agencies do not duplicate each other. Monitoring activities stemming from specific responsabilities of these two agencies provide a broad outline of the atmosphere status and its impact on human health.

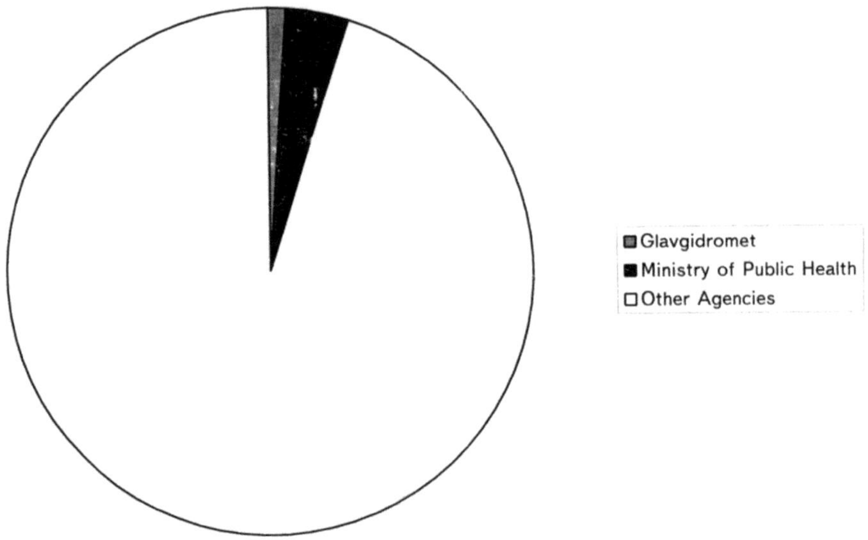

Figure 1. The contribution of various agencies to the total volume of air quality observations.

Organizational structure of the agencies is based on similar hierarchical principles. The central bodies are located in the capital and perform administrative and methodological guidance, data processing and national databanks management. Laboratories of the regional centres supervise the network of district and urban divisions.

Atmospheric pollution observations are also made at some plant laboratories. However, their share in obsevations, made outside industrial zones, is insignificant.

In this paper we will not concern ourselves with the polluters control, exerted by the Ministry of Natural Resources and Environment Protection.

Urban air is monitored in Belarus at steady-operating and mobile stations. Steady-operating stations represent specially equipped pavillions for obtaining long-term data on pollutants content and meteorological information at the fixed urban points. Steady-operating stations are the main building block of the Glavhydromet monitoring network. Mobile stations are specially fitted vehicles, used for route, "beneath-the-stack" and expeditionary observations.
Route observations are made at fixed points outside the stationary network and are used either as a complementary, or a substituting data source. Besides, this type of observations is used to substantiate the location of a stationary network and to check the representativeness of the operated observing points. "beneath-the-stack" observations are made beneath the smoke (gas) stacks of industrial enterprises at fixed intervals from an emission. Being principal for the Ministry of Public Health, these two methods of observations are used by Glavhydromet to a lesser extent.
The adequate number and location of observing stations at the area under control is a central problem in the design and refinement of a monitoring system. The most adequate and objective information on distribution and temporal dynamics of the pollutants concentration fields can be obtained by means of a reasonably dense network of observing stations. By modern standards the interval between stations is to be 2-3 km to ensure restoration of pollutants concentration fields to an accuracy of 0.2 of the maximum concentration limit. Under these conditions some 50 stations will be required in Minsk, the capital of Belarus.However, the absence of funds and material resource has dictated the development of criteria for obtaining reasonably objective meteorological information from a minimum number of stations.
By modern standards the number of stations is calculated using such a formalized index as urban population. For example, a town with a population of 50 000 people needs one station; a town with a population from 50 000 to 100 000 people needs two stations, and so forth (Figure 2)

Industrial potential, extent of polluters concentration, intricate topography, availability of unique natural and historic objects etc. serve as additional criteria for the increase of the number of stations.

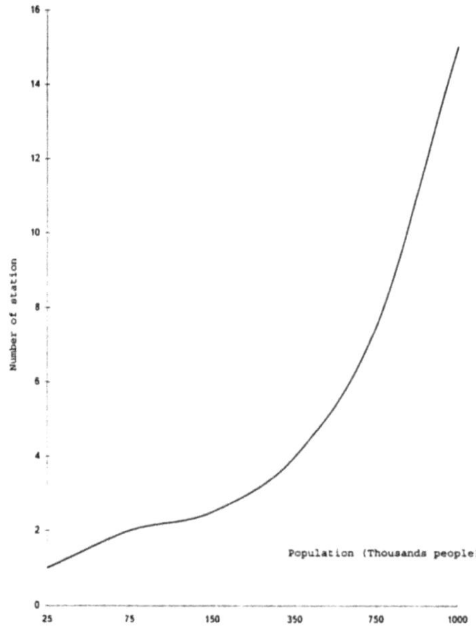

Figure 2. Relationship between the number of steady-operating monitoring stations and population.

For practical reasons the network has not been designed to obtain data adequate to construct polluters concentration fields in every Belarus town. Location of stations has been governed by the need for securing generalized data on air quality in the most typical and vital for people residential, administrative, business and overloaded with traffic areas of industrial cities. Monitoring stations layout incorporates computation of polluters dispersal fields, estimates of microclimatic conditions and orographic features of intended locations, as well as expeditionary data on circumstantial evidences of air basin pollution (precipitation, snow cover, soils). Layout of urban stations is agreed with the Ministry of Public Health and the Department for Architecture.
One of the monitoring stations, the data of which is taken as background, is set up in the most ecologically clean district, as

a rule, in an amenity forested area on the wind side of the town. Other stations are located in administrative, residential and industrial urban districts. Since these stations are meant to obtain averaged air pollution data on large areas, they are located in well-ventilated sites outside the reach of individual polluters. The stations designed to monitor air pollution in the areas of intensive traffic are located close to highways.

The same concepts are used in the layout of expeditionary and route monitoring stations. It is imperative that steady-operating and route stations should determine four principal pollutants (suspended matter, sulfur dioxide, carbon oxide, nitrogen dioxide), constituting 63% of total observations, and specific pollutants typical for the station area (Figure 3).

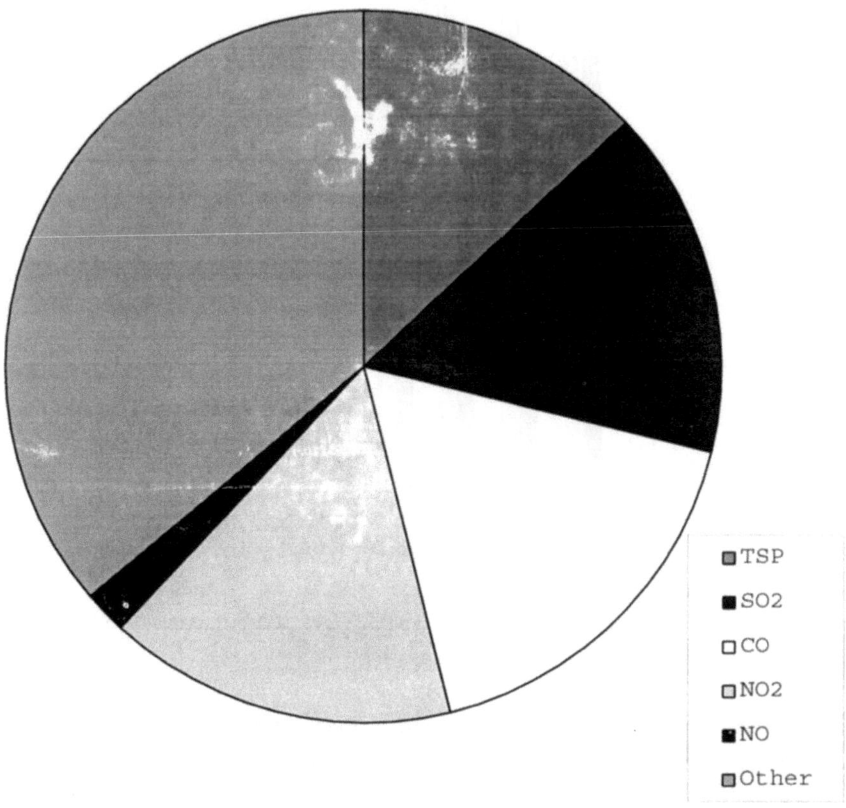

Figure 3. Relationship among pollutants monitored by Glavhydromet network.

The list of pollutants for "beneath-the-stack" observations is dictated just by polluters specificity. Priority lists of specific pollutants are compiled and regularly adjusted for each industrial town of Belarus. The lists are compiled by a special procedure with regard to such indicators as total pollutant emission, hazard class and quality standards, disperssion coefficient etc. The total number of specific pollutants, determined in Belarus, is limited by 25 due to the absence of analitical (Figure 4).

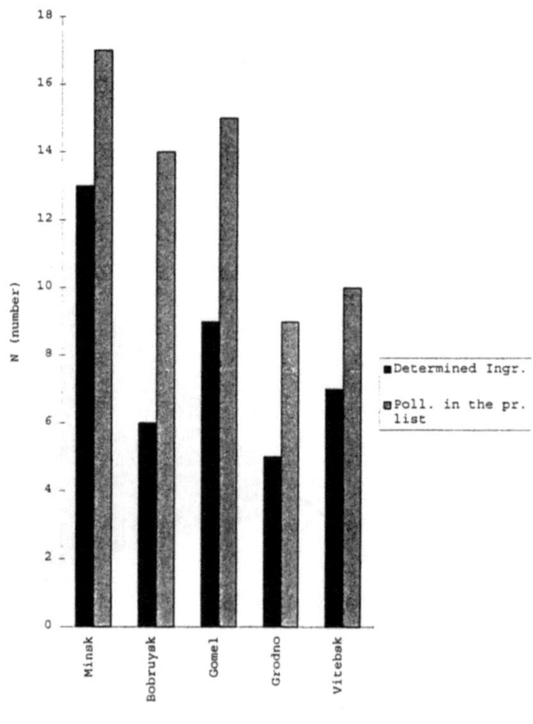

Figure 4. A correlation between priority pollutants and determined ingredients in some towns of Belarus.

The monitoring programme at steady-operating stations includes 3 to 4 daily air samplings, taken manually or automatically within 20 to 30 minutes. Modern gas analyzers are actually absent at the operated monitoring network. In this connection air samples are taken, as a rule, by aspiration method using absorbing solutions, hard sorbents and filters. The collected samples are transported to chemical laboratories.

Pollutants concentrations are measured by the unified procedures, used in the CIS countries. The major part of the procedures for determining inorganic and certain organic matters is based upon photometric analysis with a use of serial photocolorimeters and photospectrometers. The domination of these methods in the laboratories of monitoring agencies is dictated by availability and relatively high capacity of the domestic equipment. Heavy metals concentrations are determined mainly by atomic absorption spectrophotometry equipped with flame ionization. Content of most organic matters is measured by gas and liquid chromatography.

The absence of modern and reliable laboratory and analytical equipment, is a traditional problem for all laboratories of the monitoring agencies. The current lack of consumables aggravated this situation and resulted in the reduction of the number of determined components and on-line control degradation. Initial data on pollutants content and meteorological parameters is received by information centres of the monitoring agencies, subjected to control and statistical processing using special software and is entered to the agencies databanks.

The accumulated data is used to estimate and timely predict air pollution with due regard to weather forecasts and polluters parameters; to make recommendatins on air basin improvement and to evaluate performance of nature-protection measures.

Along with direct air basin observations Glavhydromet is widely uses circumstantial methods of air monitoring. The study of chemical composition of precipitation is an important costituent of the monitoring. Chemical composition of precipitation is regularly observed at 14 sites of Belarus. Acidity, main salt components and heavy metals are determined in monthly-combined precipitation samples.

Snow cover is monitored in the areas, where atmospheric pollution is measured at a limited number of stations, or is not measured at all, and in the areas of substantial emissions of stationary pollution. Snow samples are taken in industrial urban areas by a regular network in the period of maximum snow accumulation. Future development of urban monitoring network involves solution of some technical and methodological issues. The amount and layout of existing steady-operating stations are to be refined using mathematical simulation. A substantial increase of representative and expeditious results can be realized only in passing from discrete air samling at stations to the continuous cycle of analytical control of priority components, which entails complete reequipment of the existing network.

The present methodological base of the monitoring is to be revised with a view to refine and unify the methods, to intercalibrate the procedures. Mathematical simulation methods, estimates of precipitation and snow cover pollution, remote laser methods are to be widely used to enhance air monitoring. At present and for the future monitoring network remains unique experimental tool for air pollution assessment, establishment of long-term trends and causes of pollution variations.

Exposure Assessment Methodologies for Humans and Ecosystems

Daniel A. Vallero
National Exposure Research Laboratory
U.S. Environmental Protection Agency
Research Triangle Park, NC 27711
USA

Introduction

The principal components of exposure are the magnitude of the pollutant concentration, the duration of the exposure, and the frequency of the exposure (Sexton and Ryan, 1988). Human exposure assessments of these three components include measuring pollutant concentrations (outdoor, indoor, microenvironmental, personal), but also characterizing personal exposure scenarios, such as activity patterns and ventilation rates. Ecological exposure is the expression of the magnitude, duration, and frequency of contact between an ecological resource and a "stressor;" i.e., a physical, chemical, or biological entity that can induce an adverse response" (Risk Assessment Forum, 1992).

Risk assessment experts and other scientists are developing approaches to estimate exposure. Ecological scientists are exploring methodologies for estimating the exposure of ecosystems and their subdivisions to environmental stresses, while human risk assessment analysts are investigating approaches for estimating exposures to contaminants that could affect human health. Exposure assessment methods vary significantly, depending upon factors, such as the scale of the exposure, the measurement focus, and whether the measurements are actual expressions of exposure or part of an algorithm to indicate exposure. This paper compares ecological and human exposure assessment methodologies concerning the types and scales of monitoring and sampling designs, the availability of models to simulate and estimate exposure, and the components necessary to calculate exposure. The National Research Council

NATO ASI Series, Partnership Sub-Series, 2. Environment – Vol. 8
Urban Air Pollution
Edited by Ivo Allegrini and Franco De Santis

of the National Academy of Sciences (1983) developed a risk assessment paradigm with four separate steps: hazard identification; dose-response assessment; exposure assessment; and risk characterization. Regulatory agencies, like the U.S. Environmental Protection Agency (EPA), have applied this paradigm to human health risks. Each of the four steps has been subdivided further. For exposure assessment, EPA applies the steps shown in Figure 1: source characterization; transport, transformation, and fate; pathways; environmental concentrations; and exposure measurements.

Lipton et al (1993) have questioned the appropriateness of applying the NAS paradigm to ecological risk assessment, since several "intrinsic distinctions" can be drawn between human health and ecological risk assessments. Ecological target receptors may be unknown or ambiguous and the level of biological organization is variable. Exposure assessment, however, is similar for ecological and human risk assessments (Figure 2).

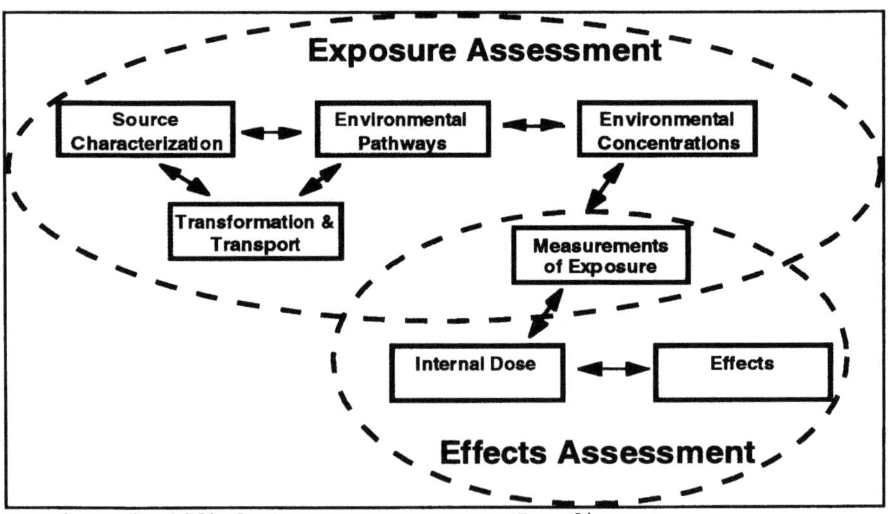

Figure 1: Simplified Exposure Assessment Paradigm.

Exposure Assessment Methodologies

Routes, magnitude, duration, and frequency of exposure are important considerations for both human and ecosystem exposure assessments. While both paradigms include measurements of pollutant concentration, the major difference rests in measurements of behavior. For humans, exposure is a function of concentration, activity pattern, and uptake (ventilation, consumption, and absorption). For ecosystems, exposure is a function of pollutant concentration in the abiotic and biotic environment, and ecological function and structure (e.g., species migration, bioaccumulation and sequestration rates, bioenergetics, succession, and nutrient cycling). This complicates ecological exposure assessments because changes in function and structure are expressions of both exposure and effect; i.e., they change as a result of the exposure. All pathways, e.g., ingestion, dermal, or inhalation, must be included to express exposure fully. A single-species exposure assessment (e.g., for an endangered species) can be very similar to the multipathway, human exposure assessment described on the left side of Figure 2.

Multiple pathway field studies are designed to measure concentrations of pollutants in various environmental media. Temporal and spatial distributions of these measurements give an indication of the frequency, magnitude and duration of the exposure. The level of temporal (continuous, hourly, 12-hour, 24 hour, monthly, annual average) and spacial precision of these measurements varies depending upon the field study objectives.

EPA developed the Total Exposure Assessment Methodology (Wallace, 1987) to estimate exposure in human microenvironments (e.g., homes, automobiles, restaurants). Results from these TEAM studies indicate that a person's activities and behavior greatly affect one's actual exposure. Even when ambient concentrations are similar, activity variables, e.g., cleaning, cooking, smoking (direct and environmental exposure), time spent indoors versus outdoors, and transportation, can introduce considerable variability for most contaminants.

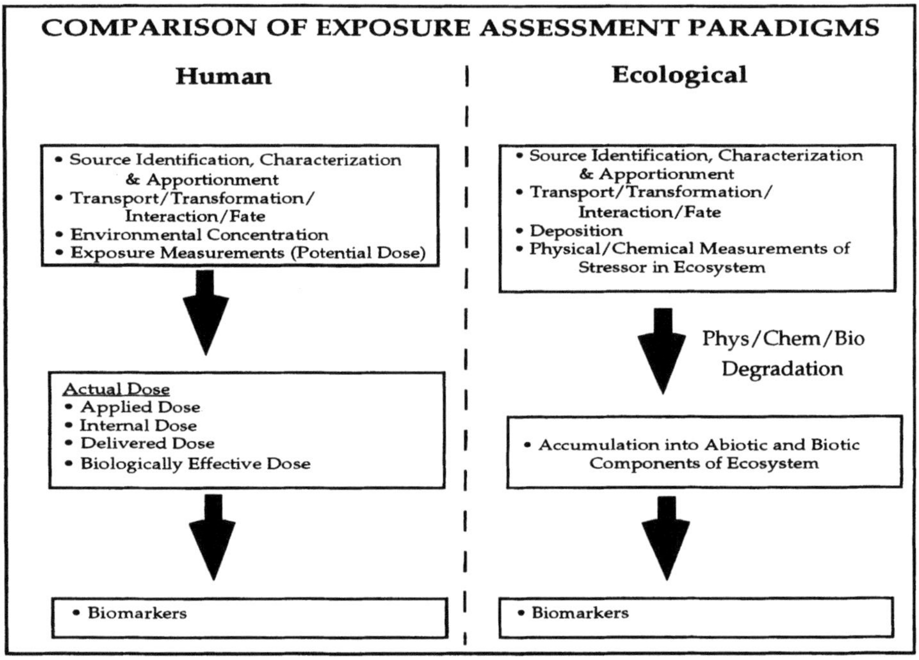

Figure 2: Exposure components of risk paradigms are similar for humans and ecosystems.

The U.S. EPA's (1990) Nonoccupational Pesticide Exposure Study (NOPES) illustrates the necessary methods to measure and estimate exposure from multiple pathways. NOPES was a multi-season study of pesticides commonly used in and around the home. The study households were selected from stratified random population samples in two urbanized areas. An embedded nine-home pilot study conducted in Jacksonville, Florida found that household dust may be a significant pathway for exposure to previously used pesticides ; e.g., Chlopyrifos, Propoxur, and Chlordane. NOPES confirmed previous studies which found indoor environmental exposures to be considerably higher than outdoor exposures. Other pathways, such as diet and drinking water, can be more significant pathways for other pollutants. NOPES was successful in estimating exposure levels for populations of two urban areas of the United States, assessing the relative importance of each exposure pathway to the overall level of exposure; characterizing the components of variability in the observed exposure levels, and, in beginning to model the relationships between exposure levels, rates of use, activity patterns, and other factors that could contribute to variation in exposure levels. These results demonstrated that the multi-pathway approach can be applied to nonoccupational exposures through inhalation. The study's probability-sampling design also allowed for inferences about the distribution of exposures for populations.

The objectives tracked well with the approaches recommended by the NAS (1991) for assessing human exposure to airborne pollutants (Figure 3), illustrating the need for data from direct measurements (personal and biomarker monitoring) and from indirect approaches (especially to gain knowledge about activities). NOPES characterized exposure, including seasonal variations, by monitoring and comparing outdoor, indoor, and personal air concentrations. The study also demonstrated that questionnaire-based models may be practical for particular analytes; e.g., certain termiticide concentrations were related to use and application history, age of home, and household inventory of the pesticides.

Scale of Exposure

Exposure studies can range from subcellular exposure to global. Methods for assessing exposure for an individual organism (e.g., one human being) differ from methods used to assess population exposure. Likewise, estimating exposures for a single ecosystem component; e.g., a lake or wetland, will be different from a large-scale exposure assessment of region or biome.

In the case of the small-scale assessment (residential, occupational, farms), a researcher may be able to determine signals of exposure for a wide array of contaminants, and provide detailed and specific information about a subject's activity patterns. Often, however, scientists are asked to estimate exposure of entire populations or target groups, wherein gathering detailed and specific information about the exposure of each individual in a population is scientifically and economically infeasible. Moreover, in the case of ecosystems, detailed information about individuals may have less importance than the interrelationships and diversity of a larger ecological community; true to the adage, "not seeing the forest for the trees." The hypothesis or study objective determines the scale of an exposure assessment.

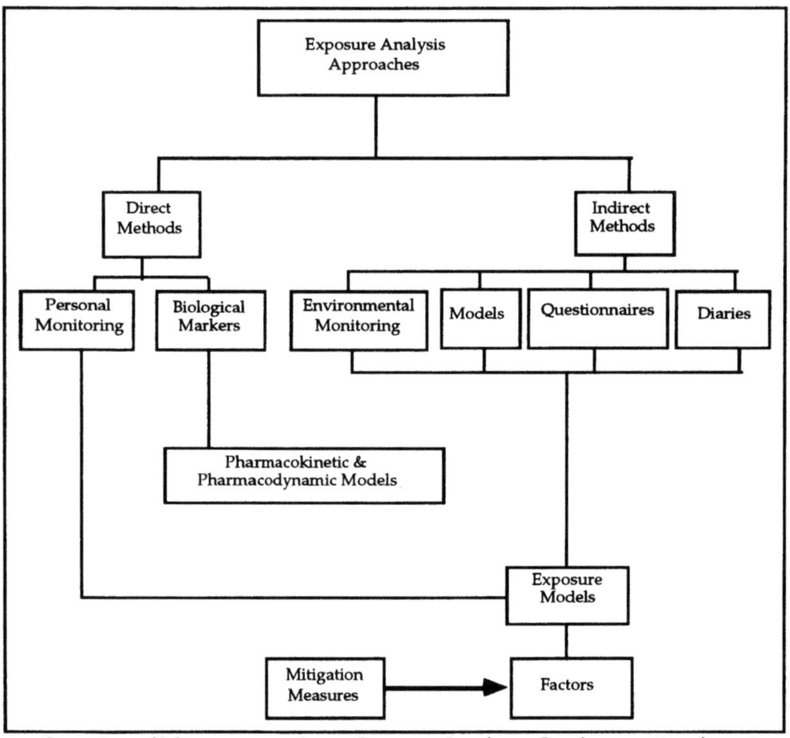

Figure 3 Possible approaches for analysis of air contaminant exposures (National Academy of Sciences, 1991).

Focus of Exposure Measurements

Human and ecosystem exposure measurement and assessment methods vary. Studies may be conducted to estimate the exposure of one type of receptor to a single pollutant; e.g., blood lead levels in children. A single pollutant exposure assessment can be conducted for a number of receptors; e.g., bone-lead concentrations in urban and rural school-aged children, lead concentrations in lawns bordering highways, and translocated lead in leafy vegetables downwind from an industrial source. The process is far more complex when a number of pollutants and receptors are included in risk assessment; e.g., ecological and human exposure to dioxins and toxic metals near potential agricultural, industrial, and transportation sources.

The measurement focus varies considerably among exposure assessments. Ambient outdoor, indoor, and personal exposures are directly measured or input into models. "Direct" measurements are usually inferences, since the measurement is a value at one point for one time period. Stationary monitoring devices provide outdoor and indoor measurements. Passive (diffusion) and active (constant flow) sampling devices are used for personal and microenvironmental measurements. Recently, researchers have deployed these devices to enhance ambient monitoring data and to provide average environmental exposure estimates for ecosystems, especially for forest stands.

Source-Receptor Assessment

Determining source characteristics and the transport and transformation of pollutants is similar for human and ecological exposure assessments. Various methods for identifying and apportioning the sources are available, including emission inventories, source-receptor models, and actual measurements (e.g., stack tests, remote sensing, and continuous emission monitoring). Emission inventories are often derived from calculations of fuel or feedstock and the manufacturing processes taken from emission forms completed by the operator; e.g., incinerator operators provide information about the type of fuel; amount and type of feedstock; a description of the combustion processes; and the types of stacks and vents at the facility, which is used to generate the emission inventory.

This information can be highly uncertain and is not sufficiently specific to characterize potential pollutant sources.

Stack tests, such as dilution samplers, are much more reliable than emission inventories, but are costly and require on-site access. Actual measurements of stack emissions are necessary to apportion the sources of pollutants to which a receptor is exposed (Figure 3). Temporally and spatially precise measurements are needed at the source (i.e., for "source signatures") to be coupled with ambient measurements of chemical species that are "markers" of particular sources. For example, the U.S. EPA's chemical mass balance model (CMB 7.0), developed by Watson et al (1990) is used by the Agency to "identify and the presence of and to quantify source contributions to receptor concentrations." Dispersion models are also useful exposure tools which require emission rates be estimated and combined with meteorology, and transformation algorithms to estimate the relative contribution of sources to measurements of pollutant concentrations at a receptor. If variability and uncertainty is high for emissions, as is common for source information derived from inventories, the dispersion model-derived source-receptor relationship is also highly uncertain and variable.

Some promising chemical markers and their associated source categories are shown in Table 1. The total, upper-bound contribution of the potential source on the measured ambient concentration can be obtained by multiplying the measured ambient concentration of the marker species by the characteristic factor (i.e., the reciprocal of the marker's abundance in the source's emission (i.e., listed in "Source Profiles," such as the U.S. EPA's VOC/Particulate Matter Speciation Data System, Version 1.4, EPA-450/4-91-027). For example, acetylene is one of the common volatile organic compounds (VOCs) found in motor vehicle tailpipe emissions. On average, acetylene represents 4 $\pm$ 2 % of total VOCs in exhaust in the U.S. If ambient acetylene is measured to be 3 $\mu g/m^3$, then the upper bound estimate = 3 $\mu g/m^3$ (25 $\pm$ 12) = 75 $\pm$ 36 µg/m3. Therefore, if total ambient VOCs = 150 $\mu g/m^3$, the greatest possible motor vehicle contribution is about one-half (75/150) of all VOC sources at this ambient site.

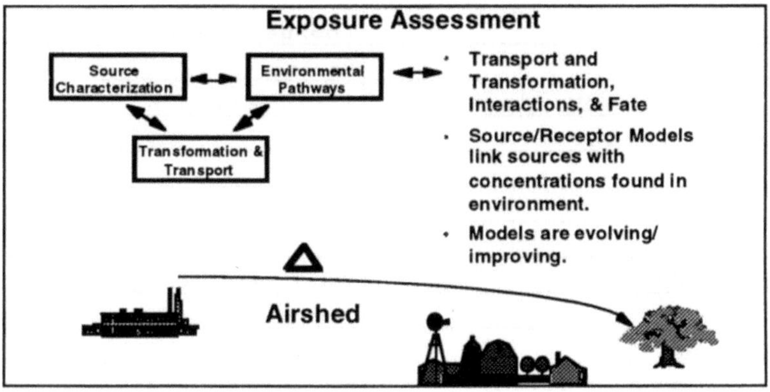

Figure 4 : After emissions are released, they undergo physical and chemical transformation before being deposited. Receptors can be human or ecological. The level of biological organization can be subcellular to regional. Human exposure assessments are often conducted at the population or subpopulation level (e.g., cancer risk per million in the United States). Ecological exposure assessments are conducted at many different levels, but regulatory and natural resources agencies often are interested in community level risk (e.g., loss of biological diversity in forest stands or wetlands).

Human exposure is defined as contact between a "contaminant and the human body" (Sexton and Ryan, 1988). By extension, ecological exposure is the contact between the contaminant and an ecosystem or its components (e.g., communities, species, or individual organisms). "Routine screening of certain indicators" in ecosystems provides an estimate of "the actual threats to the condition" of those ecosystems (Messer, 1990). Such screening for the presence of pollutants can be an indication of ecosystem exposure; however, the chemical and physical characteristics of a contaminant can ultimately determine actual exposure. For example, outdoor concentrations of fine particles near a home can be similar to fine particle concentrations inside the home, but ozone (O_3) concentrations may be much lower inside, because O_3 readily absorbs on surfaces. Aerosol acidity may be lower indoors due to higher concentrations of ammonia that buffer the acid.

Aerosols	**Dominant Source**
Na, Cl	Marine
K (soil-corrected), Cl, ^{14}C	Wood Combustion
Al, Si, K, Ca, Ti, Fe	Soil
Zn, Pb, Sn, Sb, Cl	Incinerators
V	Electric Utility Oil Combustion
Gases	
CO, various VOCs	Motor Vehicles
xylene	Industrial Solvents
ethane, propane	Natural Gas
isoprene, a-pinene, b-pinene, ^{14}C	Biogenic Emissions

Table 1: Selected examples of presently available chemical marker species. The dominant source is airshed dependent; i.e., in addition to indicating a dominant source, measurements of marker concentrations in ambient air may represent products of transformation or background concentrations. For example, Na and Cl-rich particles not near marine water bodies may be indicators of extraction or transportation activities that emit salt. High concentrations of Fe and Al may not be re-entrained dust, but may be indications of smelting activities. Therefore, an inventory of source types in the airshed should complement the receptor modeling.

The physical and chemical characteristic must also be considered when determining potential sources of measured ambient contaminant concentrations. Figure 5 illustrates three different idealized bimodal distributions for particles. The distributions can provide weight-of-evidence for whether the particles are anthropogenic or natural in origin.

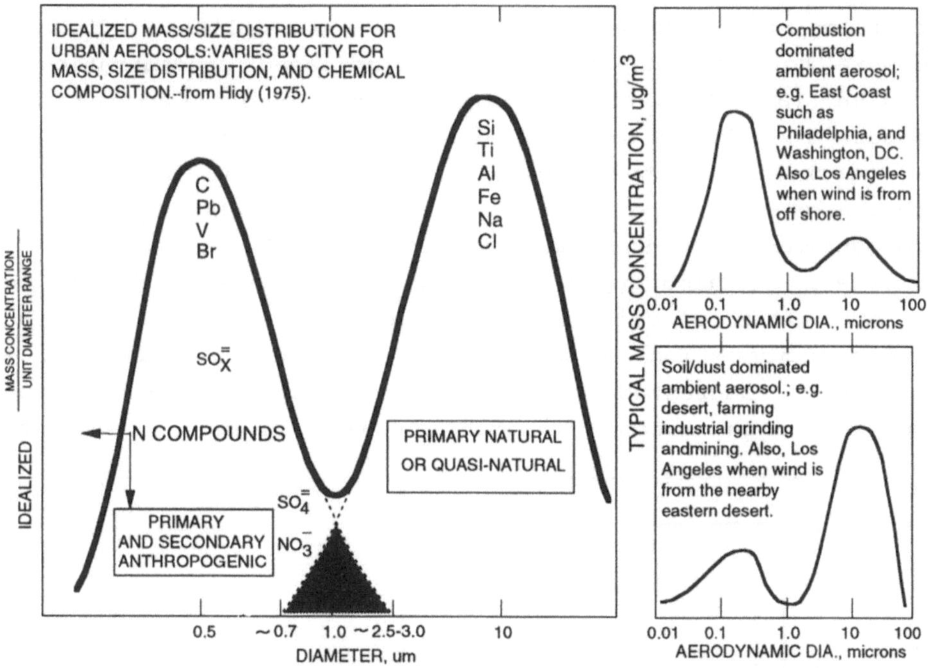

Figure 5: Particles often display a bimodal distribution by mass, originate from multiple sources, show dynamic growth and reactivity, and are carriers of other pollutants (Hidy, 1975). The upper right mass distribution is typical for an area dominated by anthropogenic (combustion) sources, while the bottom distribution is typical for areas where particles are generated from noncombustion sources (e.g., re-entrained soil and mining activities).

Dose

The nexus between exposure and effect is "dose." For human exposure assessments, the U.S. Environmental Protection Agency (1992) classifies dose as: potential dose (D_P); applied dose (D_A); internal dose (D_I); delivered dose (D_D); and biologically effective dose (D_{BE}). Figure 6 shows the pathway from an organism's first contact with a substance (D_P) to its intake, absorption, and metabolism (D_A, D_I and D_D) to its effect on the target organ (D_{BE}).

Measurements of D_P can often provide a reasonable estimate of exposure; i.e., the concentration of a contaminant around an organism. For airborne contaminants, D_A is a function of concentration, time, and ventilation. It is difficult or impossible to measure D_{BE} directly, so D_A, D_I and D_D are most often expressed by biomarkers, i.e., "indicators of changes or either be the contaminant itself or metabolites indicating exposure to the contaminant; e.g., increased concentration of cotinine (a metabolite of nicotine)in blood resulting from exposure to tobacco smoke. Similarly, biomarkers in ecosystems are "biochemical, physiological, or histological indicators of either exposure to or effects of xenobiotic chemicals at the suborganismal or organismal level" (Huggett, et al, 1992). Biomarkers can also apply to ecological exposure, although they are not often classified as measures of dose ("biotic and abiotic accumulation" in Figure 2). For example, Hunsaker, et al (1990) suggested measuring cholinesterase levels and porphyrin accumulation to indicate the level of ecosystem exposure.

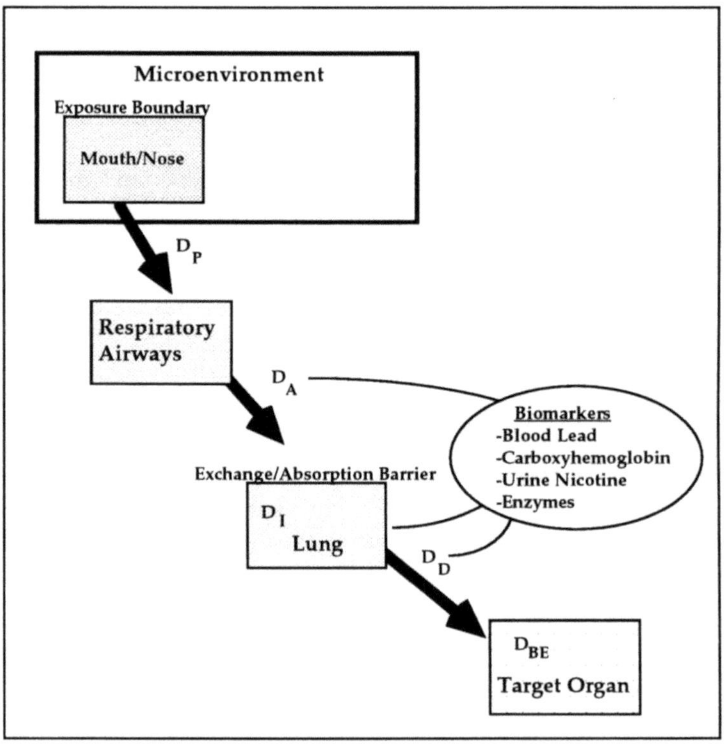

Figure 6: U.S. Environmental Protection Agency's Schematic of Dose and Exposure for Airborne Substances (Modified by McCurdy, Draft in Process). Biomarkers can be substances to which the organism is exposed or metabolites (e.g., enzymes) indicating exposure.

Comparison of Human and Ecological Assessments

A major difference between human and ecological exposure paradigms is their level of biological organization; i.e., population exposure for one species (human) versus community (several species), association, and population exposure for ecological risk assessments. Human risk assessments provide an expression of the likelihood that an adverse outcome will result from a given hazard; e.g., 10^{-6} chance of ovarian cancer in a population exposed to a particular pollutant. Ecological risk assessments are also expressions of the likelihood of an adverse outcome, but the expression depends upon the "environmental value" of concern; e.g., biological diversity, sustainability, and aesthetics (Environmental Monitoring and Assessment Program, 1993). Scientists are currently debating the usefulness of ecological risk assessments, with many instead favoring ecological benefits assessments. That is, benefits, can be gained or lost, depending on regulatory, management, and other decisions. Both risk and benefit assessments, however, will require exposure assessments.

A number of similarities exist between human and ecological exposure assessments. Both are often concerned with sensitive subpopulations, many pollutants are both human and ecological stressors, and ambient measurements for some pollutants can be indicators of both human and ecosystem exposure (e.g., ozone).

Passive monitors may improve useful data for both human and ecosystem exposure assessments, since they provide an inexpensive means of gaining coverage over large areas with reasonable accuracy for several gaseous pollutants ($\pm$ 20% for nitric oxide, ozone, and sulfur dioxide). The use of passive devices may even provide greater potential for ecosystems than for human exposure, since the need for more temporal precision may often be less for ecosystems than for human; i.e., if accumulation and degradation of a contaminant are the major areas of concern, a weekly average may be sufficient, whereas, hourly averages may be critical for human exposure assessments.

Name of Model	Keep Time Series?	V_E?	Outdoor µe Conc.	Indoor µe Conc.	Human Activity V_E	Validated Model?	User Friendly?	Citations
CARBON MONOXIDE								
Convolution	No	No	3	5	B	No	No	Duan (1989)
CO/Regress	No	No	3	5	A	No	Yes	Schwab (1989)
SHAPE	No	No	3	5	B	Limited	No	Ott (1984)
pNEM/CO	Yes	Yes	2,3	7	F	Limited	No	Johnson, et al (1992)
NITROGEN DIOXIDE								
SIMSYS	No	No	3	5	B	No	No	Ryan (1986)
REHEX	No	Yes	2	6	D	No	No	Lurmann, et al (1989)
NO2/Regress	No	No	3	5	A	No	Yes	Drye, et al (1989)
OZONE								
SAI/NEM	Yes	Yes	1	7	D	No	No	Hayes, et al (1984)
REHEX	No	Yes	2	7	D	No	No	Lurmann, et al (1989)
pNEM/O3	Yes	Yes	2	7	D	Limited	No	Johnson, et al (1993)
EPEM	No	Yes	2	-	D	No	No	Johnson, et al (1992)
LEAD								
Pb-NEM	No	No	4	6	A	No	No	OAQPS (1989)
IEUBK	No	No	4	6	A	No	Yes	Lead Work-group (1994)
RESPIRABLE PARTICULATES								
THEM	Yes	No	2	7	C	No	No	Klepeis (1994)
SULFUR DIOXIDE								
SO2-NEM	No	Yes	1	-	A	No	No	Biller et al (1986)
HAZARDOUS POLLUTANTS								
HEM	No	No	4	6	A	No	Yes	Radian (1985)
HAPEM	No	No	2	6	C	No	No	Johnson (1992)
AERAM	No	Yes	4	-	-	No	No	Eschenroder et al (1985)
SHEAR	No	No	4	-	-	No	No	Anderson & Lundberg (1983)
BEAM	No	Yes	1	6	-	No	Yes	Behar et al (1994)

Table 2: Attributes of selected air exposure models (After McCurdy Draft in Process). **Notes:** OUTDOOR µe: 1 = Use fixed site values as a surrogate. 2 = Use "adjusted" fixed site µe values. 3 = Monitor outdoor µe concentrations. 4 = Model outdoor µe concentrations. INDOOR µe: 5 = Measure indoor µe concentrations. 6 = Use indoor/outdoor ratios + indoor sources (if any). 7 = Use mass-balance model that includes indoor sources (if any). 8 = Use regression equations developed from indoor µe measurements. HUMAN ACTIVITY/ VENTILATION: A = Use of aggregate data and/or V_E. B = Simulate transitions; ignore V_E C = Sample from activity data; ignore V_E. D = Sample from joint activity/;V_E data.

Exposure Models

Both ecosystem and human exposure assessments can benefit from the use of models, although modeling ecosystem exposure pathways can be highly complex and includes much uncertainty. However, an increased understanding of fluxes and cycling of nutrients and contaminants, bioenergetics, and bioaccumulation will improve the application of ecosystem models.

Exposure models vary by scale (personal, microenvironmental, indoor, site-specific, regional), and type. Table 2 compares 20 human exposure models insofar as they incorporate ventilation rates, outdoor and indoor microenvironmental concentrations, and human activity patterns. Presently, new models are being used for carbon monoxide, oxides of nitrogen, ozone, lead, particulate matter, sulfur dioxide, and hazardous pollutants. The table illustrates that many have not yet been validated, or have been validated for limited microenvironments; e.g., within an automobile.

The application of exposure models to various ecosystems and human microenvironments is expanding rapidly, while their predictive and assessment reliability is continuing to improve.

Conclusions

Exposure assessment can be similar for humans and ecosystems, although the level of biological organization is often different for the two types of receptors. New methods for measuring, modeling, and assessing exposure are presently being developed. These scientific advances are enhancing the body of knowledge beyond simple ambient measurements, so that exposure assessments presently incorporate personal and indoor monitoring, and are increasingly making use of more reliable microenvironmental and pharmacokinetic models.

Acknowledgments

Several persons in EPA's National Exposure Research Laboratory provided invaluable information and advice to the author when preparing this report: Gerald Akland (biomarkers and multipathway methods); Tom McCurdy (models); Charles Lewis, Shaibal Mukerjee and Robert Stevens (source apportionment); Larry Purdue and Robert Burton (particles and ozone); Jim Mulik (personal monitors); Deborah Mangis (ecosystem receptors), and Andy Bond (agricultural exposure studies).

References

Anderson, G.E. and Lundberg, G.W. (1983). *User's Manual for SHEAR*. Research Triangle Park: U.S. Environmental Protection Agency.

Behar, J.V., Thomas, J., Pandian, M., and Tsang, A. (1994). "National Benzene Exposure Study Using Benzene Exposure Assessment Model (BEAM)." Paper presented at the Fourth Conference for Exposure Analysis; September.

Biller, W. B. (1986). *Analysis of Short-Term SO_2 Population Exposures in the Vicinity of Power Plants*. Durham: U.S. Environmental Protection Agency.

Drye, E., Ozkaynak, H., Burbank, R., Billick, I.H., Spengler, J.D., Ryan, P.B., Baker, P.E., and Humble, C. (1989). "Development of models for predicting the distribution of nitrogen dioxide concentrations." *J. Air Poll. Cont. Assoc.* 39: 1169-1177.

Duan, N. (1989). "Estimation of microenvironmental concentration distributions using integrated exposure measurements," pp. 15-1 to 15-14 in: T. Starks (ed.), *Proceedings of the Research Planning Conference on Human Activity Patterns*. Las Vegas: U.S. Environmental Protection Agency.

Environmental Monitoring and Assessment Program (1993). *Environmental Monitoring and Assessment Program: Master Glossary*. EPA/620/R-93/013. Research Triangle Park, NC: U.S. Environmental Protection Agency, Office of Research and Development.

Eschenroeder, A.Q., Magil, G.C., and Woodruff, C.R. (1985). *Assessing the Health Risks of Airborne Carcinogens*. Palo Alto: Electric Power Research Institute.

Hayes, S.R., Seigneur, C., and Lundberg, G.W. (1984). *Numerical Modeling of Ozone Population Exposure: Application to a Comparison of Alternative Ozone Standards*. San Rafael: Systems Applications, Inc.

Hidy, G.M. (1975). Summary of the California Aerosol Characterization Experiment. *J. Air Poll. Cont. Assoc.* 25: 1106-1114.

Hunsaker, C.T., MacCarthy, J.F., Shugart, L.R., and O'Neill, R.V. (1990). Indicators Relevant to Multiple Resource Categories. In *Ecological Indicators for the Environmental Monitoring and Assessment Program.*, eds. C.T. Hunsaker and D.E. Carpenter, 2-1 - 2-18. EPA 600/3-90/060. Research Triangle Park, NC: U.S. Environmental Protection Agency, Office of Research and Development.

Johnson, T., Paul, R.A., and Capel, J.E. (1992). *Application of the Hazardous Air Pollutant Exposure Model (HAPEM) to Mobile Source Pollutants*. Durham: IT Technology.

Johnson, T.R., Capel, J. and McCoy, M. (1993). *Estimation of Ozone Exposures Experienced by Urban Residents Using a Probabilistic Version of NEM and 1990 Population Data*. Durham: IT Technology.

Johnson, T.R., Capel, J., Olaguar, E., and Wijnberg, L. (1992). *Estimation of Ozone Exposures Experienced by Urban Residents Using a Probabilistic Version of NEM*. Research Triangle Park: U.S. Environmental Protection Agency.

Johnson, T.R., Capel, J., Paul, R., and Wijnberg, L. (1992). *Estimates of Carbon Monoxide Exposures and Associated Carboxyhemoglobin Levels in Denver Residents Using a Probabilistic Version of NEM.* Research Triangle Park:

Klepeis, N.E., Ott, W.R., and Switzer, P. (1994). "A Total human exposure model (THEM) for respirable suspended particulates." Paper 94-WA75A.03 presented at the Annual Meeting of the Air & Waste Management Association; June.

Lead Working Group (1994). *Guidance Manual for the Integrated Exposure Uptake Biokinetic Model for Lead in Children.* (The IEUBK Model.) Washington, DC: U.S. Environmental Protection Agency (EPA-450/R-93-081).

Lipton, J., Galbraith, H., Burger, J., and Wartenberg, D. (1993). A Paradigm for Ecological Risk Assessment. *Environmental Management.* Vol. 17, No. 1, 1-5.

Lurmann, F.W., Weiner, A.M., and Colome, S.D. (1989). "Development of a new regional human exposure model (REHEX) and its application to the California South Coast Air Basin." Paper presented at the Annual Meeting of the Air Pollution Control Association; June.

McCurdy, T. (Draft in Process). Modeling the "Dose Profile" in Human Exposure Assessments: Ozone as an Example. Research Triangle Park, NC: U.S. Environmental Protection Agency, Office of Research and Development.

Messer, J.J. (1990). EMAP Indicator Concepts. In *Ecological Indicators for the Environmental Monitoring and Assessment Program.*, eds. C.T. Hunsaker and D.E. Carpenter, 2-1 - 2-18. EPA 600/3-90/060. Research Triangle Park, NC: U.S. Environmental Protection Agency, Office of Research and Development.

National Academy of Sciences, National Research Council (1991). *Human Exposure Assessment for Airbone Pollutants: Advances and Opportunities.* Washington, DC: National Academy Press.

OAQPS Staff (1989). *Review of the National Ambient Air Quality Standards for Lead: Exposure Analysis Methodology and Validation.* Research Triangle Park: U.S. Environmental Protection Agency (EPA-450/2-89-011).

Ott, W.R. (1984). "Exposure estimates based on computer generated activity patterns." Clin. Toxicol. 21: 97-128.

Radian Corporation (1985). *A Study of Feasible Modeling Alternatives for Simulating Human Exposure and Risk Resulting from Airborne Pollutants.* Research Triangle Park: Radian.

Risk Assessment Forum (1992). *Framework for Ecological Risk Assessment.* EPA/600/R-92/001. Washington, DC: U.S.

Environmental Protection Agency Office of Research and Development.

Ryan, P.B., Spengler, J.D., and Letz, R. (1986). "Estimating personal exposures to NO_2." *Environ. Inter.* 12: 394-400.

Schwab, M. (1989). "The influence of daily activity patterns on differential exposure to carbon monoxide among social groups,"pp. 18-1 to 18-21 in T.H. Starks (ed.), *Proceedings of the Research Planning Conference on Human Activity Patterns.* Las Vegas: U.S. Environmental Protection Agency.

Sexton, K. and Ryan, P.B. (1988). Assessment of Human Exposure to Air Pollution: Methods, Measurements, and Models. In *Air Pollution, the Automobile, and Public Health.*, 208. Washington, DC: Health Effects Institute. National Academy Press.

U.S. Environmental Protection Agency (1990). *Nonoccupational Pesticide Exposure Study (NOPES).* EPA/600/3-90/003. Washington, DC: U.S. Environmental Protection Agency Office of Research and Development.

U.S. Environmental Protection Agency (1992). *Federal Register.* Guidelines for Exposure Assessment. Vol. 57. No. 104. FRL-4129-5.
U.S. Environmental Protection Agency.

Wallace, L.A. (1987). *The Total Exposure Assessment Methodology (TEAM) Study: Summary and Analysis*, Vol. 1. EPA/600/6-87/002a. Washington, DC: U.S. Environmental Protection Agency Office of Research and Development.

Watson, J.G, Robinson, N.F., Chow, J.C., Henry, R.C., Kim, B.M., Pace, T.G., Meyer, E.L., and Nguyen, Q. (1990). "The USEPA/DRI Chemical Mass Balance Receptor Model, CMB 7.0" *Environ. Software.*

Disclaimer:

This paper has been reviewed in accordance with the United States Environmental Protection Agency's peer and administrative review policies and approved for presentation and publication. Mention of trade names or commercial products does not constitute endorsement or recommendation for use.

Air Pollution Warning System - Critical Points of Decision Algorithm

Jacek Walczewski
Institute of Meteorology and Water Management
ul. P. Borowego 14
30-215 Cracow, Poland

Introduction

A specific feature of air pollution, especially in cities, is short-lasting, comparatively high maxima of concentrations, called episodes. They may last from one to several days, and they should be discerned from "momentary" peaks of several hours duration, and from the "background" variability of monthly and seasonal character.

The episodes are connected with a combination of meteorological factors - poor ventilation or - in some cases - unfavourable transport and transformation processes.

These air pollution episodes are often divided into "winter episodes", connected mainly with an increase of SO_2, and suspended particulate matter (SPM) concentrations, and "summer episodes" connected with the production of O_3.

The characteristics of episodes can be defined as having short duration and very high pollution concentrations. During episodes extraordinary measures can be applied to protect the population: active measures - emergency reduction of local emissions; and passive measures for a reduction of health hazards - avoiding physical exercise in the open air, taking special care of sick people, etc.

Both kinds of countermeasures, to be applied in an effective way, need information on the advent of an episode. This information should be supplied as early as possible, to reserve enough time for the organization of countermeasures. The forecast of episodes makes it possible to give this information in advance, before the episode happens.

The informative systems may be used for setting-up 2 kinds of preventive actions: warnings and alarms. Warnings are (in most

NATO ASI Series, Partnership Sub-Series, 2. Environment – Vol. 8
Urban Air Pollution
Edited by Ivo Allegrini and Franco De Santis

cases) for information only and alarms are connected with enforcement of compulsory actions. In the case of alarms the responsibility of an informative system is higher, because the operation of emergency emission-reduction system is expensive and may be troublesome for citizens (e.g. traffic limitations). Air pollution warnings and alarm systems have been developed and are currently operating in many countries (Breiling and Alcamo 1992; Breiling 1993).

Decision algorithm

We will consider two versions of a decision algorithm: for warnings, based on air-pollution forecast (Fig.1), and for alarms, based on actual measured concentration values (Fig.2). To make our discussion more concrete, we will take into consideration the case of winter episodes.

In the scheme of Fig.1 the actual concentration and meteo data, supplemented by meteo forecast, are used to prepare the air pollution forecast for the next day. It is assumed that air pollution is characterized by two "leading" substances "S" and "D" (e.g., SO_2, and fine dust).

For concentrations of both substances warning thresholds SW and DW are determined. The forecasted mean diurnal concentrations are compared with warning thresholds.

If forecasted values exceed warning thresholds, the meteorological forecast has to be checked. It should be mentioned, that the use of meteorological forecast is of a different nature at the beginning of the decision-making cycle ("Actual and forecasted meteo data") and at the point of "Meteo forecast check".For instance, the statistical models of air pollution forecast often make use of forecasted mean diurnal temperature and wind speed for some areas; then two numbers are simply introduced in the model computation. At the point "Meteo forecast check" a more thorough analysis of forecasted dispersion conditions should be made, taking into account the synoptic situation on the basis of local climatic conditions. If the analysis confirms the high probability of pollution accumulation, then a warning should be issued.

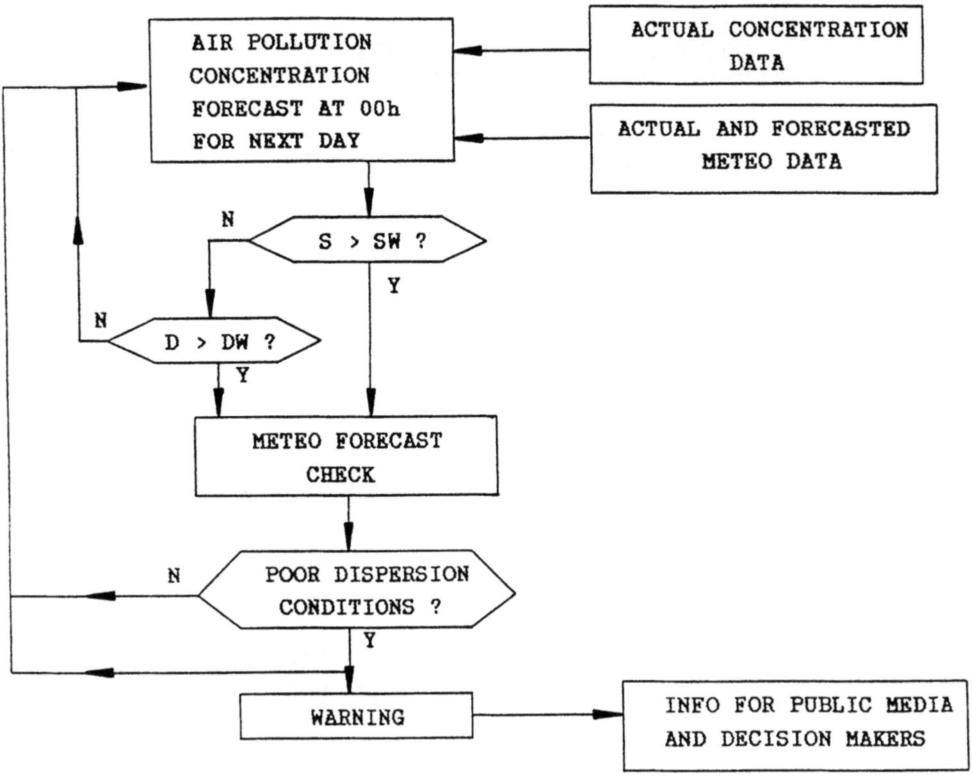

Fig.1. Scheme of the decision algorithm for air-pollution warning in winter. Air pollution is characterized by 2 leading substances: S and D. S, D - mean 24-h concentrations of pollutants S and D;
SW, DW - warning thresholds for substances S and D.

To define this decision algorithm for practical use, the following parameters must be determined:

(a) period and area for which pollution forecast is prepared;

(b) leading pollutants;

(c) warning thresholds for leading pollutants;

(d) parameters of meteorological situation, used in meteo forecast check;

(e) time and area for which warning is issued.

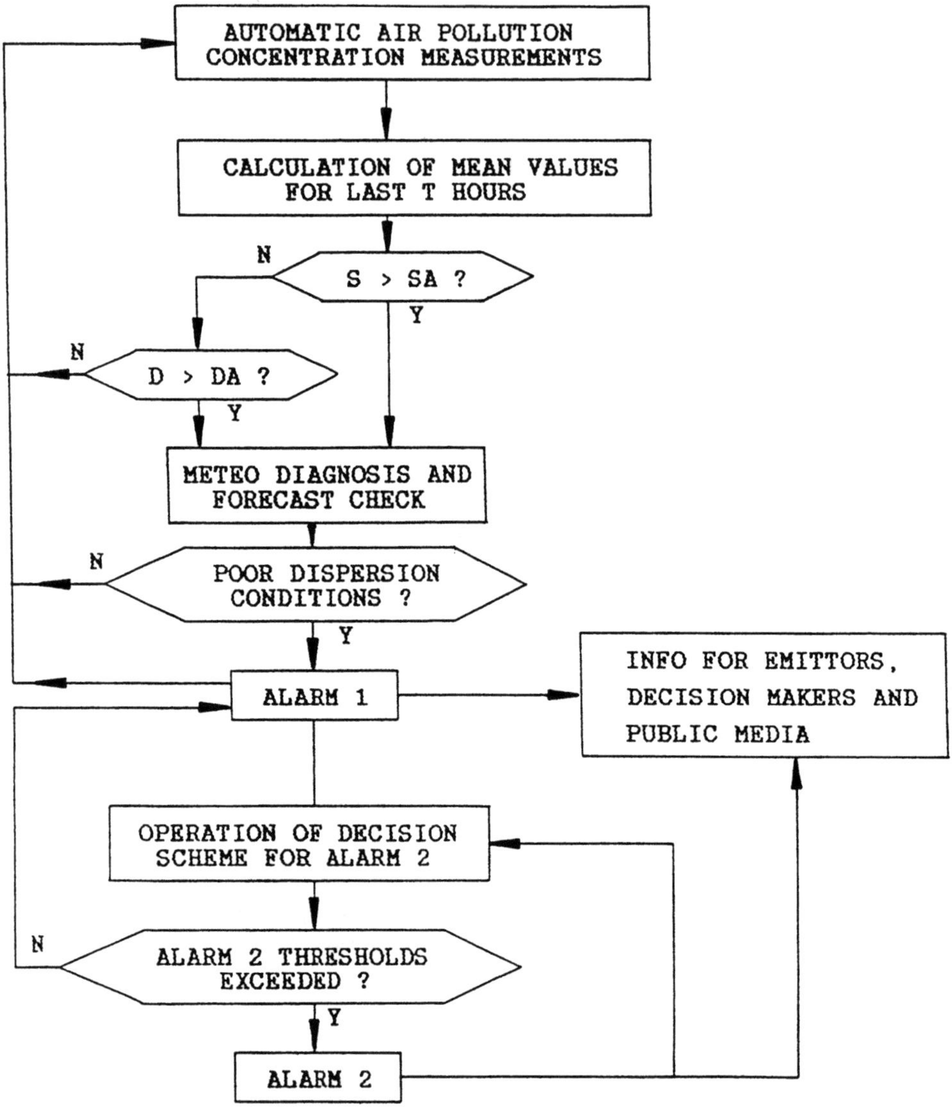

Fig.2 Scheme of the decision algorithm for air-pollution alarms in winter. Air pollution is characterized by 2 leading substances: S and D. S, D - concentrations of pollutants S and D (averages of T hours); SA, DA - warning thresholds for substances S and D.

Let us now examine the second version of decision algorithm, shown on Fig.2. This kind of algorithm, based on real concentration data, is usually preferred for alarms, because pollution forecasts with their margin of uncertainty are considered to be not sufficiently dependable as the basis for economically expensive alarm ordinances. In Fig.2 the data from automatic measuring stations are continuously processed to obtain running mean values of concentrations from the last T hours, for leading substances S and D. These values are compared with alarm thresholds SA and DA.

If threshold values are exceeded, then a meteorological forecast check is performed to confirm, that unfavourable meteo conditions will not change in the perspective of one day. If no positive change in meteorological conditions is forecasted, then an alarm signal is issued to all whom it may concern. If in the alarm system 2 stages of alarms are foreseen, a similar procedure, as for Alarm 1 is continued for Alarm 2.

The parameters of the decision algorithm are:

(a) the area (number of measuring stations) for which alarm decision procedure is applied;

(b) leading pollutants;

(c) average period for the processing of concentration values;

(d) alarm thresholds for leading pollutants;

(e) number and location of stations, for which alarm thresholds should be exceeded;

(f) parameters of meteorological situation taken into account in meteo forecast check;

(g) time and area, for which alarm is announced.

Setting the parameters of warning and alarm systems involves making difficult decisions and choices. They cannot be made correctly without the support of scientific analyses and selection of optimization factors. In the following paragraphs we will discuss briefly the points which seem to be of a critical nature for the whole decision algorithm.

Warning and alarm thresholds.

We will use the words "warning " and "alarm" in the meaning defined in the "Introduction". The general principle of setting the warning and alarm thresholds is apparently simple: warning is justified, when air pollution approaches the "dangerous" level; alarming is justified, when air pollution becomes really dangerous for human health.

In practice, the expression "dangerous pollution level" is far from exact. The health effects of air pollution are of statistical nature and they are becoming gradually more and more severe as the product of concentration and exposure time increases. For some polluting substances, like SO_2 and SPM, the World Health Organization gives definite threshold values for mean diurnal concentrations, responsible for "moderate" and "acute" health effects (World Health Organization 1990). Practical application of these figures depends on the pollution status of the air in the area where warning or alarm system is organized.

Let us consider the relationship between the concentration of some pollutants (e.g., SO_2), and the number of hours (in a year or season) in which individual values of concentrations are observed at a given site (Fig.3.). For each concentration value S we have a corresponding number of hours H. In most cases the relationship looks like that in Fig.3: lower concentration values occur more frequently than higher ones. Sometimes the highest occurrence frequency is observed for some intermediate values of concentrations, but in these cases we can also see the lowering branch of the curve at the right side of the graph.

If the air is very highly polluted, the critical ("dangerous") concentration S_k and concentrations $S > S_k$. will be present over a very high number of hours H_k, and $H > H_k$ and alarms would be announced very frequently.

This situation may not be acceptable from an economical point of view. Emergency actions must be reserved for real "emergencies": rare, but very dangerous events. If critical values of pollution concentrations become permanent, long-range pollution abatement strategies should be used (i.e flue-gas cle\aning or change of technologies).

Here we are touching the delicate problem of making the distinction between application of long-range strategies and short-range (emergency) actions.

As we see, the alarm threshold value is correlated not only with health requirements, but also with economic factors.

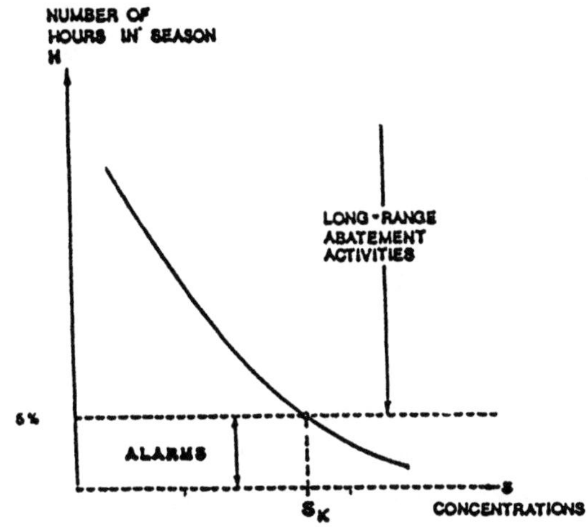

Fig.3. Scheme illustrating relationship between concentrations S of some air pollutant and numbers of hours H with concentrations S, at some site and in some period. s_k - critical value of S (alarm threshold).

This problem may be formulated as follows: what is the highest acceptable number of days in a year (season) with alarms? This number can be expressed in percent of days in the period of reference - for instance 5%, as in Fig.3. (about 9 days in a heating season). Fig.4 summarizes our discussion. S_k is here a generally accepted alarm threshold (lower limit of acute health effects, or a little lower value).

H_t is the total number of hours in a season. H_M is a fewer number of hours, for which concentrations $S < S_k$ are considered to be not essentially harmful. H_A is the economically accepted maximum number of hours for alarms in the season.

Thus we have

for $H(S_k) \leq H_M$ - low pollution situation; alarms not necessary (dashed area in Fig.4.);

for $H_M \leq H(S_K) \leq H_A$ - high pollution situation; alarms to be organized within acceptable time limits;

for $H(S_K) > H_A$ _ very high pollution situation, intensive long-range emission abatement activities necessary; alarms to be organized, but alarm hreshold has to be raised up to S_K, that does not protect the population sufficiently.

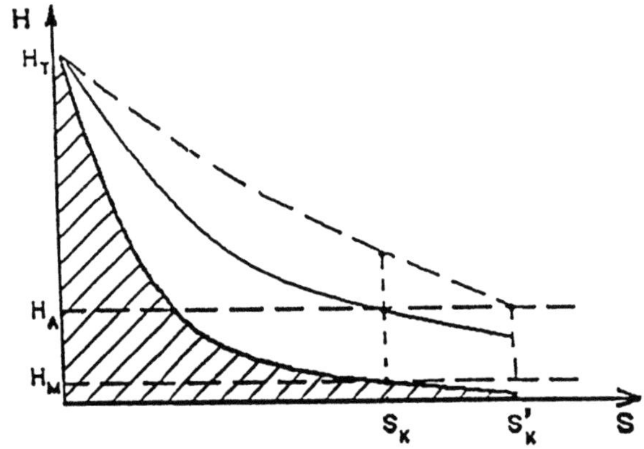

Fig.4. Scheme illustrating the identification of cases of low, high and very high air pollution, with use of curves defined as in Fig.3.

Solving the problems presented above needs research in different fields of science. The variety of schemes applied in different countries indicates the necessity to coordinate research efforts.

Averaging time for concentration values

Fig.5. presents an example of the analysis of occurrence frequency of concentration maxima of different duration. The figure (which refers to Cracow) presents the numbers of cases n in which concentration 200 $\mu g/m^3$ of SO_2, was exceeded, depending on duration (number of hours) of these events.

A considerable part (80%) of all recorded events was in the limit of the 5 hour duration. They were short-lasting maxima, mainly due to a fumigation effect. It is clear that they should not be taken into account in any warning system, because they disappear in a natural way and they cannot be affected by any emission abatement action.

This example illustrates the role of averaging the measured concentration values in an alarm system: the averaging period should be selected in such a way as to filter-out short-lasting pollution events and to signal real pollution episodes, in the

proper time. If the averaging period is too-long, signalling of the episode will be delayed too much. The proper choice is a question of optimization analysis, taking into account the alarm thresholds accepted for the given area.
In the case of warnings, based on pollution forecasts, the characteristic time period for concentration analysis will be determined mostly by the period, for which forecast may be given, using available prediction methods.

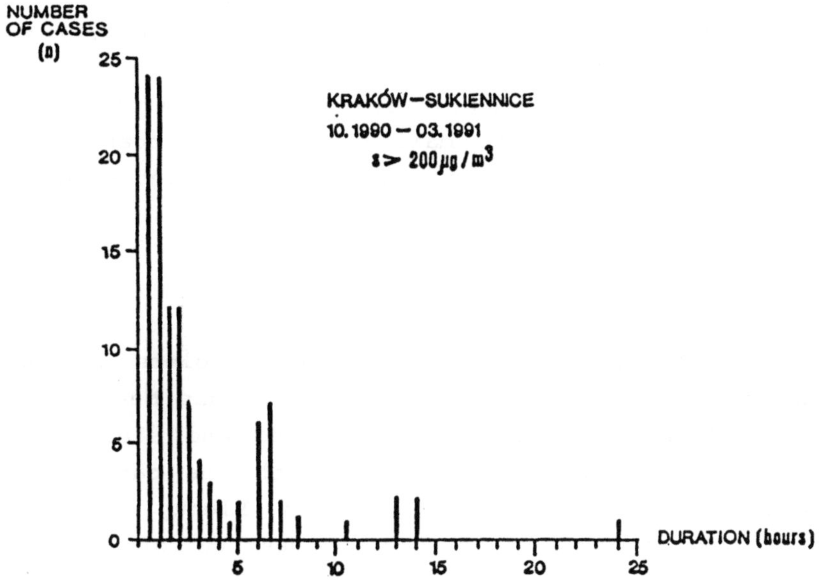

Fig.5. Numbers of observed concentration values of SO_2 a higher than 200 μg/m3, of different duration. Site: Center of Cracow, Poland. Period: winter season 1990/1991.

For instance, the statistical forecast models used in many cities usually give forecasted 24-hour-mean values of concentrations, which may be conveniently compared with standards.

The role of meteorological data

The meteorological situation is usually predictable with high probability at least for the next 24 hours and this fact is of essential importance for all methods of emergency air protection. Some air pollution warning systems only use meteorological forecasts and current meteorological data to assess the dispersion conditions, and - in this indirect way - to predict or assess the pollution situation. It is known that some synoptic situations, like stationary high pressure system in winter, cause a rise in air pollution. Others, like a front passage, with rain and strong winds, are connected with the removal of pollution from the air. Nevertheless, local dispersion conditions depend not only on large-scale (synoptic) factors, but also on meso - or even micro-scale factors (like terrain configuration, flow channels, surface roughness and albedo etc.). Thus, boundary-layer research must be combined with synoptic analyses to make pollution forecasts reliable, and general rules must be combined with knowledge of specific site factors.

Concluding remarks.

This paper covers only part of the problems connected to emergency air protection decisions. As it has been shown, such problems, like determination of warning and alarm thresholds need an interdisciplinary approach with the participation of medical, physical and economic sciences. Plenty of work remains to be done in this area. The necessity of international cooperation seems to be obvious. The forms of such cooperation range from exchange of experience to joint research projects.

References

Breiling M, Alcamo J (1992) Emergency air protection: A survey of Smog Alarm Systems. IIASA Working Paper WP-92-52, Laxemburg (Austria).

Breiling M (1993) Emergency air protection: Implementing Smog Alarm Systems in Central and Eastern Europe. IIASA, Laxemburg (Austria)

World Health Organization (1990) Acute effects on health of smog episodes. WHO Regional Publications, European Series, No.43.

Author index

The Partnership Sub-Series incorporates activities undertaken in collaboration with NATO's Cooperation Partners, the countries of the CIS and Central and Eastern Europe, in Priority Areas of concern to those countries.

The volumes published as a result of these activities are:

Vol. 1: **Clean-up of Former Soviet Military Installations.** Edited by R. C. Herndon, P. I. Richter, J. E. Moerlins, J. M. Kuperberg, and I. L. Biczó. 1995

Vol. 2: **Cleaner Technologies and Cleaner Products for Sustainable Development.** Edited by H. M. Freeman, Z. Puskas, and R. Olbina. 1995

Vol. 3: **Remediation and Management of Degraded River Basins.** Edited by V. Novotny and L. Somlyódy. 1995

Vol. 4: **Earthquakes Induced by Underground Nuclear Explosions.** Edited by R. Console and A. Nikolaev. 1995

Vol. 5: **Transportation Infrastructure.** Edited by R. M. Gutkowski and J. Kmita. 1996

Vol. 6: **Sustainable Development of the Lake Baikal Region.** Edited by V. A. Koptyug and M. Uppenbrink. 1996

Vol. 7: **Transboundary Water Resources Management.** Edited by J. Ganoulis, L. Duckstein, P. Literathy, and I. Bogardi. 1996

Vol. 8: **Urban Air Pollution.** Edited by I. Allegrini and F. De Santis. 1996

MIX
Papier aus verantwortungsvollen Quellen
Paper from responsible sources
FSC® C105338

If you have any concerns about our products,
you can contact us on
ProductSafety@springernature.com

In case Publisher is established outside the EU,
the EU authorized representative is:
Springer Nature Customer Service Center GmbH
Europaplatz 3, 69115 Heidelberg, Germany

Printed by Libri Plureos GmbH
in Hamburg, Germany